AF412567

Plasmon Resonances in Nanoparticles

Volume **6**

World Scientific Series in **Nanoscience and Nanotechnology**

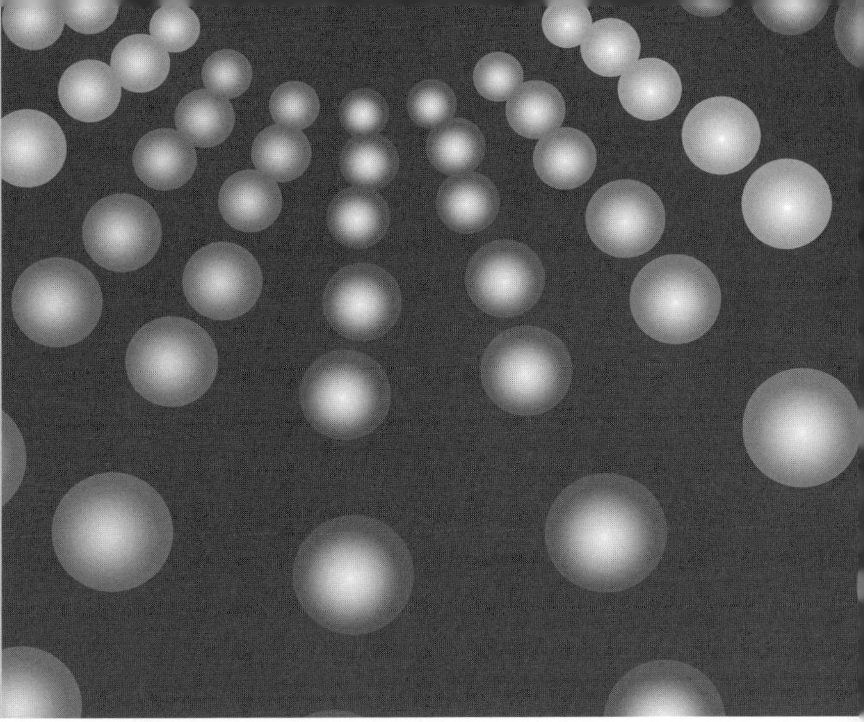

Plasmon Resonances in Nanoparticles

Isaak D Mayergoyz

Department of Electrical and Computer Engineering
UMIACS
Center for Applied Electromagnetics
University of Maryland
College Park

NEW JERSEY · LONDON · SINGAPORE · BEIJING · SHANGHAI · HONG KONG · TAIPEI · CHENNAI

Published by

World Scientific Publishing Co. Pte. Ltd.

5 Toh Tuck Link, Singapore 596224

USA office: 27 Warren Street, Suite 401-402, Hackensack, NJ 07601

UK office: 57 Shelton Street, Covent Garden, London WC2H 9HE

British Library Cataloguing-in-Publication Data
A catalogue record for this book is available from the British Library.

World Scientific Series in Nanoscience and Nanotechnology — Vol. 6
PLASMON RESONANCES IN NANOPARTICLES

ISBN 978-981-4350-65-5

Printed in Singapore by World Scientific Printers.

To my wife Doba

Preface

You have in your hands a book on plasmon resonances in metallic nanoparticles. The study of these resonances is a very active area of research in nano-science with many promising technological applications. It is emphasized in the book that plasmon resonances in nanoparticles are, by and large, electrostatic in nature and that electrostatics of particles with negative permittivity is the foundation for the understanding of these resonances. The modal approach to the analysis of plasmon resonances in nanoparticles is developed in the book. It is demonstrated that the analysis of plasmon resonance modes in electrostatic approximation can be mathematically formulated as an eigenvalue problem for specific boundary integral equations. General properties of plasmon spectrum and plasmon modes are then studied along with radiation corrections to the electrostatic approximation. The issues of excitation of specific plasmon modes, i.e., their coupling to incident optical radiation and the time-dynamics of their growth and dephasing are discussed in the book as well. Many analytical and numerical examples of plasmon modes in various nano-structures are given in the book and compared (where it is possible) with available experimental data. Finally, some selective applications of plasmon resonances in nanoparticles are presented.

The exposition of the material in the book is largely based on the publications of the author and his collaborators that have appeared over the past ten years. No attempt has been made in the book to refer to all relevant publications. For this reason, references given at the end of each chapter are not exhaustive but rather suggestive.

It is believed that this book is unique as far as its style of exposition, scope and conceptual emphasis are concerned. The topics discussed in the book will be of interest to the broad audience of electrical engineers, material scientists, physicists, applied mathematicians and numerical analysts involved in the development of novel nano-technology.

In conclusion, I would like to thank Drs. D. Fredkin, C. Krafft, G. Miano, O. Rabin, N. Gumerov, R. Duraiswami, Z. Zhang, P. McAvoy, S. Tkachuk and my current students L. Hung and G. Lang for their collaboration. The assistance of Dr. P. McAvoy in the preparation and improvement of the manuscript of this book is greatly appreciated. I gratefully acknowledge the financial support for my research on plasmon resonances by the National Science Foundation and the Office of Naval Research.

Contents

Chapter 1

Introduction

1.1 What are Plasmon Resonances?

The purpose of this introductory chapter is to discuss the physical origin of plasmon resonances in metallic nanoparticles, to describe their basic properties and to outline the approach to the study of plasmon resonances adopted in this book. The presentation in this chapter is mostly descriptive and avoids as much as possible mathematical technicalities which are provided in the subsequent chapters.

To start the discussion, consider a macroscopic piece of metal (gold or silver, for instance) subject to optical radiation (see Figure 1.1). In this situation, no unique physical phenomena of distinction or long remembrance occur; this macroscopic piece of metal is "lifeless." However, if the dimensions of this metallic piece are made smaller and smaller and are eventually reduced to nanoscale, the resulting metallic nanoparticle may come to life while being subject to optical radiation. It may glow, it may resonate and it may become a very powerful, nanoscale localized source of light. This is, in descriptive terms, the essence of the phenomena of plasmon resonances in metallic nanoparticles. These powerful localized sources of light are useful in different areas of science and technology which include scanning near-field optical microscopy [1, 2], nano-lithography [3], biosensor applications [4, 5], surface enhanced Raman scattering (SERS) [6]-[9], nanophotonics [10]-[13], optical and magnetic data storage [14, 15], etc.

The question can be immediately asked: "What is the physical mechanism of plasmon resonances?" In other words: "What is there to resonate?"

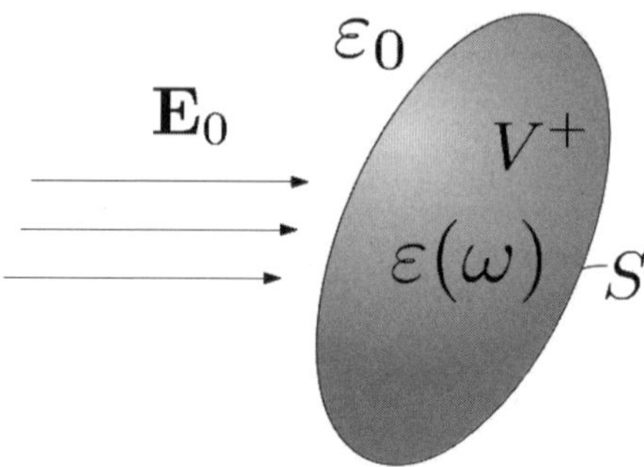

Figure 1.1

Experiments show that metallic (gold and silver) nanoparticles may exhibit resonance behavior at certain frequencies at which the following two conditions are satisfied: **1) the nanoparticle dielectric permittivity is negative and 2) the free-space wavelength is large in comparison with the nanoparticle dimensions**. The latter condition clearly suggests that these resonances are electrostatic in nature. When dielectric permittivity is negative, the uniqueness theorem of electrostatics is not valid. For this reason, source-free electrostatic fields may appear for certain negative values of dielectric permittivity. This is the manifestation of resonances, and the corresponding source-free electrostatic fields are resonant plasmon modes.

It is important to stress that plasmon resonances in metallic nanoparticles are intrinsically nanoscale phenomena. This is because the two resonance conditions (negative dielectric permittivity and smallness of the particle dimensions in comparison with free-space wavelength) can be simultaneously and naturally realized at the nanoscale.

The question can be asked why it is possible and relevant to speak of dielectric permittivity of metallic nanoparticles subject to optical radiation. The reason is that conduction electrons in metallic nanoparticles are pinned by optical radiation and execute tiny back-and-forth oscillations around some "equilibrium" positions. In this sense, these conduction electrons are indistinguishable from bound charges in dielectrics. That is why metallic nanoparticles behave at optical frequencies as dielectric particles with dispersion. The latter means that dielectric permittivity depends on frequency. It turns out that for a certain frequency range its real part may assume negative values. As discussed later in this chapter, this frequency range for metals is near their

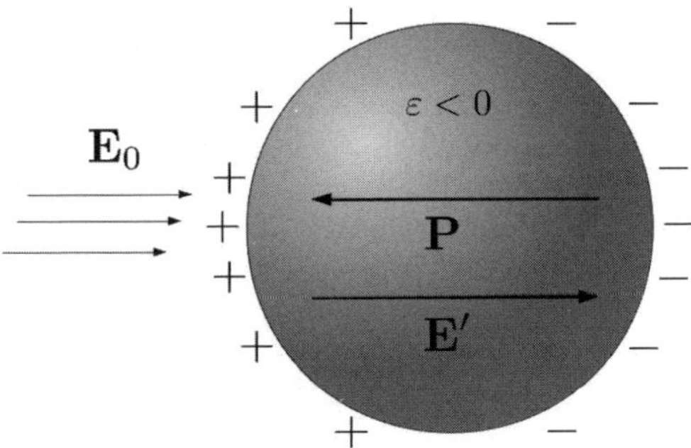

Figure 1.2

plasma frequencies, where the dispersion relations $\varepsilon(\omega)$ are fully governed by the interaction between electromagnetic radiation (light) and the conduction electrons. For good conductors such as silver and gold, plasma frequencies are in the visible frequency range, and this explains why silver and gold nanoparticles are usually employed in plasmon resonance studies and applications. It is worthwhile to remark that plasmon resonances may occur not only in metallic nanoparticles but in any nanoparticle whose permittivity exhibits dispersion and whose real part may assume negative values. One example is the silicon carbide (SiC) material whose negative permittivity is not due to the interaction of conduction electrons with light but rather due to specific properties of lattice vibrations in polar crystals.

It has been already mentioned that the second condition of smallness of particle dimensions in comparison with free-space wavelength reveals the physical nature of plasmon resonances in nanoparticles as electrostatic resonances. Indeed, due to this condition, time-harmonic electromagnetic fields within the nanoparticles and in their vicinity vary almost with the same phase. As a result, at any fixed instant of time these fields look like electrostatic fields. The electrostatic nature of plasmon resonances in nanoparticles and their occurrence for negative values of dielectric permittivity immediately suggest the enhancement of local electric fields inside nanoparticles and their vicinities. To illustrate this fact, consider an example of a spherical nanoparticle with negative permittivity ε subject to uniform external field $\mathbf{E}_0$ (see Figure 1.2). Since $\varepsilon < 0$, the polarization vector $\mathbf{P}$ has the direction opposite to $\mathbf{E}_0$ and this results in surface electric charges which create the "depolarizing" field $\mathbf{E}'$ with the same direction as $\mathbf{E}_0$. This naturally

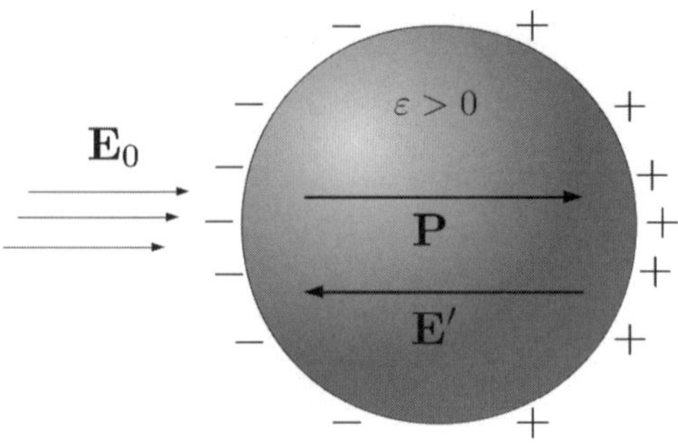

Figure 1.3

leads to the enhancement of the total electric field $\mathbf{E} = \mathbf{E}_0 + \mathbf{E}'$ inside the spherical nanoparticle*. To better appreciate this fact, it is instructive to consider the case of a spherical nanoparticle with positive permittivity (Figure 1.3) where the depolarizing field $\mathbf{E}'$ results in the attenuation of the external field $\mathbf{E}_0$.

In our study of plasmon resonances in nanoparticles, we shall follow the traditional approach when all losses are first neglected and resonance modes and resonance frequencies are first found for lossless systems. A similar approach is used, for instance, in the study of resonance modes in metallic cavities. For such cavities, the resonance mode problem is mathematically formulated as an eigenvalue problem for specific differential equations derived from the Maxwell equations. It will be demonstrated in this book that the problem of plasmon resonance modes can be also mathematically formulated as an eigenvalue problem which is posed, however, not for differential equations but rather for specific boundary integral equations. There is another important difference between plasmon resonances in metallic nanoparticles and resonances in metallic cavities. In the latter case, the resonance frequencies depend on the shape and dimensions of the metallic cavities. For instance, in the case of rectangular resonant cavities (see Figure 1.4), the

---

*It is interesting to mention that a similar phenomenon of enhancement of interior magnetic fields occurs in type-I superconductors, which are ideal diamagnets. For this reason, the magnetic fields inside type-I superconductors may exceed a critical field, while applied exterior magnetic fields remain below this field. This leads to the formation of the Landau "intermediate state" of type-I superconductors when normal and superconducting regions ("domains") coexist inside type-I superconductors.

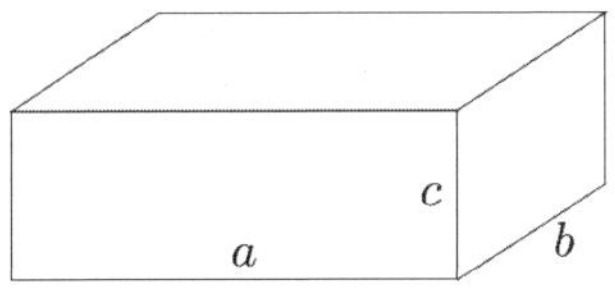

Figure 1.4

resonance frequencies are given by the formula

$$\omega_{mnk} = \frac{\pi}{\sqrt{\mu_0 \varepsilon_0}} \sqrt{\frac{m^2}{a^2} + \frac{n^2}{b^2} + \frac{k^2}{c^2}}, \tag{1.1}$$

which clearly reveals their dependence on cavity dimensions a, b and c. It will be demonstrated in this book that in the case of plasmon resonances in metallic nanoparticles, resonance frequencies are scale-invariant. This means that these frequencies depend only on particle shapes but not on their dimensions, provided that these dimensions are appreciably smaller than resonance free-space wavelengths. This scale invariance implies, among other things, that in the case of ensembles of almost self-similar (of the same shape but different dimensions) metallic nanoparticles, they may resonate at practically the same wavelength. Consequently, plasmon resonances can be simultaneously excited in all these nanoparticles.

It turns out that the eigenvalue formulation of the plasmon resonance problem has another important feature. This eigenvalue formulation leads to the direct calculation of the negative values of dielectric permittivity at which plasmon resonances may occur. These negative values of dielectric permittivity can then be used for any dispersion relation of nanoparticle to determine the corresponding resonance frequencies. In this way, the properties of plasmon resonances which depend on nanoparticle shapes can be fully separated from those which depend on the material properties of nanoparticles which define their dispersion relations. In other words, the solution of the plasmon resonance eigenvalue problem for a nanoparticle of specific shape can be used for different materials of this nanoparticle to find resonance frequencies.

Having found plasmon resonance modes and their resonance frequencies for lossless metallic nanoparticles through the solutions of the appropriate eigenvalue problem, the next step is to study the excitation of these plasmon modes by the incident radiation and the effect of ohmic and radiation losses. This approach is adopted in the book and it reveals that, by and

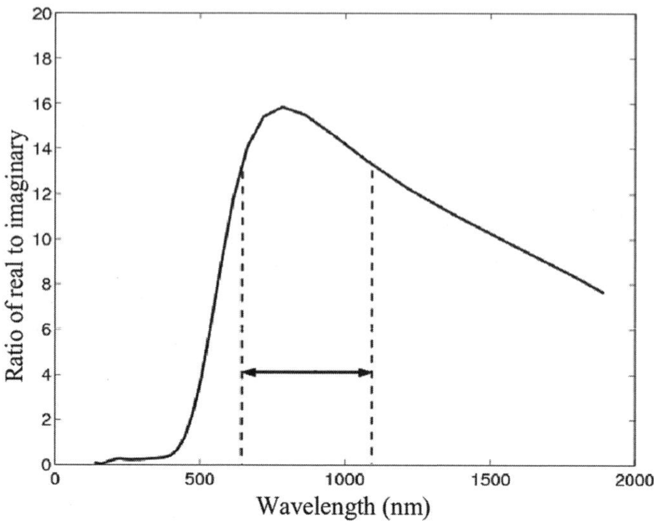

Figure 1.5

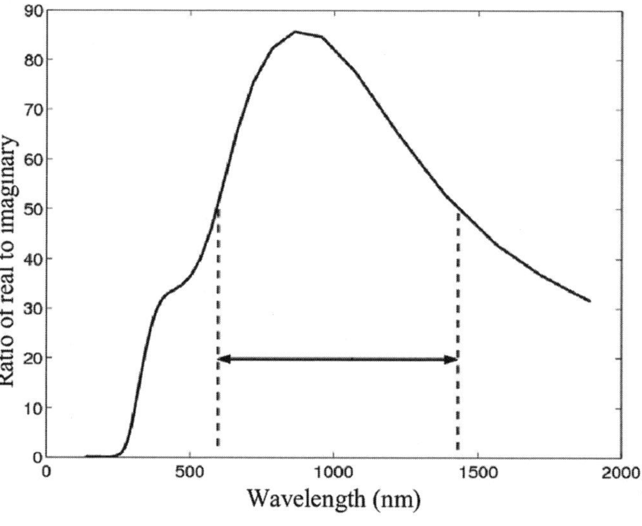

Figure 1.6

large, dipole plasmon modes are excited by incident optical radiation created by remote sources. This is because the electric field of this radiation is practically uniform over the nanoparticle region due to the smallness of nanoparticle dimensions in comparison with the wavelength of the incident radiation. It also turns out that the incident radiation is most efficiently coupled to a plasmon mode if the dipole moment of this mode is parallel to the direction of electric field of the incident radiation. The plasmon mode excitation analysis also reveals that the quality of plasmon resonances, i.e., the local enhancement of the incident optical radiation, is controlled by the ratio of the real part of dielectric permittivity to its imaginary part at the resonance frequency. For gold and silver, this ratio is most appreciable when the free-space wavelength is within the ranges of 650-1000 nm and 600-1400 nm (see Figures 1.5 and 1.6, respectively, which are based on experimental data of P. B. Johnson and R. W. Christy [16]). Therefore, plasmon resonances in gold and silver nanoparticles can be most efficiently excited in the corresponding frequency ranges. It is also worthwhile to observe that the ratio of real and imaginary parts of dielectric permittivity is appreciably higher for silver than for gold. Thus, as far as the quality of plasmon resonances is concerned, silver is "gold" and gold is "silver." This fact has long been appreciated in the area of surface enhanced Raman scattering (SERS) research where silver nanoparticles have been predominantly used in experiments. Of course, silver oxidation presents some experimental difficulties that must be dealt with.

1.2 Dispersion Relations

It is clear from the presented discussion that the dispersion relation $\varepsilon(\omega)$ of a metallic nanoparticle is instrumental for the analysis of plasmon resonances. For this reason, it is worthwhile to briefly review the simplest analytical models for the dispersion relations. We start with the case of the free-electron plasma dispersion relation, when all electron collisions are neglected. In addition, since the force on electrons arising from interaction with the magnetic field of optical radiation is typically much smaller (several orders of magnitude) than the electric force, the former force is neglected as well. Then, the equation of motion for electrons can be written as follows:

$$m\frac{d^2\mathbf{r}_i}{dt^2} = -e\mathbf{E}(t), \qquad (1.2)$$

where $\mathbf{r}_i$ is the position vector of the i-th electron.

When the Cartesian components of vector $\mathbf{E}(t)$ are time-harmonic functions of angular frequency ω, the last equation can be written in the phasor form

$$-\omega^2 m\hat{\mathbf{r}}_i = -e\hat{\mathbf{E}}, \tag{1.3}$$

where $\hat{\mathbf{r}}_i$ and $\hat{\mathbf{E}}$ are the notations for the phasors of $\mathbf{r}_i(t)$ and $\mathbf{E}(t)$, respectively. From the last equation, we find

$$\hat{\mathbf{r}}_i = \frac{e}{m\omega^2}\hat{\mathbf{E}}. \tag{1.4}$$

Now, by using the definition of the polarization vector $\hat{\mathbf{P}}$, we obtain

$$\hat{\mathbf{P}} = -\sum_{i=1}^{N} e\hat{\mathbf{r}}_i = -\frac{e^2 N}{m\omega^2}\hat{\mathbf{E}}, \tag{1.5}$$

where N is the electron density, i.e., the number of electrons per unit volume.

By using formula (1.5), we derive

$$\hat{\mathbf{D}} = \varepsilon_0\hat{\mathbf{E}} + \hat{\mathbf{P}} = \varepsilon_0\left(1 - \frac{e^2 N}{\varepsilon_0 m\omega^2}\right)\hat{\mathbf{E}}, \tag{1.6}$$

which implies that

$$\hat{\mathbf{D}} = \varepsilon(\omega)\hat{\mathbf{E}}(\omega), \tag{1.7}$$

with

$$\varepsilon(\omega) = \varepsilon_0\left(1 - \frac{\omega_p^2}{\omega^2}\right). \tag{1.8}$$

In the last formula, ω_p is the plasma frequency defined by the formula

$$\omega_p^2 = \frac{e^2 N}{\varepsilon_0 m}. \tag{1.9}$$

It is clear from the dispersion relation (1.8) that

$$\varepsilon(\omega) < 0 \quad \text{if} \quad \omega < \omega_p. \tag{1.10}$$

The derivation of the dispersion relation (1.8) has been based on the greatly simplified equation of motion (1.2). For this reason, it is natural to question its validity and accuracy. To test the latter, we shall compare this dispersion relation with the available experimental data for gold and silver. The

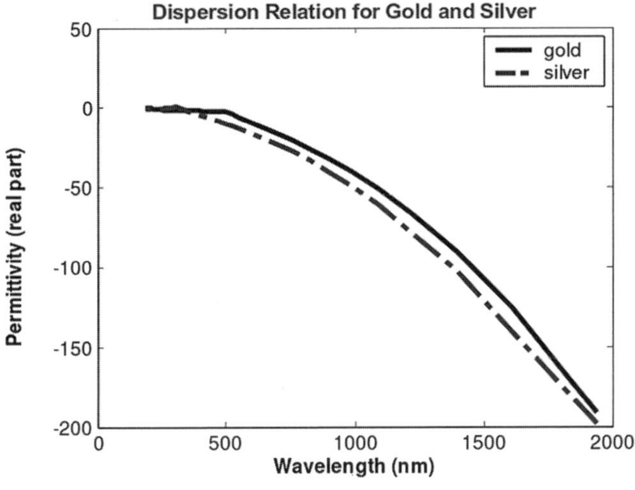

Figure 1.7

remarkable fact is that the electron density N for gold is practically the same as for silver, namely,

$$N_{\text{Au}} = 5.90 \times 10^{22} \text{ cm}^{-3}, \quad N_{\text{Ag}} = 5.86 \times 10^{22} \text{ cm}^{-3}. \tag{1.11}$$

This implies that according to formulas (1.8) and (1.9) the dispersion relation $\varepsilon(\omega)$ must be practically identical for gold and silver in the optical frequency range. By using the experimental data from P. B. Johnson and R. W. Christy [16], the dispersion relations for gold and silver are plotted in Figure 1.7, which reveals that the **real parts** of dielectric permittivities of gold and silver are indeed approximately the same for the free-space wavelength range between 500 nm and 2000 nm. In addition, the comparison between $\varepsilon(\omega)$ computed by using formulas (1.8) and (1.9) and the experimentally measured [16] real parts of dielectric permittivity of gold and silver are shown in Figures 1.8 and 1.9, respectively. These figures suggest that the dispersion relation (1.8)-(1.9) is indeed fairly accurate in the 500 nm-2000 nm wavelength range, and it is in this wavelength range that plasmon resonances in metallic nanoparticles are usually studied.

Formulas (1.8) and (1.9) suggest that the dispersion relation $\varepsilon(\omega)$ can be controlled by manipulating the electron density N. This may be especially attractive in the case of semiconductors where the conduction electron

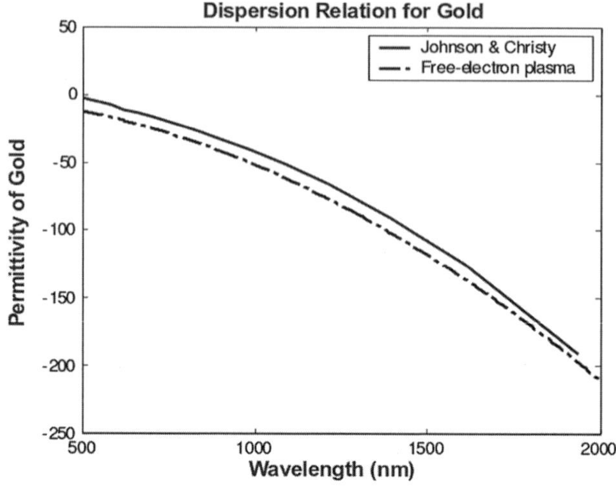

Figure 1.8

density can be manipulated optically. This may open opportunities for optical controllability (optical gating) of plasmon resonances in semiconductor nanoparticles.

The constitutive relation (1.7) is local in the frequency domain. However, due to the dispersion of dielectric permittivity, the constitutive relation between $\mathbf{D}(t)$ and $\mathbf{E}(t)$ is non-local in the time domain. It is instructive and somewhat interesting to find the non-local-in-time constitutive relation for $\mathbf{D}(t)$ and $\mathbf{E}(t)$ which corresponds to the dispersion relation (1.8)-(1.9). The simplest way to do this is to integrate twice the equation of motion (1.2) and to perform the integration by parts in the resulting double integral. This leads to the following expression:

$$\mathbf{r}_i(t) = -\frac{e}{m} \int_0^t (t - \tau)\mathbf{E}(\tau)d\tau. \tag{1.12}$$

Next, by using the same definitions of $\mathbf{P}$ and $\mathbf{D}$ as in formulas (1.5) and (1.6), we arrive at the following constitutive relation:

$$\mathbf{D}(t) = \varepsilon_0 \left[\mathbf{E}(t) + \omega_p^2 \int_0^t (t - \tau)\mathbf{E}(\tau)d\tau \right]. \tag{1.13}$$

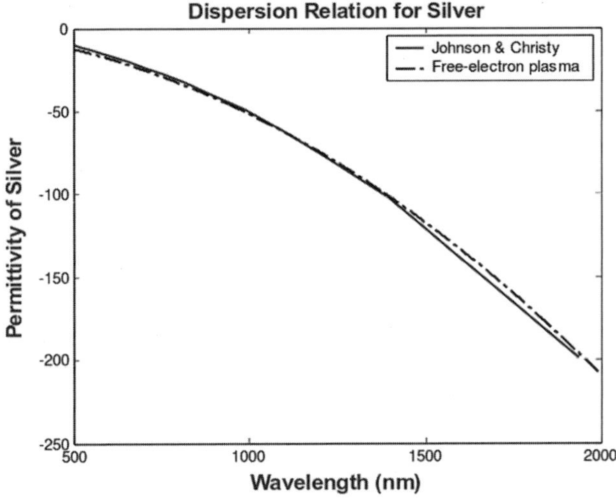

Figure 1.9

In general, non-local time-domain relations between $\mathbf{D}(t)$ and $\mathbf{E}(t)$ can be quite complicated. Mathematically, these non-local-in-time constitutive relations can be treated as pseudo-differential operators [17] and the experimentally measured dispersion relations $\varepsilon(\omega)$ as their symbols. This pseudo-differential operator interpretation of dispersion may be especially promising for the study of spatially dispersive media, but this is beyond the scope of this book.

The greatly simplified equation of motion (1.2) leads to the real-valued dielectric permittivity $\varepsilon(\omega)$ given by formulas (1.8) and (1.9). Experiments reveal that the actual dielectric permittivity of metals in the optical frequency range has an imaginary part as well which accounts for energy losses. It turns out that the equation of motion (1.2) can be modified to obtain the complex-valued dispersion relation $\varepsilon(\omega)$. Indeed, if the effect of electron collisions is modeled by introducing a "friction" term representing some average loss of electron momentum, then this results in the following equation of motion:

$$m\frac{d^2\mathbf{r}_i}{dt^2} - \gamma m\frac{d\mathbf{r}_i}{dt} = -e\mathbf{E}(t). \qquad (1.14)$$

Now, by using the same line of reasoning as in the derivation of formulas (1.8) and (1.9), we arrive at the following dispersion relation

$$\varepsilon(\omega) = \varepsilon_0 \left[1 - \frac{\omega_p^2}{\omega(\omega + j\gamma)} \right], \qquad (j = \sqrt{-1}), \qquad (1.15)$$

which is quite often referred to as the Drude model for dielectric permittivity.

From the last formula, the following expressions for the real ε' and imaginary ε'' parts of dielectric permittivity can be easily derived,

$$\varepsilon' = \varepsilon_0 \left[1 - \frac{\omega_p^2}{\omega^2 + \gamma^2} \right], \qquad (1.16)$$

$$\varepsilon'' = \varepsilon_0 \frac{\omega_p^2 \gamma}{\omega(\omega^2 + \gamma^2)}. \qquad (1.17)$$

In the typical (for plasmon resonances) case when $1 \ll \omega_p^2/(\omega^2 + \gamma^2)$, from the last two formulas we find

$$\frac{\varepsilon'}{\varepsilon''} \simeq -\frac{\omega}{\gamma}. \qquad (1.18)$$

The Drude model will be extensively used in this book for the analytical study of time-dynamics of excitation and dephasing (decay) of plasmon modes. In particular, it will be shown that $1/\gamma$ can be identified as the decay time for the light intensity of plasmon modes in the absence of optical excitation.

The imaginary part ε'' of the dielectric permittivity can be used for computation of power losses in metallic nanoparticles. As discussed before, plasmon resonances result in appreciable enhancement of electric fields inside nanoparticles. This, in turn, results in substantial increase of power losses, which leads to the peaks in extinction (total or forward-scattering) cross section. These peaks of extinction cross sections are used in experiments to identify plasmon resonances and the corresponding resonance frequencies.

As discussed, the real part of dielectric permittivity for dispersive media may assume negative values ($\varepsilon'(\omega) < 0$). This makes the classical formula

$$\bar{w}_e = \frac{1}{4} \varepsilon' \left| \hat{\mathbf{E}} \right|^2 \qquad (1.19)$$

for the time-average stored electric energy density meaningless. It turns out that the last formula can be properly modified to be valid for slightly lossy (transparent) dispersive media [18, 19]. The modified formula is

$$\bar{w}_e = \frac{1}{4} \frac{d\left(\omega\varepsilon'(\omega)\right)}{d\omega} \left| \hat{\mathbf{E}} \right|^2. \qquad (1.20)$$

It is apparent that in the case of non-dispersive media the last formula is reduced to formula (1.19). It is also clear that in accordance with the last formula $\bar{w}_e > 0$ in the case of the dispersion relation described by formula (1.8) as well as by the Drude model (1.16) under the transparency condition $\omega > \gamma$.

1.3 Overview of Book Contents

It is apparent from the previous discussion that the approach to the study of plasmon resonances in metallic nanoparticles adopted in this book is the eigenmode approach. This means that the problem of resonant plasmon modes (and the corresponding resonance frequencies) in lossless metallic nanoparticles is first framed as the eigenvalue problem for specific boundary integral equations. After this problem is solved and plasmon modes are identified, the excitation of these plasmon modes by incident optical radiation is studied along with the radiation corrections. This approach is quite distinct and different from techniques frequently used in scientific literature. Indeed, resonances in metallic nanoparticles are often found experimentally and numerically by using a "trial-and-error" approach, i.e., by probing metallic nanoparticles of complex shapes with optical radiation of various frequencies and polarizations [20]-[23]. The numerical analysis is typically performed by using the finite-difference time-domain (FDTD) technique. It seems to us that the eigenmode technique adopted in this book has important advantages over FDTD and other techniques.

First, the eigenmode technique leads to the direct calculation of the resonance values of dielectric permittivity which can then be used for any dispersion relation of a metallic nanoparticle to immediately find the corresponding resonance frequency. In this way, the properties of plasmon resonances which depend on nanoparticle shapes are clearly separated from those properties which depend on dispersion relations. In contrast, separate FDTD computations have to be performed for different dispersion relations.

Second, the eigenmode technique has analytical capabilities for the time-domain analysis of plasmon resonance modes. As discussed in this book, the eigenmode technique explicitly reveals the coupling conditions between specific plasmon modes and the incident radiation. It leads to analytical formulas for time dynamics of plasmon modes (their excitation and dephasing). It results in analytical expressions for steady-state amplitudes of plasmon resonance modes and for steady-state amplitudes in the case of off-resonance excitations; the latter reveal the sharpness of plasmon resonances. In contrast, FDTD is a purely numerical technique without analytical capabilities.

Furthermore, FDTD requires discretization of three-dimensional space, while the technique advocated in this book is a surface integral equation technique, which requires discretization of two-dimensional boundaries of the nanoparticles. Since only a finite region of three-dimensional space can be discretized and used in computations, FDTD requires the introduction of artificial external boundaries. Special absorbing boundary conditions must be posed on these boundaries to minimize distortions and errors caused by the introduction of artificial boundaries. In the eigenmode technique, no such artificial boundaries are required.

Finally, plasmon resonances occur in dispersive media with non-local-in-time (convolution-type) constitutive relations between electric displacement and electric field. These non-local constitutive relations lead to finite difference schemes with the electric field coupling at all previous time steps. This past history-coupling of the electric field may diminish the effectiveness of FDTD for plasmon resonance computations.

The book consists of four chapters. The detailed review of the book content is given below, chapter by chapter. The review is presented without invoking complicated mathematical formulas, but rather emphasizing the physical aspects of the matter.

Chapter 2 deals with the modal analysis of plasmon resonances in metallic nanoparticles. This analysis is first framed as an eigenvalue problem for boundary integral equations with respect to surface electric charges distributed over nanoparticle boundaries. These virtual (fictitious) charges are introduced on S in free space in order to mimic the boundary conditions which occur for electric fields of plasmon modes on the nanoparticle surface. As a result, these surface charges create in free space the same electric fields **E** which exist for actual plasmon modes in the presence of metallic nanoparticles. This replication of the boundary conditions for **E** leads to the eigenvalue problem for a specific boundary integral equation with respect to surface electric charges which has nonzero solutions only for specific negative values of dielectric permittivity. In practice, metallic nanoparticles are placed on dielectric substrates. These substrates can be accounted for in the mathematical formalism by using the appropriate Green functions in the kernel of the boundary integral equations. In the case of flat substrates, this Green function can be found by using the method of images. Some analytical results for plasmon modes in spherical and ellipsoidal nanoparticles are presented here as well. In particular, it is demonstrated that for any negative value of dielectric permittivity an ellipsoidal nanoparticle with the appropriate aspect ratio can be found that will resonate for this value of permittivity and, consequently, for the corresponding value of frequency (wavelength) of optical radiation. This suggests that ellipsoidal nanoparticles can be used (at least in principle) for the solution of the tunability problem for plasmon resonances.

It turns out that there exists a dual eigenvalue approach to the study of plasmon modes in metallic nanoparticles. In this approach, virtual (fictitious) double layers of electric charges are introduced on S in free space to reproduce on S the same boundary conditions for electric displacement $\mathbf{D}$ which exist for actual plasmon modes in the presence of metallic nanoparticles. This eventually leads to the eigenvalue problem for boundary integral equations for double layer densities. These integral equations are adjoint to the integral equations for simple (single) layers of electric charges.

The detailed study of general properties of the spectrum of the derived boundary integral equations (which is the plasmon spectrum) is then followed. It is demonstrated that for any shape S of nanoparticles the spectrum is discrete and real, and that, as expected, the corresponding values of dielectric permittivity are negative. It is also shown that the spectrum is scale-invariant. Since the kernels of the corresponding integral equations are not symmetric, the eigenfunctions are not orthogonal on S. However, the corresponding electric fields of plasmon modes are orthogonal. Moreover, strong orthogonality conditions hold for plasmon mode electric fields. The term "strong" means that these fields are orthogonal not only in the entire space but they are also separately orthogonal inside and outside the nanoparticles.

Unique spectral properties of plasmon resonances occur for metallic nanowires, i.e., in the two-dimensional case. In this case, for any cross-sectional shape of nanowires, the spectrum of the corresponding integral equations consists of pairs (couples) of positive and negative eigenvalues of the same magnitude. In other words, each positive eigenvalue has its twin (counterpart) of negative value and the same magnitude. This phenomenon of twin spectrum is due to unique symmetry properties of the mathematical formulation which appear in the two-dimensional case due to the existence of the stream function which is conjugate to the electric potential. The twin spectrum phenomenon results in two distinct bands of plasmon resonances with relative dielectric permittivities which are reciprocal to one another for different bands.

It is apparent that the tunability of plasmon resonances may be of value in many applications. It turns out that a wide range of tunability can be achieved by using metallic nanoshells [24, 25] and controlling the plasmon resonance frequencies via adjustment of the shell thickness. It is demonstrated in the chapter that the plasmon mode analysis in metallic nanoshells can be reduced to a generalized eigenvalue problem for specific boundary (integral) equations and the detailed analysis of plasmon spectrum in metallic nanoshells is carried out.

The chapter is concluded with the discussion of interesting relation of eigenvalue treatment of plasmon resonances to the Riemann hypothesis. Namely, it is pointed out that for any sufficiently regular shapes of nanowire cross sections Fredholm determinants $D(\lambda)$ of integral equations used in the calculation of plasmon spectrum form the class of entire functions with properties that have been conjectured or proved for the Riemann xi-function $\xi(\lambda)$. For this reason, the problem can be posed to find such curved boundary L of nanowire cross section that

$$\xi(\lambda) = D(\lambda), \tag{1.21}$$

which will prove the Riemann hypothesis. The last formula is consistent with a spectral interpretation of the Riemann hypothesis asserted in the Hilbert-Pólya conjecture. Finally, a promising approach to the proof of the Riemann hypothesis based on the Grommer theorem (and suggested by formula (1.21)) is discussed as well.

Chapter 3 deals with analytical and numerical analysis of plasmon resonances in metallic nanoparticles. Before proceeding with the discussion of numerical issues, analytical solutions for plasmon modes in certain nanostructures are presented. There are two reasons why these analytical solutions are of importance. First, these analytical solutions are used in the book for testing the accuracy of numerical computations. Second, these analytical solutions for plasmon modes are of interest in their own right because they are derived for nanostructures that have appeared (or will appear) in various applications of plasmon resonances. These analytical solutions are obtained by using the method of separation of variables in various coordinate systems. It is known that in this method possible solutions are expressed as products of functions, each of which depends only on one of the variables of the coordinate system used. These product solutions are plasmon modes, and they are realized for specific negative values of dielectric permittivity of metallic nanoparticles or nanowires. These specific negative (resonance) values of permittivity are found from the interface boundary conditions.

Next, the numerical issues of the implementation of the analysis of plasmon resonances in metallic nanoparticles presented in Chapter 2 are discussed. The discussion starts with the description of the special discretization technique for the eigenvalue problem for boundary integral equations with respect to single layers of electric charges. This discretization technique completely circumvents the difficulties associated with (weak) singularities of the kernels of the integral equations as well as with singularities of surface electric charge densities on edges and corners of nanoparticle boundaries.

Discretizations of boundary integral equations used in the analysis of plasmon resonances result in fully populated matrices. For this reason, these discretized equations are computationally expensive to solve. This is especially true when many nanoparticles are involved in the design of metallic nanostructures. Fortunately, since the fully populated matrices are generated through discretizations of integral operators with $1/r$-type kernels, this computational problem can be considerably alleviated by using the fast multipole method [26] introduced by V. Rokhlin and L. Greengard. This method greatly speeds up the matrix-vector multiplications, resulting at most in $O(N)$ computational cost for $N \times N$ fully populated matrices. The central idea of the fast multipole method is to split the computations for near-field and far-field regions and then utilize factorized representations for kernels in these regions in terms of spherical harmonics which follow from "addition theorem" expansions.

The theoretical discussion of discretization techniques is illustrated in the chapter by numerous computational examples. Many of these computations are carried out for metallic nanoparticle arrangements that have been already studied experimentally due to their physical interest or possible technological applications. The comparison of computational results with available experimental data reveals the coincidence which is mostly within five to seven percent. This agreement with experimental data is quite good and, by and large, within the accuracy of measuring techniques. This agreement can also be construed as some justification for using the macroscopically measured dispersion relation $\varepsilon(\omega)$ at the nanoscale, i.e., in the constitutive relation for metallic nanoparticles.

The chapter also contains the discussion of a numerical technique for the solution of inhomogeneous boundary integral equations with singular kernels. Such integral equations arise in the analysis of extinction cross sections. The unique features of this numerical technique are 1) unique solvability of discretized equations for any mesh and any surface geometry of metallic nanoparticles and 2) guaranteed convergence of the numerical technique which is valid for any singularity in the kernels of the integral equations. The unique solvability of discretized equations, convergence of the approximate solutions to the exact solutions and the rate of this convergence are established under only one natural condition of unique solvability of the inhomogeneous boundary integral equations for any right-hand sides (any "forcing" terms).

The chapter is concluded with the discussion of exact absorbing boundary conditions for finite-difference time-domain (FDTD) analysis of scattering problems. It may seem that this discussion is somewhat out of place in this book which advocates alternatives to the FDTD technique. Nevertheless, it is included in the book for two reasons: its distinctness and the extensive current use of the FDTD technique in the study of plasmon resonances. The central point of the presented discussion is to use the time-domain versions of Kirchhoff or Stratton-Chu formulas as Dirichlet-type "absorbing" boundary conditions that can be posed on **any** artificial boundary that encloses a scattering nanoparticle. These boundary conditions are exact and they are updated as computations proceed. The physical foundation for this type of absorbing boundary conditions is the retardation phenomena. The Stratton-Chu-type boundary conditions can be naturally coupled with the Yee finite-difference scheme, while the Kirchhoff-type boundary conditions can be naturally coupled with the standard explicit finite-difference scheme for scalar wave equations written for each Cartesian component of electric (or magnetic) field.

Chapter 4 deals with several topics. The first one is the radiation corrections to the electrostatic plasmon mode analysis presented in Chapter 2. These radiation corrections are mathematically treated as perturbations with respect to small parameters β, which are the ratios of particle dimensions (their diameters) to the free-space wavelengths. To compute these corrections, homogeneous Maxwell equations are written in a scaled (perturbative) form which explicitly includes the small parameter β. Then, source-free solutions of these Maxwell equations and the values of dielectric permittivities for which they occur are expanded in power series with respect to the small parameter β. In this way, the appropriate boundary value problems are derived for zero-, first- and second-order terms of these power expansions. It turns out, as expected, that the zero-order approximation coincides with the electrostatic formulation of plasmon modes studied in detail in Chapter 2. Then, by using the principle of normal solvability of integral equations, the first- and second-order radiation corrections with respect to β are studied. Namely, it is demonstrated that the first-order radiation corrections to the resonance value of dielectric permittivity (hence, to resonance frequencies) and to plasmon mode electric fields are always equal to zero for any shape of metallic nanoparticle. Meanwhile, the first-order radiation corrections for the magnetic fields of plasmon modes are not equal to zero, and the explicit analytical expressions for these corrections are found. Finally, the explicit second-order radiation corrections are derived for resonance values of dielectric permittivity of plasmon modes in terms of the zero-order electrostatic solutions for these modes. It is demonstrated that in the particular case of spherical nanoparticles these second-order radiation corrections coincide with the radiation corrections obtained from the classical Mie theory.

The chapter also contains the analysis of extinction cross section of metallic nanoparticles subject to optical radiation. This analysis is performed by using the same perturbation technique that has been employed for the calculation of radiation corrections. Namely, it is demonstrated that the main (zero-order) term in the power expansion of the solution of the scattering problem is electrostatic in nature, and it can be efficiently computed by using inhomogeneous boundary integral equations that have the same kernels as the integral equations used in Chapter 2 for the analysis of plasmon modes. Then the algorithm for the calculation of first- and second-order terms in β for scattered electric and magnetic field is presented. Once the scattered fields are found, the extinction cross section of metallic nanoparticles can be computed by using the optical theorem. This theorem relates the extinction cross sections to the far electric field scattered in the forward direction. It is worthwhile to remark that calculations of extinction cross section are helpful in the analysis of experimental data because the plasmon resonances in metallic nanoparticles are usually studied through the measurements of this cross section.

The analysis of extinction cross section is concluded with the discussion of nanoparticle-structured plasmon waveguides of light. These waveguides consist of arrays of metallic nanoparticles with their plasmon resonance frequency in the range of optical waveguiding. These nanoparticle-structured waveguides are quite promising for light guiding and bending at the nanoscale. In the chapter, a technique for calculations of extinction cross sections as well as guiding (resonance) frequencies of these waveguides is presented. This technique is based on the boundary integral equation method and, in this sense, it is the further extension of the techniques already developed in the book.

The next discussion in the chapter deals with temporal analysis of plasmon resonances in metallic nanoparticles, which is the least studied area of plasmonics. The purpose of this analysis is to develop techniques for the quantitative characterization of time dynamics of excitation and dephasing (decay) of plasmon resonance modes. The central mathematical element of this analysis is the biorthogonal expansion of bound electric charges induced on nanoparticle boundaries by incident radiation. This expansion is given in terms of eigenfunctions of boundary integral equations which are surface electric charges corresponding to specific plasmon modes. Conceptually, this analysis is quite similar to the traditional approach to the study of time dynamics of excitation of resonant cavities. In the above biorthogonal expansion, the expansion coefficients depend on time. It turns out that simple analytical expressions can be derived for Fourier transforms of these coefficients in terms of dispersion relation $\varepsilon(\omega)$ and the Fourier transform of the electric field of incident radiation. These analytical expressions for the expansion coefficients reveal that the incident radiation is efficiently coupled only to dipole plasmon modes when the incident electric fields have the same directions as the plasmon mode dipoles. It is also observed that the Fourier transforms of the plasmon mode expansion coefficients exhibit resonance behavior at plasmon resonance frequencies. This fact is used to derive the analytical formulas for the steady-state amplitudes of plasmon modes in terms of the real and imaginary parts of dielectric permittivity, amplitude of incident field and its spatial orientation with respect to the dipole moments of the plasmon modes. These formulas reveal that the quality of plasmon resonances (i.e., the local enhancement of incident electric fields and light intensity) is fully controlled by the ratio of the real to imaginary parts of dielectric permittivity at resonance frequencies of plasmon modes. The analytical expressions for the steady-state amplitudes of plasmon modes for off-resonance excitations are derived as well. These expressions are presented in terms of real and imaginary parts of dielectric permittivity evaluated at off-resonance excitation frequencies. The obtained formulas are instrumental for the assessment of the width (sharpness) of plasmon resonances.

The chapter further contains the extensive analytical study of time dynamics of excitation and dephasing of specific plasmon modes. This study has been carried out by using the Drude model for the permittivity dispersion relation. The analytical calculations reveal the nearly linear growth of the amplitude of resonance plasmon modes at the beginning of the excitation process. This growth eventually saturates due to ohmic losses accounted for by the imaginary part of dielectric permittivity. This type of excitation dynamics is generic for slightly lossy resonance systems. The performed analysis also suggests that the reciprocal of the Drude damping factor can be identified with the dephasing time for the light intensity of plasmon modes. In particular, for gold and silver nanoparticles the dephasing time is in the range of 5-12 femtoseconds, which is consistent with the available experimental data. Other comparisons of obtained results with the published experimental data are presented as well.

Finally, the chapter deals with several selective applications of plasmon resonances and some of these applications may be novel. First, plasmon resonance enhancement of Faraday rotation in thin garnet films is discussed. It is known that on the macroscopic level magnetic garnets act as gyrotropic media which discriminate between right-handed and left-handed polarizations of light. This results in the Faraday rotation. However, on the microscopic level, magneto-optic effects are controlled by spin-orbit interaction (coupling) whose Hamiltonian depends on local electric fields. These fields can be optically induced by exciting plasmon resonances in metallic nanoparticles embedded in garnets, and these induced fields may eventually lead to the enhancement of magneto-optic effects. In this way, the plasmon resonances in garnet-embedded nanoparticles can be utilized for the enhancement of the Faraday effect as well as for the probing of the origin of this effect on the fundamental microscopic (quantum mechanical) level. Some encouraging experimental results in this direction are presented in the chapter. Namely, it is reported that garnet films have been grown by liquid phase epitaxy over (100)-oriented gadolinium gallium substrates populated with gold nanoparticles and the noticeable enhancement of Faraday rotation in such films has been observed.

The applications of plasmon resonances to heat-assisted and all-optical magnetic recording are next discussed in the chapter. Heat-assisted magnetic recording (HAMR) is currently the focus of considerable research and technological interest. A central issue of high density HAMR is the development of optical sources of high intensity and nanometer resolution. The proper profiling (shaping) of optical spots is also very important in order to reduce the collateral heating of adjacent recorded bits. It is demonstrated in the chapter that plasmon resonances in metallic nanoparticles and perforated nanofilms can be efficiently used for nano-localization of high-intensity optical radiation and proper profiling of optical spots.

It has been recently demonstrated [27] that magnetization reversal can be consistently achieved by using only circularly polarized laser pulses, that is, without applying any external magnetic fields. In this all-optical switching, the direction of reversed magnetization is controlled by the helicity of circularly polarized light, which acts as an "effective" dc magnetic field aligned with the light propagation direction. The performed experiments [27] have demonstrated femtosecond magnetization reversals of 100 μm spots on magnetic media. This all-optical magnetization switching will be technologically feasible for magnetic recording only if the techniques for delivery of nanoscale-focused circularly polarized light are developed. It is demonstrated in the book that the focusing of light at the nanoscale and, at the same time, the preservation of its circular polarization can be simultaneously achieved by exciting circularly polarized plasmon modes in metallic nanoparticles with uniaxial (rotational) symmetry. This rotational symmetry leads to the existence of such circularly polarized plasmon modes and their effective coupling to circularly polarized optical radiation.

Next, the discussion of the electromagnetic mechanism of surface enhanced Raman scattering (SERS) is presented in the chapter. The essence of the electromagnetic mechanism of SERS can be briefly outlined as follows. The incoming optical radiation excites a desired plasmon resonance mode in a metallic (usually silver) nanoparticle or a cluster of nanoparticles. This results in strong electric field on the particle boundaries. This strong electric field causes molecules adsorbed on metal surfaces to radiate at the Raman-shifted frequency. This molecule radiation excites in turn a resonance plasmon mode in the nanoparticles at the (slightly shifted from resonance) Raman frequency, which may significantly enhance the overall Raman scattering. It is apparent from the presented description of the electromagnetic mechanism of SERS that the fine-tuning of the following conditions must be performed to achieve very strong SERS enhancement: a) the incident optical radiation matches the resonance frequency and the polarization of the desired plasmon resonance mode, b) adsorbed molecules are in the region where the plasmon mode electric field is the strongest, and c) molecule radiation at the Raman-shifted frequency closely matches the resonance frequency of the plasmon mode and is efficiently coupled to it. Various ways to achieve the fine-tuning of the above three conditions are discussed in the chapter along with supporting computational results.

The chapter is concluded with the discussion of optical controllability of plasmon resonances in semiconductor and metallic nanoparticles and the plausible plasmon resonance mechanism for ball lightning formation. These applications of plasmon resonances are speculative in nature at the current state of affairs; nevertheless, they are interesting and quite promising. The optical controllability of plasmon resonances is attractive because it may eventually lead to the development of controllable nanoscale light switches and all-optical nanotransistors. In such devices, one light beam can be used to generate conduction electrons in semiconductor nanoparticles and, in this way, to properly manipulate the dispersion relation $\varepsilon(\omega)$ and drive semiconductor nanoparticles into conditions where the desired plasmon mode can be resonantly excited by another light beam. Another possibility of optical controllability of plasmon resonances in metallic nanoparticles is to control in time the polarization of incident optical radiation and, consequently, its coupling to a desired plasmon mode. Finally, it is discussed that the theory of plasmon resonances and, particularly, their scale invariance, may shed light on the nucleation and growth of ball lightning, the physical mechanism of accumulation of electromagnetic energy in this lightning, as well as provide an explanation for its "ball" shape.

References

[1] T.J. Silva and S. Schultz, Review of Scientific Instruments 67, 715 (1996).

[2] R.M. Stockle, Y.D. Suh, V. Deckert, and R. Zenobi, Chemical Physics Letters 318, 131 (2000).

[3] J.C. Hulteen, D.A. Treichel, M.T. Smith, M.L. Duval, T.R. Jensen, and R.P. Van Dyune, Journal of Physical Chemistry B 103, 3854 (1999).

[4] R. Elghanian, J.J. Storhoff, R.C. Mucic, R.L. Letsinger, and C.A. Mirkin, Science 277, 1078 (1997).

[5] S. Schultz, D.R. Smith, J.J. Mock, and D.A. Schultz, Proceedings of the National Academy of Science U.S.A. February 1, 97, 996 (2000).

[6] M. Moskovits, Review of Modern Physics 57, 783 (1985).

[7] S. Nie and S.R. Emory, Science 275, 1102 (1997).

[8] H. Xu, E.J. Bjerneld, M. Käll, and L. Börjesson, Physical Review Letters 83, 4357 (1999).

[9] J.P. Kottman, O.J.F. Martin, D.R. Smith, and S. Schultz, Chemical Physics Letters 341, 1 (2001).

[10] J.R. Krenn, A. Dereux, J.C. Weeber, E. Bourillot, Y. Lacroute, J.-P. Goudonnet, G. Schider, W. Gotschy, A. Leitner, F.R. Aussenegg, and C. Girard, Physical Review Letters 82, 2590 (1999).

[11] J.C. Weeber, A. Dereux, C. Girard, J.R. Krenn, and J.P. Goudonnet, Physical Review B 60, 9061 (1999).

[12] T. Yatsui, M. Kourogi, and M. Ohtsu, Applied Physics Letters 79, 4583 (2001).

[13] J. Tominaga, C. Mihalcea, D. Büchel, H. Fukuda, T. Nakano, N. Atoda, H. Fuji, and T. Kikukawa, Applied Physics Letters 78, 2417 (2001).

[14] J. Tominaga, T. Nakano, and N. Atoda, Proceedings of the SPIE 3467, 282 (1998).

[15] L. Men, J. Tominaga, H. Fuji, Q. Chen, and N. Atoda, Proceedings of the SPIE 4085, 204 (2001).

[16] P.B. Johnson and R.W. Christy, Physical Review B 6, 4370 (1972).

[17] M.A. Shubin, Pseudodifferential Operators and Spectral Theory, Springer Verlag (2001).

[18] L.D. Landau and E.M. Lifshitz, Electrodynamics of Continuous Media, Pergamon Press, Oxford (1960).

[19] G. Diener, Annalen der Physik 7, 639 (1998).

[20] M.R. Pufall, A. Berger, and S. Schultz, Journal of Applied Physics 81, 5689 (1997).

[21] E. Prodan and P.J. Nordlander, Journal of Chemical Physics 120, 5444 (2004).

[22] W.M. Saj, Optics Express 13, 4818 (2005).

[23] H. Wang, D.W. Brandt, F. Le, P. Nordlander, and N.J. Halas, Nano Letters 6, 827 (2006).

[24] R.D. Averitt, S.L. Westcott, and N.J. Halas, Journal of the Optical Society of America B 16, 1824 (1999).

[25] E. Prodan, C. Radloff, N.J. Halas, and P. Nordlander, Science 302, 419 (2003).

[26] L. Greengard and V. Rokhlin, Journal of Computational Physics 73(2), 325 (1987).

[27] C.D. Stanciu, F. Hansteen, A.V. Kimel, A. Kirilyuk, A. Tsukamoto, A. Itoh, and Th. Rasing, Physical Review Letters 99, 047601 (2007).

Chapter 2

Modal Analysis of Plasmon Resonances in Nanoparticles

2.1   Plasmon Resonances as an Eigenvalue Problem

In the study of plasmon resonances in nanoparticles, we shall follow the traditional approach when all losses are first neglected and resonance frequencies and resonance modes are found for lossless and sourceless systems. In the case when free-space wavelengths are appreciably larger than nanoparticle dimensions, this approach leads to the consideration of plasmon resonances in the electrostatic limit (see, for instance, [1]-[7]). In this limit, all radiation losses as well as "ohmic" losses (due to the imaginary part $\varepsilon''(\omega)$ of dielectric permittivity) are neglected. Radiation corrections to the electrostatic theory of plasmon resonances are studied in Chapter 4.

To start the discussion, consider a nanoparticle of arbitrary shape with permittivity ε in free space with permittivity ε_0 (see Figure 2.1). We are interested to find such negative values of ε for which source-free electrostatic fields may exist. These fields satisfy the following equations inside (V^+) and outside (V^-) the particle, respectively:

$$\operatorname{div} \mathbf{E}^+ = 0, \tag{2.1}$$

$$\operatorname{curl} \mathbf{E}^+ = 0, \tag{2.2}$$

$$\operatorname{div} \mathbf{E}^- = 0, \tag{2.3}$$

$$\operatorname{curl} \mathbf{E}^- = 0, \tag{2.4}$$

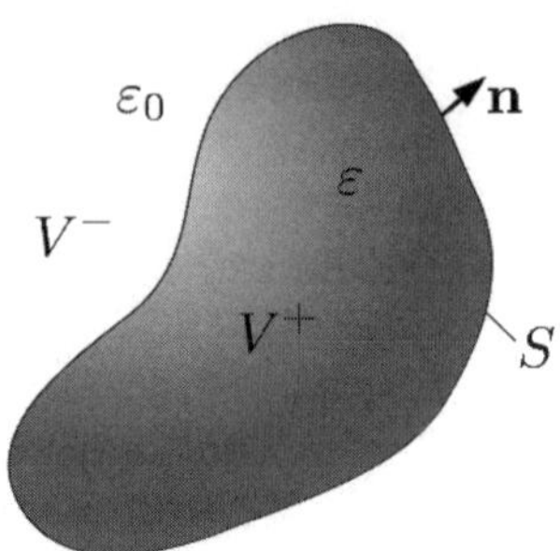

Figure 2.1

as well as the following interface conditions on the particle boundary S:

$$\mathbf{n} \times \mathbf{E}^+ = \mathbf{n} \times \mathbf{E}^-, \tag{2.5}$$

$$\varepsilon \mathbf{n} \cdot \mathbf{E}^+ = \varepsilon_0 \mathbf{n} \cdot \mathbf{E}^-, \tag{2.6}$$

where $\mathbf{n}$ is a unit outward normal to S.

Finally, the electric field vanishes at infinity,

$$\mathbf{E}^- = 0. \tag{2.7}$$

Equations (2.2) and (2.4) are satisfied when plasmon fields $\mathbf{E}^\pm$ are described in terms of electric potential φ,

$$\mathbf{E}^\pm = -\operatorname{grad} \varphi^\pm. \tag{2.8}$$

Furthermore, equations (2.1) and (2.3) as well as the boundary conditions (2.5)-(2.6) and the condition at infinity (2.7) will be satisfied if the potential φ is the solution of the following homogeneous boundary value problem:

$$\nabla^2 \varphi^+ = 0 \ \text{ in } V^+, \tag{2.9}$$

$$\nabla^2 \varphi^- = 0 \ \text{ in } V^-, \tag{2.10}$$

$$\varphi^+ = \varphi^- \ \text{ on } S, \tag{2.11}$$

$$\varepsilon \frac{\partial \varphi^+}{\partial n} = \varepsilon_0 \frac{\partial \varphi^-}{\partial n} \ \text{ on } S, \tag{2.12}$$

and

$$\varphi^-(\infty) = 0. \tag{2.13}$$

Next, we shall demonstrate that nonzero solutions of boundary value problem (2.9)-(2.13) may exist only if the dielectric permittivity of the nanoparticle is negative,

$$\varepsilon < 0. \tag{2.14}$$

To this end, we shall use the "divergence theorem",

$$\int_{V^\pm} \operatorname{div} \mathbf{a}^\pm dv = \pm \oint_S \mathbf{n} \cdot \mathbf{a}^\pm ds, \tag{2.15}$$

where sign "−" in front of the surface integral reflects that $\mathbf{n}$ is the inward normal for V^-.

By setting

$$\mathbf{a}^\pm = \varphi^\pm \operatorname{grad} \varphi^\pm, \tag{2.16}$$

we derive from formula (2.15) that

$$\varepsilon \int_{V^+} \left(\varphi^+ \nabla^2 \varphi^+ + \left| \operatorname{grad} \varphi^+ \right|^2 \right) dv = \varepsilon \oint_S \varphi^+ \frac{\partial \varphi^+}{\partial n} ds, \tag{2.17}$$

$$\varepsilon_0 \int_{V^-} \left(\varphi^- \nabla^2 \varphi^- + \left| \operatorname{grad} \varphi^- \right|^2 \right) dv = -\varepsilon_0 \oint_S \varphi^- \frac{\partial \varphi^-}{\partial n} ds. \tag{2.18}$$

By adding equations (2.17) and (2.18) and taking into account equations (2.9) and (2.10) as well as boundary conditions (2.11) and (2.12), we obtain

$$\varepsilon \int_{V^+} \left| \operatorname{grad} \varphi^+ \right|^2 dv + \varepsilon_0 \int_{V^-} \left| \operatorname{grad} \varphi^- \right|^2 dv = 0. \tag{2.19}$$

It is obvious from the last formula and the condition (2.13) that if $\varepsilon > 0$ then

$$\left| \operatorname{grad} \varphi^\pm \right| = 0 \ \text{ and } \ \varphi^\pm = 0. \tag{2.20}$$

This means that plasmon resonances may exist only if the permittivity of the nanoparticle is negative. It is also clear from (2.19) that the negative resonance values of ε are determined by the ratios of the Dirichlet integrals in V^- and V^+, respectively.

To find the specific negative values of ε for which source-free plasmon fields may exist, we replace the permittivity of the nanoparticle by ε_0 (see Figure 2.2) and introduce on its boundary S virtual (fictitious) surface electric charges of density $\sigma(M)$. The electric field and potential created by these

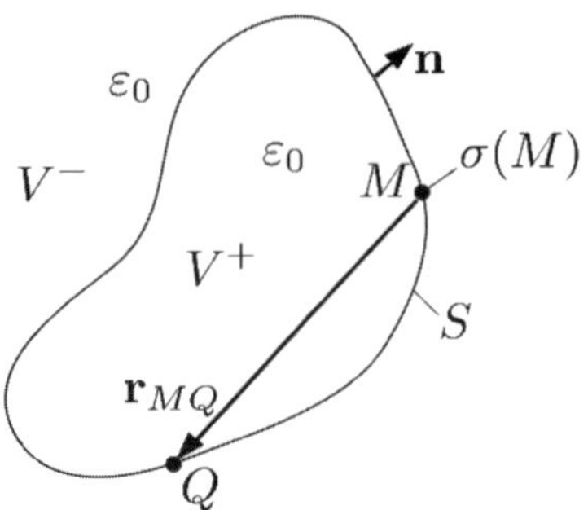

Figure 2.2

charges are respectively given by the formulas

$$\mathbf{E}(Q) = \frac{1}{4\pi\varepsilon_0} \oint_S \sigma(M)\frac{\mathbf{r}_{MQ}}{r_{MQ}^3} ds_M, \tag{2.21}$$

$$\varphi(Q) = \frac{1}{4\pi\varepsilon_0} \oint_S \frac{\sigma(M)}{r_{MQ}} ds_M, \tag{2.22}$$

where $\mathbf{r}_{MQ}$ is the vector directed from the "integration" point M to the "observation" point Q and its length r_{MQ} is equal to the distance between M and Q. It is apparent that for any distribution of surface charges $\sigma(M)$ the electric field (2.21) and the potential (2.22) respectively satisfy the equations (2.1)-(2.4) and (2.9)-(2.10) as well as boundary conditions (2.5) and (2.11) along with conditions at infinity (2.7) and (2.13).

Next, we shall discuss under what conditions on $\sigma(M)$ the boundary condition (2.6) for normal components of electric field (or, equivalently, boundary condition (2.12)) will be satisfied. In other words, we shall discuss under what conditions on surface charges $\sigma(M)$ the electric fields created by these charges in homogeneous space (Figure 2.2) may replicate (may mimic) the source-free plasmon fields of actual nanoparticles (Figure 2.1). To this end, we shall derive the formulas for the normal component of electric fields created by $\sigma(M)$ on S. To do this, we partition S into two pieces $S - \Delta S$ and ΔS; the latter contains observation point Q (see Figure 2.2). Then the electric field $\mathbf{E}(Q)$ can be represented as

$$\mathbf{E}(Q) = \mathbf{E}'(Q) + \frac{1}{4\pi\varepsilon_0} \int_{S-\Delta S} \sigma(M)\frac{\mathbf{r}_{MQ}}{r_{MQ}^3} ds_M, \tag{2.23}$$

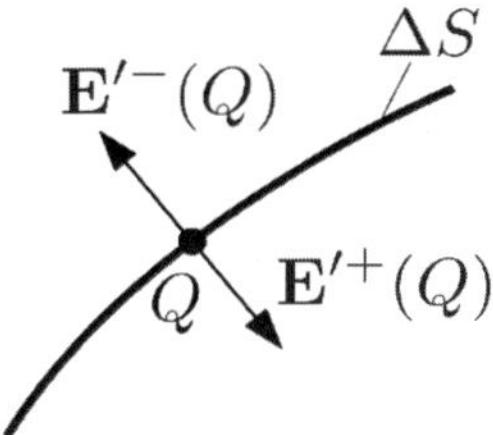

Figure 2.3

where $\mathbf{E}'(Q)$ is the electric field created by charges on ΔS. For this field, we have

$$\mathbf{n}_Q \cdot \left(\mathbf{E}'^{-}(Q) - \mathbf{E}'^{+}(Q) \right) = \frac{\sigma(Q)}{\varepsilon_0}. \tag{2.24}$$

It is clear that the normal components of the field $\mathbf{E}'(Q)$ created by charges on ΔS have opposite directions (see Figure 2.3). For sufficiently small and almost flat ΔS, the electric field is nearly odd-symmetric with respect to the interior and exterior sides of S. As a result of this symmetry, we have

$$\mathbf{n}_Q \cdot \mathbf{E}'^{-}(Q) \simeq -\mathbf{n}_Q \cdot \mathbf{E}'^{+}(Q). \tag{2.25}$$

From formulas (2.24) and (2.25) we find

$$\mathbf{n}_Q \cdot \mathbf{E}'^{-}(Q) \simeq \frac{\sigma(Q)}{2\varepsilon_0}, \tag{2.26}$$

$$\mathbf{n}_Q \cdot \mathbf{E}'^{+}(Q) \simeq -\frac{\sigma(Q)}{2\varepsilon_0}. \tag{2.27}$$

By substituting formulas (2.26) and (2.27) into equation (2.23), we derive

$$\mathbf{n}_Q \cdot \mathbf{E}^{-}(Q) \simeq \frac{\sigma(Q)}{2\varepsilon_0} + \frac{1}{4\pi\varepsilon_0} \int_{S-\Delta S} \sigma(M) \frac{\mathbf{r}_{MQ} \cdot \mathbf{n}_Q}{r_{MQ}^3} ds_M, \tag{2.28}$$

$$\mathbf{n}_Q \cdot \mathbf{E}^{+}(Q) \simeq -\frac{\sigma(Q)}{2\varepsilon_0} + \frac{1}{4\pi\varepsilon_0} \int_{S-\Delta S} \sigma(M) \frac{\mathbf{r}_{MQ} \cdot \mathbf{n}_Q}{r_{MQ}^3} ds_M. \tag{2.29}$$

The smaller ΔS, the more accurate the symmetry condition (2.25) and the more accurate formulas (2.28) and (2.29). Consequently, in the limit of infinitesimally small ΔS we obtain the exact formulas

$$\mathbf{n}_Q \cdot \mathbf{E}^-(Q) = \frac{\sigma(Q)}{2\varepsilon_0} + \frac{1}{4\pi\varepsilon_0} \oint_S \sigma(M) \frac{\mathbf{r}_{MQ} \cdot \mathbf{n}_Q}{r_{MQ}^3} ds_M, \qquad (2.30)$$

$$\mathbf{n}_Q \cdot \mathbf{E}^+(Q) = -\frac{\sigma(Q)}{2\varepsilon_0} + \frac{1}{4\pi\varepsilon_0} \oint_S \sigma(M) \frac{\mathbf{r}_{MQ} \cdot \mathbf{n}_Q}{r_{MQ}^3} ds_M. \qquad (2.31)$$

It is apparent from the presented derivation that the cause of discontinuity of the normal components of electric fields expressed by formulas (2.30) and (2.31) is the "self"-field due to the "elementary" charge distributed in the infinitesimally small and almost flat vicinity of Q. The lines of this "self"-field are almost normal to S at Q and have opposite directions on interior and exterior sides of S. It is clear from the presented discussion that formulas (2.30)-(2.31) are also valid when S consists of several closed surfaces.

Formulas (2.30) and (2.31) lead to the following expressions for the normal derivatives on S of the potential (2.22):

$$\frac{\partial \varphi^\pm}{\partial n}(Q) = \pm\frac{\sigma(Q)}{2\varepsilon_0} - \frac{1}{4\pi\varepsilon_0} \oint_S \sigma(M) \frac{\mathbf{r}_{MQ} \cdot \mathbf{n}_Q}{r_{MQ}^3} ds_M. \qquad (2.32)$$

By substituting formulas (2.31) and (2.32) into the boundary condition (2.6), we arrive at

$$\frac{\varepsilon}{2\varepsilon_0}\sigma(Q) - \frac{\varepsilon}{4\pi\varepsilon_0} \oint_S \sigma(M) \frac{\mathbf{r}_{MQ} \cdot \mathbf{n}_Q}{r_{MQ}^3} ds_M$$
$$= -\frac{\sigma(Q)}{2} - \frac{1}{4\pi} \oint_S \sigma(M) \frac{\mathbf{r}_{MQ} \cdot \mathbf{n}_Q}{r_{MQ}^3} ds_M. \qquad (2.33)$$

Now, after simple transformation we obtain

$$\sigma(Q) = \frac{\lambda}{2\pi} \oint_S \sigma(M) \frac{\mathbf{r}_{MQ} \cdot \mathbf{n}_Q}{r_{MQ}^3} ds_M, \qquad (2.34)$$

where

$$\lambda = \frac{\varepsilon - \varepsilon_0}{\varepsilon + \varepsilon_0}. \qquad (2.35)$$

Thus, the electric field created by the distribution of virtual electric charges $\sigma(M)$ over S will replicate in free homogeneous space (Figure 2.2) the source-less plasmon fields of the nanoparticle with permittivity ε (Figure 2.1) if these charges satisfy the homogeneous integral equation (2.34).* It turns out that this homogeneous integral equation has nonzero solutions $\sigma_k(M)$ only for special values of $\lambda = \lambda_k$, which are the eigenvalues, i.e.,

$$\sigma_k(Q) = \frac{\lambda_k}{2\pi} \oint_S \sigma_k(M) \frac{\mathbf{r}_{MQ} \cdot \mathbf{n}_Q}{r_{MQ}^3} ds_M. \tag{2.36}$$

In this way, the problem of analysis of sourceless plasmon fields is framed as the eigenvalue problem for integral equation (2.36). If this problem is solved and λ_k as well as $\sigma_k(M)$ are found, then by using the formula (see (2.35))

$$\lambda_k - \frac{\varepsilon_k - \varepsilon_0}{\varepsilon_k + \varepsilon_0} \tag{2.37}$$

the corresponding value ε_k of dielectric permittivity of nanoparticles for which sourceless plasmon fields may exist can be determined. Then, by using the known dispersion relation $\varepsilon(\omega)$, the frequencies ω_k at which these plasmon fields can be excited can be found by using the equation

$$\varepsilon'(\omega_k) = \varepsilon_k, \tag{2.38}$$

where $\varepsilon'(\omega)$ stands for the real part of $\varepsilon(\omega)$. Finally, by using eigenfunctions $\sigma_k(M)$ the corresponding sourceless plasmon fields can be calculated by the formula

$$\mathbf{E}_k(Q) = \frac{1}{4\pi\varepsilon_0} \oint_S \sigma_k(M) \frac{\mathbf{r}_{MQ} \cdot \mathbf{n}_Q}{r_{MQ}^3} ds_M. \tag{2.39}$$

These sourceless plasmon fields are resonance plasmon modes that can be excited at the frequency ω_k.

It will be shown in section 2.3 that for any shape S of nanoparticles all the eigenvalues of integral equation (2.36) are real and satisfy the inequality

$$|\lambda_k| > 1. \tag{2.40}$$

Since according to (2.37)

$$\varepsilon_k = \varepsilon_0 \frac{1 + \lambda_k}{1 - \lambda_k}, \tag{2.41}$$

the inequality (2.40) implies that (as expected) ε_k must be negative.

---

*Integral equation (2.34) has been used for the analysis of plasmon resonances in metallic nanoparticles in [1]-[4]. Similar integral equations have been extensively used for the analysis of static fields in [8, 9].

It is important to point out that, as evident from integral equation (2.36), resonance values ε_k of nanoparticle permittivity depend only on the nanoparticle shape. As soon as ε_k are determined through the numerical solution of integral equation (2.36), they can be used in formula (2.38) to find the corresponding resonance frequency for any material (any dispersion relation) of the nanoparticle. In this way, the properties of plasmon resonance modes that depend on nanoparticle shapes are clearly separated from those which depend on material properties.

It is worthwhile to remark that for all plasmon modes

$$\oint_S \sigma_k(M)ds_M = 0. \tag{2.42}$$

Indeed, by integrating both sides of (2.36) with respect to Q, we find

$$\oint_S \sigma_k(Q)ds_Q = \frac{\lambda_k}{2\pi} \oint_S \sigma_k(M) \left[\oint_S \frac{\mathbf{r}_{MQ} \cdot \mathbf{n}_Q}{r_{MQ}^3} ds_Q \right] ds_M. \tag{2.43}$$

By taking into account the well-known fact that for $Q \in S$

$$\oint_S \frac{\mathbf{r}_{MQ} \cdot \mathbf{n}_Q}{r_{MQ}^3} ds_Q = 2\pi, \tag{2.44}$$

from (2.43) we obtain

$$\oint_S \sigma_k(Q)ds_Q = \lambda_k \oint_S \sigma_k(M)ds_M. \tag{2.45}$$

Formulas (2.45) and (2.40) imply the equality (2.42).

Numerical techniques for the solution of integral equation (2.36) along with many numerical examples will be discussed in the next chapter.

It is instructive to demonstrate at this point that simple analytical expressions for ε_k and $\sigma_k(M)$ can be derived from the integral equation (2.34) in the case of spherical nanoparticles. Indeed, when S is a sphere of radius R, we find (see Figure 2.4)

$$\frac{\mathbf{r}_{MQ} \cdot \mathbf{n}_Q}{r_{MQ}^3} = \frac{\cos \alpha}{r_{MQ}^2} = \frac{1}{2Rr_{MQ}}. \tag{2.46}$$

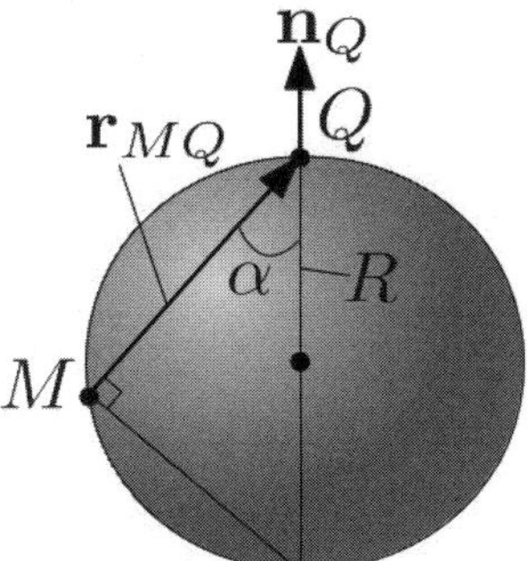

Figure 2.4

Consequently, for spherical nanoparticles the integral equation (2.34) has the form

$$\sigma(Q) = \frac{\lambda}{4\pi R} \oint_S \frac{\sigma(M)}{r_{MQ}} ds_M. \tag{2.47}$$

Now, by introducing spherical coordinates (R, θ, ϕ) and (R, θ', ϕ') for points Q and M, respectively, and by using the addition theorem relation [10]

$$\frac{1}{r_{MQ}} = \frac{4\pi}{R} \sum_{k_0=0}^{\infty} \sum_{m_0=-k_0}^{k_0} \frac{1}{2k_0 + 1} Y_{k_0 m_0}(\theta, \phi) Y_{k_0 m_0}^*(\theta', \phi'), \tag{2.48}$$

the integral equation (2.47) can be transformed as follows:

$$\sigma(\theta, \phi) =$$
$$\lambda \int_0^{2\pi} \left[\int_0^{\pi} \sigma(\theta', \phi') \sum_{k_0=0}^{\infty} \sum_{m_0=-k_0}^{k_0} \frac{1}{2k_0 + 1} Y_{k_0 m_0}(\theta, \phi) Y_{k_0 m_0}^*(\theta', \phi') \sin \theta' d\theta' \right] d\phi',$$
$$\tag{2.49}$$

where symbol Y_{km} is used for the notation of spherical harmonics. These spherical harmonics satisfy the following well-known orthogonality conditions:

$$\int_0^{2\pi} \left[\int_0^{\pi} Y_{km}(\theta', \phi') Y_{k_0 m_0}^*(\theta', \phi') \sin \theta' d\theta' \right] d\phi' = \delta_{kk_0} \delta_{mm_0}, \tag{2.50}$$

where δ_{kk_0} and δ_{mm_0} are Kronecker deltas. By using the above orthogonality conditions and by substituting

$$\boxed{\sigma_{km}(\theta, \phi) = Y_{km}(\theta, \phi), \quad (m = -k, ..., 0, ..., k)} \tag{2.51}$$

into integral equation (2.49), we find

$$\sigma_{km}(\theta, \phi) = \frac{\lambda}{2k + 1} \sigma_{km}(\theta, \phi). \tag{2.52}$$

This suggests that $\sigma_{km}(\theta, \phi)$ defined by formula (2.51) are the eigenfunctions corresponding to eigenvalues

$$\boxed{\lambda_k = 2k + 1.} \tag{2.53}$$

This implies that plasmon resonances occur for the following values of dielectric permittivity of the spherical nanoparticle:

$$\boxed{\varepsilon_k = -\varepsilon_0 \frac{k + 1}{k}.} \tag{2.54}$$

The last result is consistent with the classical Mie theory [11, 12].

It is apparent from formula (2.51) that eigenvalues λ_k have $(2k + 1)$-"geometric" multiplicity, i.e., for each λ_k there are $2k+1$ linearly independent eigenfunctions. This fact is the consequence of geometrical symmetry of spherical nanoparticles, and it is a particular case of the general situation when geometric symmetry of nanoparticles results in geometric multiplicity of eigenvalues and resonance values ε_k of dielectric permittivity. It follows from formulas (2.51) and (2.54) that for

$$\varepsilon_1 = -2\varepsilon_0 \tag{2.55}$$

there are three linearly independent plasmon modes corresponding to the eigenfunctions $\sigma_{1m}(\theta, \phi) = Y_{1m}(\theta, \phi)$. It is easy to see that for these plasmon modes the electric fields are spatially uniform and mutually orthogonal inside the spherical nanoparticles. These plasmon modes have nonzero dipole moments and they are excited at about 360 nm and 500 nm wavelengths of incident radiation for silver and gold nanoparticles, respectively.

The fact that the electric fields of these plasmon modes are spatially uniform inside the spherical nanoparticles suggests that these modes can be

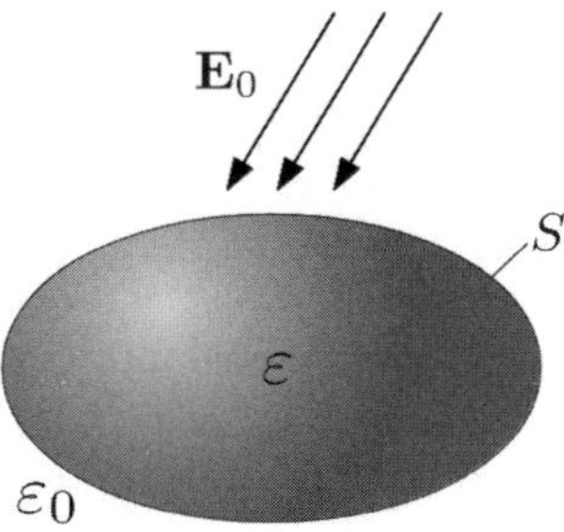

Figure 2.5

found without using integral equations (2.36). Indeed, we shall demonstrate now that this can be done for the more general problem of ellipsoidal nanoparticles. To this end, we consider an ellipsoidal nanoparticle with depolarizing coefficients N_k, $(k = 1, 2, 3)$ subject to spatially uniform electric field $\mathbf{E}_0$ (see Figure 2.5). The electric field $\mathbf{E}^+$ inside the nanoparticle is the superposition of the applied field $\mathbf{E}_0$ and the depolarizing field $\mathbf{E}'$, which can be expressed in terms of polarization vector $\mathbf{P}$ and depolarizing coefficients as

$$\mathbf{E}' = \frac{1}{\varepsilon_0}\hat{N}\mathbf{P}, \tag{2.56}$$

where

$$\hat{N} = \begin{pmatrix} N_1 & 0 & 0 \\ 0 & N_2 & 0 \\ 0 & 0 & N_3 \end{pmatrix}. \tag{2.57}$$

This leads to the formula

$$\mathbf{E}^+ + \frac{1}{\varepsilon_0}\hat{N}\mathbf{P} = \mathbf{E}_0. \tag{2.58}$$

By using relation

$$\mathbf{D} = \varepsilon\mathbf{E}^+ = \varepsilon_0\mathbf{E}^+ + \mathbf{P}, \tag{2.59}$$

we find

$$\mathbf{P} = (\varepsilon - \varepsilon_0)\mathbf{E}^+. \tag{2.60}$$

By substituting formula (2.60) into equation (2.58), we end up with

$$\mathbf{E}^+ + \frac{\varepsilon - \varepsilon_0}{\varepsilon_0}\hat{N}\mathbf{E}^+ = \mathbf{E}_0. \tag{2.61}$$

The last equation can be written in the matrix form as follows:

$$\begin{pmatrix} 1 + \frac{\varepsilon - \varepsilon_0}{\varepsilon_0} N_1 & 0 & 0 \\ 0 & 1 + \frac{\varepsilon - \varepsilon_0}{\varepsilon_0} N_2 & 0 \\ 0 & 0 & 1 + \frac{\varepsilon - \varepsilon_0}{\varepsilon_0} N_3 \end{pmatrix} \mathbf{E}^+ = \mathbf{E}_0. \tag{2.62}$$

From the last equation we find that the source-free and spatially uniform electric fields $\mathbf{E}^+$ inside the ellipsoidal nanoparticle must satisfy the homogeneous equations

$$\begin{pmatrix} 1 + \frac{\varepsilon - \varepsilon_0}{\varepsilon_0} N_1 & 0 & 0 \\ 0 & 1 + \frac{\varepsilon - \varepsilon_0}{\varepsilon_0} N_2 & 0 \\ 0 & 0 & 1 + \frac{\varepsilon - \varepsilon_0}{\varepsilon_0} N_3 \end{pmatrix} \mathbf{E}^+ = 0. \tag{2.63}$$

Nonzero solutions of the last homogeneous equations exist for such values of ε that the determinant of the diagonal matrix in (2.63) is equal to zero. This will be the case if at least one of the following three equations is satisfied:

$$1 + \frac{\varepsilon_k - \varepsilon_0}{\varepsilon_0} N_k = 0, \quad (k = 1, 2, 3). \tag{2.64}$$

It follows now that spatially uniform plasmon resonance modes may exist only for the special values of permittivity of the ellipsoidal nanoparticle given by the formula

$$\boxed{\varepsilon_k = \varepsilon_0 \left(1 - \frac{1}{N_k} \right), \quad (k = 1, 2, 3).} \tag{2.65}$$

It is clear from equations (2.63) that for each ε_k given by formula (2.65) the corresponding nonzero solution of homogeneous equations (2.63) has the form

$$\mathbf{E}_k^+ = \begin{pmatrix} \delta_{1k} E_{x1}^+ \\ \delta_{2k} E_{x2}^+ \\ \delta_{3k} E_{x3}^+ \end{pmatrix}, \quad (k = 1, 2, 3), \tag{2.66}$$

where δ_{ik} are Kronecker deltas. This means that plasmon modes $\mathbf{E}_k^+$ corresponding to different values of ε_k are mutually orthogonal inside ellipsoidal nanoparticles.

It is known that

$$N_1 + N_2 + N_3 = 1. \tag{2.67}$$

For this reason, $N_k < 1$ and according to formula (2.65)

$$\varepsilon_k < 0, \quad (k = 1, 2, 3). \tag{2.68}$$

In applications, it may be of interest to design nanoparticles that will resonate at a specified wavelength (or frequency) of incident radiation. This is the tuning problem. It turns out that this problem can be solved by using ellipsoidal nanoparticles. Indeed, by using the dispersion relation $\varepsilon'(\omega)$ for ellipsoidal nanoparticle material and the desired resonance frequency ω_k we find the desired resonance value of dielectric permittivity,

$$\varepsilon_k = \varepsilon'(\omega_k). \tag{2.69}$$

By using this value of ε_k in formula (2.65), we obtain

$$N_k = \frac{\varepsilon_0}{\varepsilon_0 - \varepsilon_k} = \frac{\varepsilon_0}{\varepsilon_0 - \varepsilon'(\omega_k)}. \tag{2.70}$$

Thus, an ellipsoidal nanoparticle of appropriate aspect ratio (appropriate N_k) can be always found that will resonate for the desired frequency ω_k (or wavelength).

In the case of spherical nanoparticles, $N_1 = N_2 = N_3$, which implies according to formula (2.67) that

$$N_k = \frac{1}{3}, \quad (k = 1, 2, 3). \tag{2.71}$$

By using the last formula in equation (2.65), we find

$$\varepsilon_k = -2\varepsilon_0, \quad (k = 1, 2, 3), \tag{2.72}$$

which is consistent with the previous result (see (2.55)). In the case of dispersion relation (see section 1.2)

$$\varepsilon'(\omega) = \varepsilon_0 \left(1 - \frac{\omega_p^2}{\omega^2} \right), \tag{2.73}$$

it follows from formula (2.70) that

$$N_k = \frac{\omega_k^2}{\omega_p^2}, \quad (k = 1, 2, 3). \tag{2.74}$$

By substituting the last expressions into formula (2.67), we find

$$\sum_{k=1}^{3} \omega_k^2 = \omega_p^2. \tag{2.75}$$

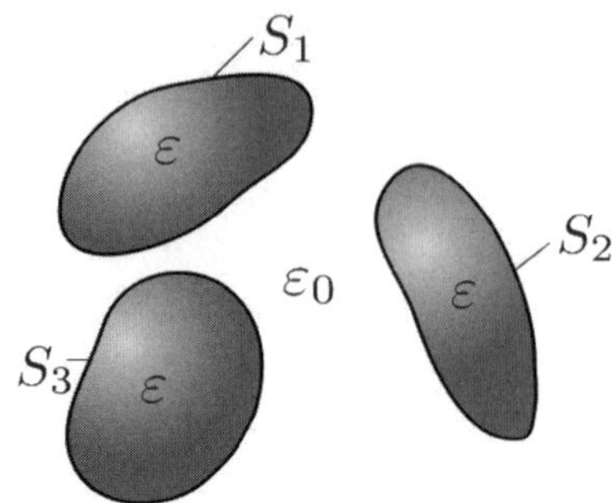

Figure 2.6

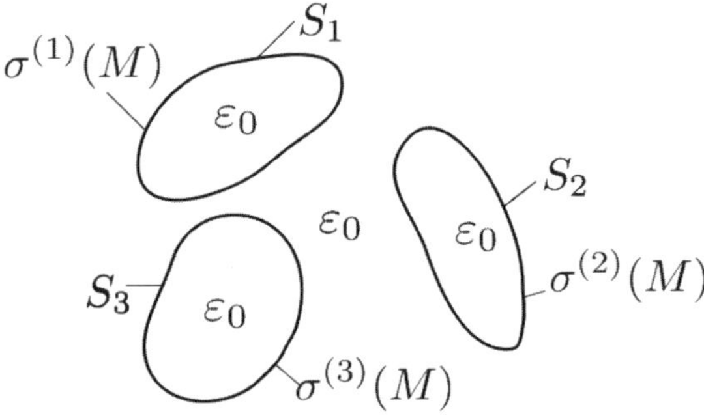

Figure 2.7

Now, it is appropriate to return to the discussion of integral equation (2.34) and consider its two important extensions. The first extension deals with plasmon mode analysis in systems of several nanoparticles (see Figure 2.6, where only three nanoparticles are shown). Plasmon resonances in multiple nanoparticles are utilized in SERS research. As before, we replace permittivity ε of the nanoparticles by ε_0 (see Figure 2.7) and introduce on their boundaries S_i virtual surface electric charges of densities $\sigma^{(i)}(M)$. The electric field and potential created by these charges are respectively given by the formulas

$$\mathbf{E}(Q) = \frac{1}{4\pi\varepsilon_0} \sum_{i=1}^{n} \oint_{S_i} \sigma^{(i)}(M) \frac{\mathbf{r}_{MQ}}{r_{MQ}^3} ds_M, \tag{2.76}$$

$$\varphi(Q) = \frac{1}{4\pi\varepsilon_0} \sum_{i=1}^{n} \oint_{S_i} \frac{\sigma^{(i)}(M)}{r_{MQ}} ds_M. \tag{2.77}$$

By using exactly the same line of reasoning as in the derivation of equation (2.34), it can be shown that the electric field created by distributions of virtual electric charges $\sigma^{(i)}(M)$ in free homogeneous space will replicate source-free plasmon fields of actual nanoparticles with permittivity ε if these charges satisfy the following homogeneous integral equations:

$$\sigma^{(j)}(Q) = \frac{\lambda}{2\pi} \sum_{i-1}^{n} \oint_{S_i} \sigma^{(i)}(M) \frac{\mathbf{r}_{MQ} \cdot \mathbf{n}_Q}{r_{MQ}^3} ds_M, \quad (j = 1, 2, ..., n), \tag{2.78}$$

where as before

$$\lambda = \frac{\varepsilon - \varepsilon_0}{\varepsilon + \varepsilon_0}. \tag{2.79}$$

Thus, the problem of analysis of plasmon modes in multiple nanoparticles is reduced to the eigenvalue problem for a system (2.78) of integral equations. Homogeneous equations (2.78) have nonzero solutions $\sigma_k^{(i)}(M)$ only for special values of $\lambda = \lambda_k$, which are the eigenvalues, i.e.,

$$\boxed{\sigma_k^{(j)}(Q) = \frac{\lambda_k}{2\pi} \sum_{i=1}^{n} \oint_{S_i} \sigma_k^{(i)}(M) \frac{\mathbf{r}_{MQ} \cdot \mathbf{n}_Q}{r_{MQ}^3} ds_M, \quad (j = 1, 2, ..., n),} \tag{2.80}$$

$$\boxed{\lambda_k = \frac{\varepsilon_k - \varepsilon_0}{\varepsilon_k + \varepsilon_0}.} \tag{2.81}$$

When the eigenvalues λ_k and eigenfunctions $\sigma_k^{(i)}(M)$ are found, then the electric field of the corresponding plasmon modes can be computed by using the formula

$$\boxed{\mathbf{E}_k(Q) = \frac{1}{4\pi\varepsilon_0} \sum_{i=1}^{n} \oint_{S_i} \sigma_k^{(i)}(M) \frac{\mathbf{r}_{MQ}}{r_{MQ}^3} ds_M,} \tag{2.82}$$

while the resonance frequencies are determined by using formula (2.38). Finally, by using the known formula

$$\oint_{S_i} \frac{\mathbf{r}_{MQ} \cdot \mathbf{n}_Q}{r_{MQ}^3} ds_M = \begin{cases} 0, & \text{if } Q \notin S_i \\ 2\pi, & \text{if } Q \in S_i \end{cases} \tag{2.83}$$

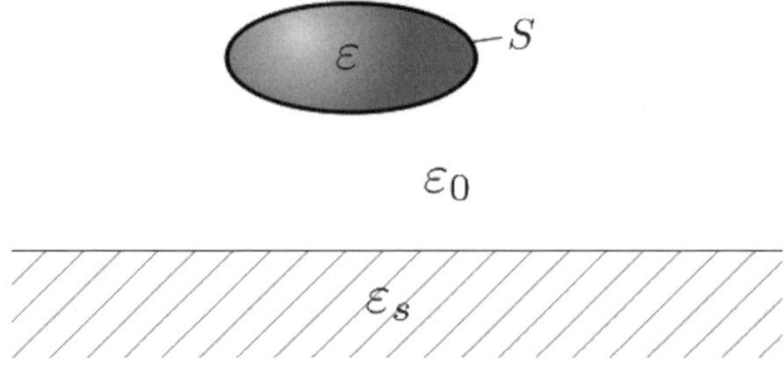

Figure 2.8

and by integrating both sides of integral equations (2.80) with respect to Q over S_j $(j = 1, 2, ..., n)$, we derive

$$\oint_{S_i} \sigma_k^{(i)}(M)ds_M = 0, \quad (i = 1, 2, ..., n), \tag{2.84}$$

which is the generalization of formula (2.42) to the multi-particle case.

Now, we proceed to the discussion of the second extension of integral equation (2.34) to the case when nanoparticles are located not in free space but on some substrate with dielectric permittivity ε_s (Figure 2.8). This is a typical case in many applications, and the substrate thickness is usually much larger than the geometric dimensions of the nanoparticles. As a result, it can be assumed with sufficient accuracy that the thickness of the dielectric substrate is infinite. Under this assumption, the field of the electric charges above the substrate can be evaluated by using the method of images. This means that if we replace the permittivity ε of the nanoparticle by ε_0 and introduce virtual electric charges $\sigma(M)$ distributed over the nanoparticle boundary S (see Figure 2.9), then the electric field created by these charges above the flat substrate is given by the formula

$$\mathbf{E}(Q) = -\frac{1}{4\pi\varepsilon_0} \oint_S \sigma(M)\,\mathrm{grad}_Q G(Q, M)ds_M. \tag{2.85}$$

Here, according to the method of images, the Green function $G(Q, M)$ is computed as

$$G(Q, M) = \frac{1}{r_{MQ}} - \frac{\varepsilon_s - \varepsilon_0}{\varepsilon_s + \varepsilon_0}\frac{1}{r_{M'Q}}, \tag{2.86}$$

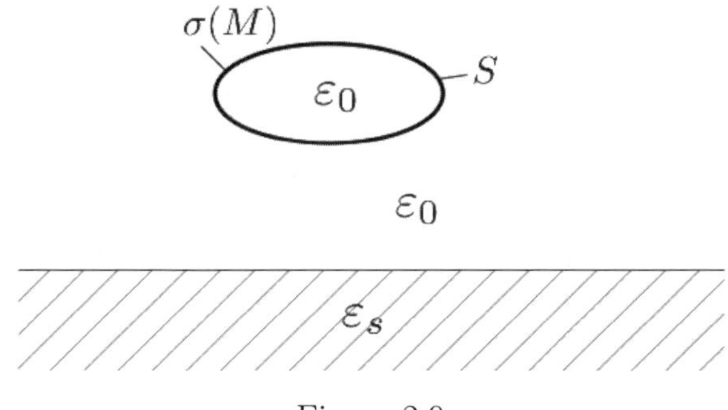

Figure 2.9

with M' being the mirror image of M with respect to the flat substrate boundary.

By using the same line of reasoning as in the derivation of integral equation (2.34), it is straightforward to show that electric charges $\sigma(M)$ create the electric field which replicates the source-free plasmon electric field of the nanoparticle on the dielectric substrate if $\sigma(M)$ are the eigenfunctions of the integral equation

$$\boxed{\sigma_k(Q) = -\frac{\lambda_k}{2\pi} \oint_S \sigma_k(M)\mathbf{n}_Q \cdot \mathrm{grad}_Q G(Q, M) ds_M,} \qquad (2.87)$$

where as before

$$\lambda_k = \frac{\varepsilon_k - \varepsilon_0}{\varepsilon_k + \varepsilon_0}. \qquad (2.88)$$

In the case of multiple nanoparticles placed on a dielectric substrate, equations (2.85) and (2.87) are naturally modified as follows (see (2.76) and (2.80)):

$$\mathbf{E}(Q) = -\frac{1}{4\pi\varepsilon_0} \sum_{i=1}^{n} \oint_{S_i} \sigma^{(i)}(M)\, \mathrm{grad}_Q G(Q, M) ds_M, \qquad (2.89)$$

$$\sigma_k^{(j)}(Q) = -\frac{\lambda_k}{2\pi} \sum_{i=1}^{n} \oint_{S_i} \sigma^{(i)}(M)\mathbf{n}_Q \cdot \mathrm{grad}_Q G(Q, M) ds_M, \quad (j = 1, 2, ..., n).$$

$$(2.90)$$

Up to this point, the discussion has been concerned with three-dimensional problems. However, the obtained results can be easily modified

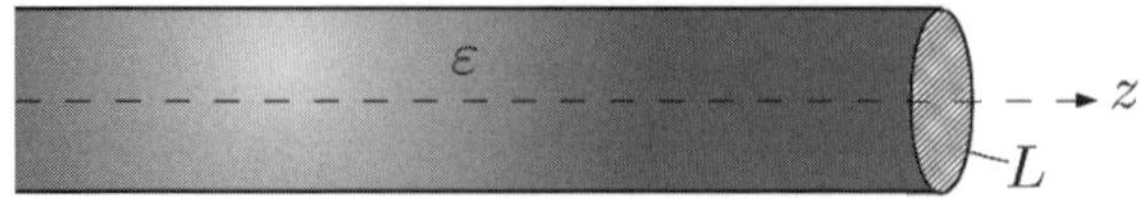

Figure 2.10

to be used for two-dimensional problems when plasmon resonance modes in nanowires are of interest. These modes are excited by plane electromagnetic waves whose directions of propagation and electric fields are parallel to the nanowire cross-sectional planes. The corresponding eigenvalue problem for boundary integral equations is stated as follows:

$$\sigma_k(Q) = \frac{\lambda_k}{\pi} \oint_L \sigma_k(M) \frac{\mathbf{r}_{MQ} \cdot \mathbf{n}_Q}{r_{MQ}^2} dl_M, \tag{2.91}$$

where L is the boundary of the nanowire cross section (see Figure 2.10) and λ_k is given as before by the formula (2.88). When $\sigma_k(M)$ and λ_k are found, then the electric field of the plasmon mode and its potential are computed by using the formulas

$$\mathbf{E}_k(Q) = \frac{1}{2\pi\varepsilon_0} \oint_L \sigma_k(M) \frac{\mathbf{r}_{MQ}}{r_{MQ}^2} dl_M, \tag{2.92}$$

$$\varphi_k(Q) = -\frac{1}{2\pi\varepsilon_0} \oint_L \sigma_k(M) \ln r_{MQ} dl_M. \tag{2.93}$$

It is worthwhile to point out that three-dimensional integral equation (2.36) (or (2.34)) has the kernel $\frac{\mathbf{r}_{MQ} \cdot \mathbf{n}_Q}{2\pi r_{MQ}^3}$ with weak (i.e., integrable in the usual sense) singularity as it is clear from formula (2.46). In contrast, the kernel $\frac{\mathbf{r}_{MQ} \cdot \mathbf{n}_Q}{\pi r_{MQ}^2}$ of two-dimensional integral equation (2.91) is a bounded and continuous function on L if L is sufficiently smooth. The latter can be beneficial in numerical computations.

It is clear from the presented discussion that the analysis of plasmon modes in nanoparticles of arbitrary shape can be framed as an eigenvalue problem for specific boundary integral equations. For this reason, it is fitting and proper to review at this point the main facts of the theory of these equations, because these facts will be extensively used throughout this book.

Consider integral equation of the second kind

$$\sigma(Q) - \lambda \oint_S \sigma(M) K(Q, M) ds_M = f(Q), \tag{2.94}$$

where $K(Q, M)$ is called the kernel of the integral equation, while $f(Q)$ is its right-hand side. This equation can be written in the operator form

$$\sigma - \lambda \hat{K} \sigma = f, \tag{2.95}$$

where $\hat{K}$ is the notation for the integral operator in (2.94), i.e.,

$$\hat{K}\sigma = \oint_S \sigma(M) K(Q, M) ds_M. \tag{2.96}$$

The homogeneous integral equation corresponding to (2.94) is

$$\sigma(Q) = \lambda \oint_S \sigma(M) K(Q, M) ds_M. \tag{2.97}$$

The nonzero solutions $\sigma_k(M)$ of this equation are called eigenfunctions and the corresponding values of $\lambda = \lambda_k$ for which these solutions exist are called eigenvalues. By comparing equations (2.34) and (2.97) we conclude that the kernel of equation (2.34) is

$$K(Q, M) = \frac{\mathbf{r}_{MQ} \cdot \mathbf{n}_Q}{2\pi r_{MQ}^3}. \tag{2.98}$$

Along with integral equation (2.94), it is customary to consider the adjoint equation

$$\tau(Q) - \overline{\lambda} \oint_S \tau(M) \overline{K(M, Q)} ds_M = g(Q). \tag{2.99}$$

Here, $\overline{\lambda}$ is the complex conjugate of λ, and the kernel $\overline{K(M, Q)}$ is obtained by permutation of "integration" and "observation" points and subsequent complex conjugation. In the operator form, the adjoint integral equation can be written as follows:

$$\sigma - \overline{\lambda} \hat{K}^* \tau = g, \tag{2.100}$$

where $\hat{K}^*$ is the adjoint operator

$$\hat{K}^* \tau = \oint_S \tau(M) \overline{K(M, Q)} ds_M. \tag{2.101}$$

The homogeneous integral equation corresponding to (2.99) is

$$\tau(Q) = \overline{\lambda} \oint_S \tau(M) \overline{K(M, Q)} ds_M. \tag{2.102}$$

It is apparent from (2.98) that

$$\overline{K(M,Q)} = \frac{\mathbf{r}_{QM} \cdot \mathbf{n}_M}{2\pi r_{QM}^3}.$$ (2.103)

In the next section, the dual approach to the analysis of plasmon modes in nanoparticles is presented. In this approach, the plasmon mode analysis is framed as the eigenvalue problem for the adjoint to (2.34) integral equation with the kernel defined by formula (2.103).

In the analysis of integral equations it is useful to introduce the Hilbert space $\mathcal{L}_2(S)$ of square-summable functions with the inner product

$$\langle \sigma, \tau \rangle = \oint_S \sigma(M)\overline{\tau(M)}ds_M.$$ (2.104)

It is easy to verify that in this space, the following relation is valid for operators $\hat{K}$ and $\hat{K}^*$:

$$\left\langle \hat{K}\sigma, \tau \right\rangle = \left\langle \sigma, \hat{K}^*\tau \right\rangle.$$ (2.105)

It is clear that operator $\hat{K}$ with the kernel (2.98) is not Hermitian (not self-adjoint). However, as will be demonstrated in section 2.3, this operator has many properties of self-adjoint operators, and it actually becomes a self-adjoint operator in the properly chosen "energy" space.

As has been discussed before, the kernel of integral equation (2.36) has weak singularity while the kernel of equation (2.91) is bounded. For such kernels, the corresponding integral operators are compact [13, 14] and, for this reason, the Fredholm theory is valid for the integral equations with these kernels. This theory can be viewed as a generalization of linear algebra to infinite dimensions. The basic facts of this theory are presented in the following theorems.

<u>Spectral Theorem.</u> There exists only a finite or countable set of eigenvalues λ_k and

$$\lim_{k \to \infty} |\lambda_k| = \infty.$$ (2.106)

This means that the spectrum of compact integral operators is discrete.

<u>Existence Theorem.</u> If the homogeneous integral equation (2.97) has only the trivial zero-solution, then there exists the unique solution of inhomogeneous equation (2.94) for any right-hand side $f(Q)$.

In other words, this theorem states that uniqueness implies existence.

<u>Zero-Index Theorem.</u> The numbers of linearly independent nonzero solutions of homogeneous equation (2.97) and its adjoint equation (2.102) are the same.

The difference between the numbers of nontrivial linearly independent solutions of a homogeneous integral equation and its adjoint is called the index. The last theorem states that for compact integral operators the index is equal to zero.

<u>Normal Solvability Theorem.</u> If λ is the eigenvalue ($\lambda = \lambda_k$), then integral equation (2.94) is solvable (i.e., its solution exists) if and only if the right-hand side $f(Q)$ of this equation is orthogonal (normal) in the sense of inner product (2.104) to all linearly independent eigenfunctions of the adjoint homogeneous integral equation (2.102) at $\overline{\lambda} = \overline{\lambda}_k$.

This normal solvability property of integral equations will be instrumental in Chapter 4 for the calculation of radiation corrections to the electrostatic theory of plasmon modes in nanoparticles.

2.2   Dual Formulation

In the previous section, the analysis of plasmon modes in nanoparticles has been framed as the eigenvalue problem for boundary integral equation (2.34). This has been achieved by replicating electric fields of plasmon modes by electric fields in homogeneous space created by surface electric charges distributed over nanoparticle boundaries. In this section, the dual formulation is developed in which electric displacement fields of plasmon modes are replicated by electric displacement fields created in homogeneous space by double layers of electric charges distributed over nanoparticle boundaries. This leads to the eigenvalue problem for the boundary integral equation which is adjoint to equation (2.34).

To start the discussion, consider source-free electric displacement fields of plasmon modes in a generic nanoparticle shown in Figure 2.1. These fields satisfy the following equations inside (V^+) and outside (V^-) the nanoparticle, respectively,

$$\operatorname{div} \mathbf{D}^+ = 0, \tag{2.107}$$

$$\operatorname{curl} \mathbf{D}^+ = 0, \tag{2.108}$$

$$\operatorname{div} \mathbf{D}^- = 0, \tag{2.109}$$

$$\operatorname{curl} \mathbf{D}^- = 0. \tag{2.110}$$

These displacement fields are also subject to the following boundary conditions on S:

$$\frac{1}{\varepsilon}\mathbf{n} \times \mathbf{D}^+ = \frac{1}{\varepsilon_0}\mathbf{n} \times \mathbf{D}^-, \tag{2.111}$$

$$\mathbf{n} \cdot \mathbf{D}^+ = \mathbf{n} \cdot \mathbf{D}^-. \tag{2.112}$$

Finally, the electric displacement vanishes at infinity:

$$\mathbf{D}^-(\infty) = 0. \tag{2.113}$$

Equations (2.108) and (2.110) are satisfied when plasmon fields $\mathbf{D}^\pm$ are described in terms of "displacement" potential:

$$\mathbf{D}^\pm = -\text{grad}\ \Phi^\pm. \tag{2.114}$$

It is easy to check out that equations (2.107) and (2.109) as well as the boundary conditions (2.111) and (2.112) and the condition at infinity (2.113) will be satisfied as well if the potential Φ is the solution of the following boundary value problem:

$$\nabla^2\Phi^+ = 0 \ \text{ in } V^+, \tag{2.115}$$

$$\nabla^2\Phi^- = 0 \ \text{ in } V^-, \tag{2.116}$$

$$\frac{\Phi^+}{\varepsilon} = \frac{\Phi^-}{\varepsilon_0} \ \text{ on } S, \tag{2.117}$$

$$\frac{\partial\Phi^+}{\partial n} = \frac{\partial\Phi^-}{\partial n} \ \text{ on } S, \tag{2.118}$$

and

$$\Phi^-(\infty) = 0. \tag{2.119}$$

By using the same line of reasoning as in section 2.1 (see formulas (2.15)-(2.20)), it can be proved that nonzero solution of the boundary value problem (2.115)-(2.119) may exist only for negative values of dielectric permittivity ε of nanoparticles. These nonzero solutions and corresponding electric displacement fields represent resonance plasmon modes. To find these specific negative values of ε and the plasmon modes, we replace the permittivity

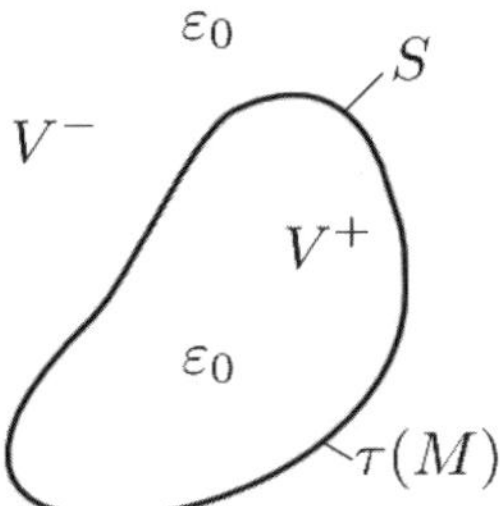

Figure 2.11

of the nanoparticle by ε_0 (see Figure 2.11) and introduce on nanoparticle boundary S a double layer of electric charges of moment density $\tau(M)$. The potential Φ created by this double layer of electric charges is given by the formula

$$\Phi(Q) = \frac{1}{4\pi} \oint_S \tau(M) \frac{\mathbf{r}_{QM} \cdot \mathbf{n}_M}{r_{QM}^3} ds_M, \tag{2.120}$$

where all notations have the same meaning as in the previous section.

It is our intention to demonstrate that the double layer density $\tau(M)$ can be chosen in such a way that the displacement fields created by the double charge layer in homogeneous space will replicate the displacement fields of the plasmon modes of nanoparticle with permittivity ε. To this end, we shall discuss the unique properties of double layer potential (2.120). We begin with brief review of the origin of formula (2.120). Consider two similar surfaces S^+ and S^- separated by a tiny distance ℓ (see Figure 2.12). These surfaces are charged in such a way that $\sigma(M^+) = -\sigma(M^-)$, where M^+ and M^- are points on S^+ and S^-, respectively, which share a common normal to these surfaces. To be specific, it is assumed that $\sigma(M^+) > 0$. The "displacement" potential Φ created by these double layers of electric charges can be written as follows:

$$\begin{aligned}
\Phi(Q) &= \frac{1}{4\pi} \left(\oint_{S^+} \frac{\sigma(M^+)}{r_{QM^+}} ds_M + \oint_{S^-} \frac{\sigma(M^-)}{r_{QM^-}} ds_{M^-} \right) \\
&\approx -\frac{1}{4\pi} \oint_S \sigma(M^+) \left[\frac{1}{r_{QM^-}} - \frac{1}{r_{QM^+}} \right] ds_M,
\end{aligned} \tag{2.121}$$

where S is the surface in the middle between S^+ and S^- and M is the "middle" point between M^+ and M^-.

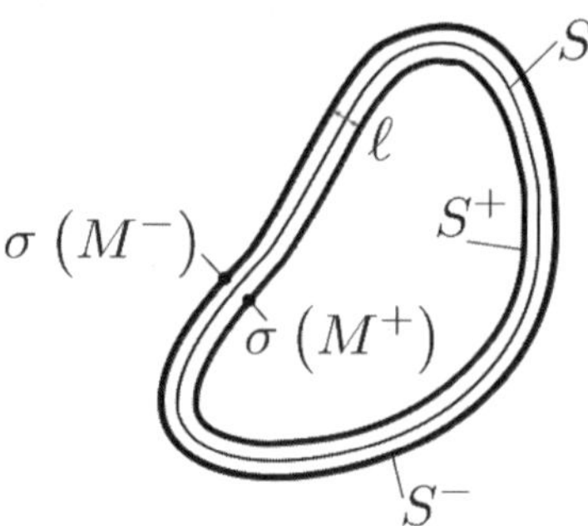

Figure 2.12

It is apparent that

$$\frac{1}{r_{QM^-}} - \frac{1}{r_{QM^+}} \approx \ell \frac{\partial}{\partial n_M}\left(\frac{1}{r_{QM}}\right) = -\ell \frac{\mathbf{r}_{QM} \cdot \mathbf{n}_M}{r_{QM}^3}. \tag{2.122}$$

By substituting formula (2.122) into (2.121) and introducing notation

$$\tau(M) = \sigma(M^+)\ell, \tag{2.123}$$

we obtain

$$\Phi(Q) \approx \frac{1}{4\pi} \oint_S \tau(M) \frac{\mathbf{r}_{QM} \cdot \mathbf{n}_M}{r_{QM}^3} ds_M. \tag{2.124}$$

In the limit when the separation ℓ between S^+ and S^- tends to zero in such a way that S^+ and S^- merge with S while $\sigma(M^+)$ tends to infinity in order to keep $\tau(M)$ being finite, then the approximate formula (2.124) is reduced to the exact expression (2.120).

It is apparent from the previous discussion that normal components of electric displacement $\mathbf{D}$ are continuous across the double layer S. This is because the jumps in normal components of $\mathbf{D}$ across S^+ and S^- completely canceled. The continuity of normal components of $\mathbf{D}$ implies the continuity of normal derivatives of Φ across the double layer S.

In contrast, the "displacement" potential Φ is discontinuous across the double layer S. This is due to the very strong (in the limit $\ell \to 0$, infinitely strong) internal field $\mathbf{D}$ between oppositely charged surfaces S^+ and S^- that form the double layer. This internal field leads to a finite jump in Φ across the double layer S. We shall next discuss this matter in detail and derive the

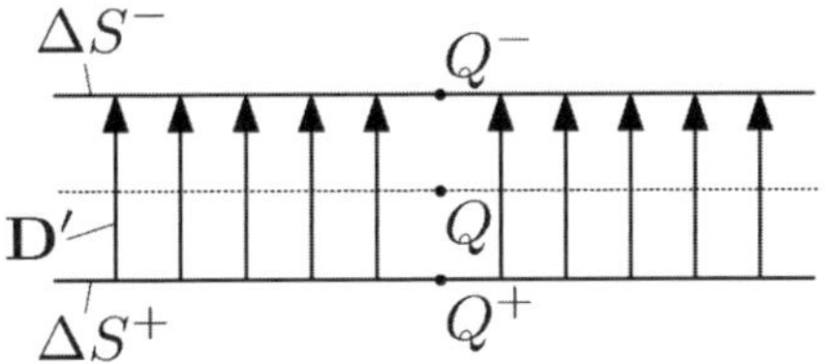

Figure 2.13

formulas for the values of Φ^+ and Φ^- on the interior and exterior sides of S, respectively. To do this, we partition S into two pieces $S - \Delta S$ and ΔS (see Figure 2.11) and rewrite formula (2.120) as follows:

$$\Phi^-(Q) \simeq \Phi'(Q^-) + \frac{1}{4\pi} \int_{S-\Delta S} \tau(M) \frac{\mathbf{r}_{QM} \cdot \mathbf{n}_M}{r_{QM}^3} ds_M, \tag{2.125}$$

$$\Phi^+(Q) \simeq \Phi'(Q^+) + \frac{1}{4\pi} \int_{S-\Delta S} \tau(M) \frac{\mathbf{r}_{QM} \cdot \mathbf{n}_M}{r_{QM}^3} ds_M, \tag{2.126}$$

where $\Phi'(Q^-)$ and $\Phi'(Q^+)$ are the values of the potential at points Q^- and Q^+, respectively, which are due to the double layer of charges on ΔS (see Figure 2.13). If the dimensions of ΔS are sufficiently small that it is almost flat and surface charges on ΔS^+ and ΔS^- are almost spatially uniform, while (at the same time) these dimensions are appreciably larger than the separation ℓ between ΔS^+ and ΔS^-, then the displacement field $\mathbf{D}$ between ΔS^+ and ΔS^- mimics with high accuracy the field in a parallel plate capacitor. This means that the potential at the midpoint Q is practically equal to zero,

$$\Phi'(Q) \approx 0, \tag{2.127}$$

and

$$\mathbf{D}'(Q) \approx \mathbf{n}_Q \sigma(Q^+), \tag{2.128}$$

where $\mathbf{D}'$ is the displacement field created by charges on ΔS^+ and ΔS^-. From formulas (2.123), (2.127) and (2.128) it follows that

$$\Phi'(Q^-) \simeq \Phi'(Q) - \frac{\sigma(Q^+)\ell}{2} = -\frac{\tau(Q)}{2}, \tag{2.129}$$

$$\Phi'(Q^+) \simeq \Phi'(Q) + \frac{\sigma(Q^+)\ell}{2} = \frac{\tau(Q)}{2}. \tag{2.130}$$

By substituting formulas (2.129) and (2.130) into relations (2.125) and (2.126), respectively, we obtain

$$\Phi^-(Q) \simeq -\frac{\tau(Q)}{2} + \frac{1}{4\pi} \int_{S-\Delta S} \tau(M) \frac{\mathbf{r}_{QM} \cdot \mathbf{n}_M}{r_{QM}^3} ds_M, \qquad (2.131)$$

$$\Phi^+(Q) \simeq \frac{\tau(Q)}{2} + \frac{1}{4\pi} \int_{S-\Delta S} \tau(M) \frac{\mathbf{r}_{QM} \cdot \mathbf{n}_M}{r_{QM}^3} ds_M. \qquad (2.132)$$

In the limit when ℓ and dimensions of ΔS tend to zero in such a way that these dimensions are appreciably larger than ℓ, the last two formulas become exact, i.e., we derive

$$\Phi^-(Q) = -\frac{\tau(Q)}{2} + \frac{1}{4\pi} \oint_{S} \tau(M) \frac{\mathbf{r}_{QM} \cdot \mathbf{n}_M}{r_{QM}^3} ds_M, \qquad (2.133)$$

$$\Phi^+(Q) = \frac{\tau(Q)}{2} + \frac{1}{4\pi} \oint_{S} \tau(M) \frac{\mathbf{r}_{QM} \cdot \mathbf{n}_M}{r_{QM}^3} ds_M. \qquad (2.134)$$

The presented derivation of formulas (2.133) and (2.134) is physically transparent. Mathematically rigorous derivations of these formulas as well as formulas (2.30)-(2.32) can be found in mathematical books on the potential theory [15, 16].

It is clear from the previous discussion that the double layer potential $\Phi(Q)$ satisfies equations (2.115) and (2.116) as well as boundary conditions (2.118) for any double layer density $\tau(M)$. Next, we shall demonstrate that the density $\tau(M)$ can be chosen to guarantee the validity of boundary condition (2.117). Indeed, by substituting formulas (2.133) and (2.134), we obtain

$$\frac{\tau(Q)}{2\varepsilon} + \frac{1}{4\pi\varepsilon} \oint_{S} \tau(M) \frac{\mathbf{r}_{QM} \cdot \mathbf{n}_M}{r_{QM}^3} ds_M = -\frac{\tau(Q)}{2\varepsilon_0} + \frac{1}{4\pi\varepsilon_0} \oint_{S} \tau(M) \frac{\mathbf{r}_{QM} \cdot \mathbf{n}_M}{r_{QM}^3} ds_M.$$

$$(2.135)$$

Now, after simple transformations we derive

$$\tau(Q) = \frac{\lambda}{2\pi} \oint_{S} \tau(M) \frac{\mathbf{r}_{QM} \cdot \mathbf{n}_M}{r_{QM}^3} ds_M, \qquad (2.136)$$

where

$$\lambda = \frac{\varepsilon - \varepsilon_0}{\varepsilon + \varepsilon_0}. \qquad (2.137)$$

Thus, the electric displacement fields created in homogeneous space by the distribution of virtual double layer charges over S (see Figure 2.11) will replicate the displacement fields of plasmon modes of nanoparticle with permittivity ε if the density $\tau(M)$ of double layers satisfies the homogeneous integral equation (2.136). It is apparent that this integral equation is adjoint to integral equation (2.34). It will be shown in the next section that the eigenvalues of integral equation (2.34) are real and satisfy inequality (2.40). By using the Fredholm theory outlined at the end of the previous section, we conclude that the same is true for the eigenvalues of integral equation (2.136). Thus, the analysis of plasmon modes is reduced to the calculation of eigenvalues λ_k and eigenfunctions $\tau_k(M)$ of integral equation (2.136), i.e.,

$$\boxed{\tau_k(Q) = \frac{\lambda_k}{2\pi} \oint_S \tau_k(M) \frac{\mathbf{r}_{QM} \cdot \mathbf{n}_M}{r_{QM}^3} ds_M,} \tag{2.138}$$

and the computation of resonance values ε_k of dielectric permittivity ε of the nanoparticle and resonance frequencies by using formulas (2.41) and (2.38), respectively. Finally, the electric displacement $\mathbf{D}_k$ of the corresponding plasmon modes can be determined by using the formula

$$\boxed{\mathbf{D}_k(Q) = -\frac{1}{4\pi} \mathrm{grad}_Q \left(\oint_S \tau_k(M) \frac{\mathbf{r}_{QM} \cdot \mathbf{n}_M}{r_{QM}^3} ds_M \right).} \tag{2.139}$$

It can be shown that eigenfunctions $\sigma_k(M)$ and $\tau_i(M)$ of integral equations (2.36) and (2.138), respectively, are biorthogonal. This means that

$$\oint_S \sigma_k(Q)\tau_i(Q)ds_Q = \delta_{ki}, \tag{2.140}$$

or in concise notation,

$$\langle \sigma_k, \tau_i \rangle = \delta_{ki}. \tag{2.141}$$

Indeed, by writing integral equations (2.36) and (2.138) in operator forms (see (2.96) and (2.101))

$$\sigma_k = \lambda_k \hat{K} \sigma_k, \tag{2.142}$$

$$\tau_i = \lambda_i \hat{K}^* \tau_i, \tag{2.143}$$

and by using formula (2.105), we find

$$\frac{1}{\lambda_k}\left\langle \sigma_k, \tau_i \right\rangle = \left\langle \hat{K}\sigma_k, \tau_i \right\rangle = \left\langle \sigma_k, \hat{K}^*\tau_i \right\rangle = \frac{1}{\lambda_i}\left\langle \sigma_k, \tau_i \right\rangle. \tag{2.144}$$

Since λ_k and λ_i are different ($k \neq i$), the last formula implies that

$$\left\langle \sigma_k, \tau_i \right\rangle = 0, \quad \text{if } k \neq i. \tag{2.145}$$

Next, we demonstrate that $\langle \sigma_k, \tau_k \rangle$ are not equal to zero for any k. To this end, we remark that from formulas (2.12) and (2.32) it follows that

$$\sigma_k(Q) = (\varepsilon_0 - \varepsilon_k)\frac{\partial \varphi_k^+}{\partial n}(Q). \tag{2.146}$$

Similarly, from formulas (2.117), (2.133) and (2.134) and taking into account that, for the same plasmon modes, potentials φ_k^+ and Φ_k^+ are related by the formula $\Phi_k^+ = \varepsilon_k \varphi_k^+$, we derive

$$\tau_k(Q) = (\varepsilon_k - \varepsilon_0)\,\varphi_k^+(Q). \tag{2.147}$$

By using formulas (2.146) and (2.147), we find

$$\left\langle \sigma_k, \tau_k \right\rangle = -(\varepsilon_k - \varepsilon_0)^2 \left\langle \frac{\partial \varphi_k^+}{\partial n}, \varphi_k^+ \right\rangle = -(\varepsilon_k - \varepsilon_0)^2 \oint_S \varphi_k^+(Q)\frac{\partial \varphi_k^+}{\partial n}(Q)ds_Q. \tag{2.148}$$

Now, by invoking the Green formula (see (2.17)) and the fact that $\nabla^2 \varphi_k^+ = 0$, we obtain

$$\left\langle \sigma_k, \tau_k \right\rangle = -(\varepsilon_k - \varepsilon_0)^2 \int_{V^+} |\text{grad } \varphi_k^+|^2 dv \neq 0. \tag{2.149}$$

The last formula means that by scaling either σ_k or τ_k or both of them, we arrive at

$$\left\langle \sigma_k, \tau_k \right\rangle = 1. \tag{2.150}$$

Thus, the biorthogonality (2.141) is established. It has been tacitly assumed in our derivation that eigenvalues λ_k are simple, i.e., that for each eigenvalue λ_k there exists only one eigenfunction. However, by using the biorthonormalization procedure, the presented proof can be extended to the case when eigenvalues are not simple.

By using the biorthogonality condition (2.140), biorthogonal expansion of any function $f(Q)$ defined on S can be developed. Indeed, consider the expansion

$$f(Q) = \sum_k a_k \sigma_k(Q). \tag{2.151}$$

Then, by multiplying both sides of (2.151) by $\tau_i(Q)$, subsequently integrating over S and by using the biorthogonality conditions (2.140) (or (2.141)), we derive

$$a_i = \oint_S f(Q)\tau_i(Q)ds_Q = \langle f, \tau_i \rangle. \tag{2.152}$$

By substituting the last formula into expression (2.151), we find

$$f(Q) = \sum_k \langle f, \tau_k \rangle \, \sigma_k(Q). \tag{2.153}$$

The biorthogonal expansion (2.153) will be extensively used in Chapter 4 in the analysis of excitation and time dynamics of specific plasmon modes. It will be demonstrated that mostly dipolar modes (i.e., plasmon modes with nonzero dipole moments) can be efficiently excited. For this reason, it is appropriate at this point to discuss the issue of dipole moments of plasmon modes and derive various equivalent expressions for these moments.

By using the "charge" model of plasmon modes discussed in the previous section, the dipole moments of plasmon modes can be defined as follows:

$$\boxed{\mathbf{p}_k = \oint_S \mathbf{r}_{OM}\sigma_k(M)ds_M,} \tag{2.154}$$

where $\mathbf{r}_{OM}$ is the radius vector directed from some origin O to the "integration" point M. It is apparent that due to the "charge-neutrality" property (2.42) of plasmon modes the dipole moment is well defined, i.e., it does not depend on the choice of origin O. The x-component of the dipole moment is given by the formula

$$p_{x,k} = \oint_S x\sigma_k(M)ds_M, \tag{2.155}$$

where x stands for the x-coordinate of $\mathbf{r}_{OM}$.

By using formulas (2.30), (2.31) and (2.6), we find

$$\sigma_k(M) = \varepsilon_0 \mathbf{n}_M \cdot \left[\mathbf{E}_k^-(M) - \mathbf{E}_k^+(M) \right] = \varepsilon_0 \left(\frac{\varepsilon_k}{\varepsilon_0} - 1 \right) \mathbf{n}_M \cdot \mathbf{E}_k^+(M). \tag{2.156}$$

From the last two formulas we obtain

$$p_{x,k} = \varepsilon_0 \left(\frac{\varepsilon_k}{\varepsilon_0} - 1 \right) \oint_S x \, \mathbf{n}_M \cdot \mathbf{E}_k^+(M) ds_M. \tag{2.157}$$

Now, by introducing the vector

$$\mathbf{a} = x\mathbf{E}_k^+, \tag{2.158}$$

taking into account that according to equation (2.1)

$$\operatorname{div} \mathbf{a} = E_{x,k}^+ \tag{2.159}$$

and finally by invoking the "divergence" theorem, from the last three equations we derive

$$p_{x,k} = (\varepsilon_k - \varepsilon_0) \int_{V^+} E_{x,k}^+ dv. \tag{2.160}$$

Now, it is apparent that

$$\boxed{\mathbf{p}_k = (\varepsilon_k - \varepsilon_0) \int_{V^+} \mathbf{E}_k^+ dv.} \tag{2.161}$$

The last formula may be useful for the identification of dipolar plasmon modes. In particular, it is proved in the next section that electric fields of various plasmon modes are orthogonal in V^+,

$$\int_{V^+} \mathbf{E}_k^+ \cdot \mathbf{E}_i^+ dv = 0. \tag{2.162}$$

From the last two formulas it is easy to conclude that in the case of spherical and ellipsoidal nanoparticles only spatially uniform plasmon modes are dipolar and only those modes are most efficiently excited by incident radiation.

We shall next derive the expression for the dipole moments of plasmon modes in terms of double layer density $\tau_k(M)$. From formula (2.160) we find

$$p_{x,k} = \frac{\varepsilon_k - \varepsilon_0}{\varepsilon_k} \int_{V^+} D_{x,k}^+ dv = \frac{\varepsilon_0 - \varepsilon_k}{\varepsilon_k} \int_{V^+} \mathbf{e}_x \cdot \operatorname{grad} \Phi_k^+ dv, \tag{2.163}$$

where $\mathbf{e}_x$ is the unit vector directed along the x-axis. It is apparent that

$$\mathbf{e}_x \cdot \operatorname{grad} \Phi_k^+ = \operatorname{div} \left(\mathbf{e}_x \Phi_k^+ \right). \tag{2.164}$$

By substituting the last formula into equation (2.163) and by using the "divergence" theorem, we derive

$$p_{x,k} = \frac{\varepsilon_0 - \varepsilon_k}{\varepsilon_k} \oint_S \Phi_k^+(M)\mathbf{e}_x \cdot \mathbf{n}_M ds_M. \tag{2.165}$$

From formula (2.147) it follows that

$$\Phi_k^+(M) = \frac{\varepsilon_k}{\varepsilon_k - \varepsilon_0} \tau_k(M). \tag{2.166}$$

From the last two formulas we find

$$p_{x,k} = -\mathbf{e}_x \cdot \oint_S \tau_k(M)\mathbf{n}_M ds_M, \tag{2.167}$$

which implies that

$$\boxed{\mathbf{p}_k = - \oint_S \tau_k(M)\mathbf{n}_M ds_M.} \tag{2.168}$$

The last formula is consistent with the interpretation of a double layer of charges as a surface layer of dipoles of density $\tau(M)$. The minus sign in formula (2.168) appears because it has been assumed in the discussion that negative charges are distributed on the exterior side of the double layer (see Figure 2.12).

Next, we shall discuss another matter. It is well known that the surface distribution of dipoles (i.e., double layer of charges) of constant density $\tau(M) = const$ does not create any electric displacement field. This is because the following well-known formula is valid:

$$\oint_S \frac{\mathbf{r}_{QM} \cdot \mathbf{n}_M}{r_{QM}^3} ds_M = \begin{cases} 4\pi, & \text{if } Q \in V^+ \\ 2\pi, & \text{if } Q \in S \\ 0, & \text{if } Q \in V^-. \end{cases} \tag{2.169}$$

Thus, there exist classes of equivalent double layers with densities $\tau_k(M)+d_k$, where d_k are arbitrary constants. It is clear that additive constants d_k can be chosen in such a way that

$$\oint_S \tau_k(M)ds_M = 0. \tag{2.170}$$

The integral equation (2.138) (or (2.136)) is valid only for one specific representative of these equivalence classes. It is possible to derive the boundary integral equations for any representative of these equivalence classes. This derivation is based on the observations that the boundary condition (2.117) follows from the continuity of the tangential components of the plasmon electric fields. However, this continuity will be also preserved if the following more general boundary condition is adopted on S:

$$\frac{\Phi^+}{\varepsilon} = \frac{\Phi^-}{\varepsilon_0} + C, \tag{2.171}$$

where C is an arbitrary constant. By substituting formulas (2.133) and (2.134) into the boundary condition (2.171), we find

$$\frac{\tau(Q)}{2\varepsilon} + \frac{1}{4\pi\varepsilon} \oint_S \tau(M) \frac{\mathbf{r}_{QM} \cdot \mathbf{n}_M}{r_{QM}^3} ds_M$$

$$= -\frac{\tau(Q)}{2\varepsilon_0} + \frac{1}{4\pi\varepsilon_0} \oint_S \tau(M) \frac{\mathbf{r}_{QM} \cdot \mathbf{n}_M}{r_{QM}^3} ds_M + C, \tag{2.172}$$

which after straightforward transformation leads to the following integral equation:

$$\tau(Q) = \frac{\lambda}{2\pi} \oint_S \tau(M) \frac{\mathbf{r}_{QM} \cdot \mathbf{n}_M}{r_{QM}^3} ds_M + \frac{2\varepsilon\varepsilon_0}{\varepsilon + \varepsilon_0} C. \tag{2.173}$$

Here, as before, λ is given by formula (2.137).

It is easy to see that if $\tilde{\tau}_k(M)$ is the nonzero solution of integral equation (2.173) for some eigenvalue λ_k and some constant C_k, i.e.,

$$\tilde{\tau}_k(Q) = \frac{\lambda_k}{2\pi} \oint_S \tilde{\tau}_k(M) \frac{\mathbf{r}_{QM} \cdot \mathbf{n}_M}{r_{QM}^3} ds_M + \frac{2\varepsilon_k\varepsilon_0}{\varepsilon_k + \varepsilon_0} C_k, \tag{2.174}$$

then the additive constant d_k can be found to guarantee that

$$\tau_k(M) = \tilde{\tau}_k(M) + d_k \tag{2.175}$$

is the solution of integral equation (2.138). Indeed, by using formula (2.169), we find

$$(1 - \lambda_k)\, d_k = \frac{2\varepsilon_k\varepsilon_0}{\varepsilon_k + \varepsilon_0} C_k. \tag{2.176}$$

For plasmon modes $|\lambda_k| > 1$ (see the next section), and, consequently, formula (2.176) establishes one-to-one correspondence between the additive constraints d_k and constants C_k in the integral equation (2.174). Thus, by varying C_k in equation (2.174) the entire equivalence class of $\tilde{\tau}_k(M)$ can be found.

Actually, the constant C_k can be chosen in such a way that the condition (2.170) is satisfied. Indeed, by integrating both sides of equation (2.174) with respect to Q over S and imposing the condition

$$\oint_S \tilde{\tau}_k(Q)ds_Q = 0, \tag{2.177}$$

we obtain

$$-\frac{\lambda_k}{2\pi S} \oint_S \tilde{\tau}_k(M) \left(\oint_S \frac{\mathbf{r}_{QM} \cdot \mathbf{n}_M}{r_{QM}^3} ds_Q \right) ds_M = \frac{2\varepsilon_k \varepsilon_0}{\varepsilon_k + \varepsilon_0} C_k. \tag{2.178}$$

By substituting formula (2.178) into equation (2.174), we derive

$$\tilde{\tau}_k(Q) = \frac{\lambda_k}{2\pi} \oint_S \tilde{\tau}_k(M) \left[\frac{\mathbf{r}_{QM} \cdot \mathbf{n}_M}{r_{QM}^3} - \frac{1}{S} \oint_S \frac{\mathbf{r}_{QM} \cdot \mathbf{n}_M}{r_{QM}^3} ds_Q \right] ds_M. \tag{2.179}$$

It is easy to check by integration of both sides of equation (2.179) that the solution of this equation satisfies the condition (2.177). The structure of the integral equation (2.179) is somewhat more complex than that of integral equation (2.138). For this reason, the actual computations can be performed by using the integral equation (2.138) and then by changing the found solution as follows:

$$\tilde{\tau}_k(M) = \tau_k(M) - \frac{1}{S} \oint_S \tau_k(M)ds_M. \tag{2.180}$$

It is apparent that this will guarantee the condition (2.177) and that $\tilde{\tau}_k(M)$ will be the solution of integral equation (2.179).

Next, we shall demonstrate that all eigenfunctions from the same equivalence class have the same dipole moment. According to formula (2.168) and the definition of equivalence class, it is sufficient to prove that

$$\oint_S \mathbf{n}_M ds_M = 0. \tag{2.181}$$

The proof proceeds as follows:

$$\mathbf{e}_x \cdot \oint_S \mathbf{n}_M ds_M = \oint_S \mathbf{e}_x \cdot \mathbf{n}_M ds_M = \int_{V+} \operatorname{div} \mathbf{e}_x dv = 0. \tag{2.182}$$

The same result can be obtained by replacing $\mathbf{e}_x$ by $\mathbf{e}_y$ and $\mathbf{e}_z$.

We shall conclude the discussion in this section by extending the integral equation (2.138) to the case of several nanoparticles and the case when these nanoparticles are placed on a dielectric substrate. These extensions are straightforward and are performed conceptually in the same way as in the previous section.

In the case of several nanoparticles (see Figure 2.6), we replace permittivity ε of these nanoparticles by ε_0 (see Figure 2.7) and introduce on their boundaries S_i double layers of electric charges of densities $\tau^{(i)}(M)$. By using exactly the same line of reasoning as in the derivation of integral equation (2.138), it can be shown that the fields of electric displacement $\mathbf{D}_k$ created by these double layers in free homogeneous space will coincide with source-free plasmon mode displacement fields of actual nanoparticles with permittivity ε if the densities $\tau_k^{(i)}(M)$ are the solutions of the following integral equations:

$$\tau_k^{(j)}(Q) = \frac{\lambda_k}{2\pi} \sum_{i=1}^{n} \oint_{S_i} \tau_k^{(i)}(M) \frac{\mathbf{r}_{QM} \cdot \mathbf{n}_M}{r_{QM}^3} ds_M, \quad (j = 1, 2, ..., n), \tag{2.183}$$

where λ_k are defined by formula (2.37). By solving the eigenvalue problem (2.183) and finding λ_k and $\tau_k^{(i)}(M)$, the ε_k and resonance frequencies can be determined by using formulas (2.41) and (2.38), respectively, while the plasmon mode displacement fields can be computed by using the formula

$$\mathbf{D}_k(Q) = -\frac{1}{4\pi} \sum_{i=1}^{n} \mathrm{grad}_Q \left(\oint_{S_i} \tau_k^{(i)}(M) \frac{\mathbf{r}_{QM} \cdot \mathbf{n}_M}{r_{QM}^3} ds_M \right). \tag{2.184}$$

In the case of multiple nanoparticles placed on a dielectric substrate, the last two formulas are naturally modified as follows:

$$\tau_k^{(j)}(Q) = -\frac{\lambda_k}{2\pi} \sum_{i=1}^{n} \oint_{S_i} \tau_k^{(i)}(M) \mathbf{n}_M \cdot \mathrm{grad}_M G(Q, M) ds_M, \quad (j = 1, 2, ..., n),$$

$$\tag{2.185}$$

$$\mathbf{D}_k(Q) = \frac{1}{4\pi} \sum_{i=1}^{n} \mathrm{grad}_Q \left(\oint_{S_i} \tau_k^{(i)}(M) \mathbf{n}_M \cdot \mathrm{grad}_M G(Q, M) ds_M \right), \tag{2.186}$$

where the Green function $G(Q, M)$ is given by formula (2.86).

Finally, in the case of two-dimensional problems when plasmon resonance modes in nanowires are of interest, the integral equation (2.183) and formula (2.184) are modified as follows:

$$\tau_k^{(j)}(Q) = \frac{\lambda_k}{2\pi} \sum_{i=1}^{n} \oint_{L_i} \tau_k^{(i)}(M) \frac{\mathbf{r}_{QM} \cdot \mathbf{n}_M}{r_{QM}^2} dl_M, \quad (j = 1, 2, ..., n),$$ (2.187)

$$\mathbf{D}_k(Q) = -\frac{1}{4\pi} \sum_{i=1}^{n} \mathrm{grad}_Q \left(\oint_{L_i} \tau_k^{(i)}(M) \frac{\mathbf{r}_{QM} \cdot \mathbf{n}_M}{r_{QM}^2} dl_M \right),$$ (2.188)

where L_i are cross-sectional boundaries of the nanowires.

It is worthwhile to point out again that in contrast with the three-dimensional case the kernels of integral equations (2.187) are not singular. They are actually continuous functions on L_i if L_i are sufficiently smooth. This can be beneficial in numerical computations.

2.3 General Properties of Plasmon Spectrum

In the previous two sections, the analysis of plasmon modes in nanoparticles has been reduced to the eigenvalue problems for specific boundary integral equations. By solving these eigenvalue problems, resonance frequencies and the electric fields of corresponding plasmon modes can be determined. In this section, we shall study the general properties of plasmon spectrum, i.e., the general properties of eigenvalues of the boundary integral equations which describe the plasmon modes in nanoparticles. The term "general" means that we shall be concerned with the properties of eigenvalues that are valid for *any* shape of nanoparticles.

We start our discussion with the following:

Property 1. For any shape of nanoparticles the corresponding plasmon spectrum is discrete.

This property immediately follows from the Fredholm theory of integral equations outlined at the end of section 2.1. According to this theory (see Spectral Theorem) there exists only a countable (denumerable) set of eigenvalues λ_k of integral equation (2.36) and $|\lambda_k|$ tends to infinity as $k \to \infty$ (see formula (2.106)). This implies that plasmon resonance modes may exist only for denumerable values ε_k of dielectric permittivity of nanoparticles and, according to (2.41),

$$\lim_{k \to \infty} \varepsilon_k = -\varepsilon_0. \tag{2.189}$$

For sufficiently high-order plasmon modes (large k), $\varepsilon_k \approx -\varepsilon_0$ and according to the dispersion relation (2.73) (see also (1.8)) we find that the following approximate formula for resonance frequencies

$$\omega_k \approx \frac{\omega_p}{\sqrt{2}} \tag{2.190}$$

is valid for any shape of nanoparticles. In other words, the values of resonance frequencies ω_k for high-order plasmon modes are universal (i.e., shape-independent). It will be shown in the next chapter (see section 3.1) that for the particular case of nanowires of circular cross sections, $\varepsilon_k = -\varepsilon_0$ for all plasmon modes.

We now proceed to

Property 2. Plasmon spectrum is scale-invariant.

This means that eigenvalues λ_k of integral equation (2.36) depend on shapes of nanoparticles but not their geometric dimensions.

To demonstrate this, consider some scaling of geometric dimensions of S by an arbitrary parameter α. This implies the following transformations:

$$r_{M'Q'} = \alpha r_{MQ}, \qquad ds'_{M'} = \alpha^2 ds_M, \tag{2.191}$$

where Q' and M' are scaled versions of Q and M, respectively. By substituting relations (2.191) into the integral equation (2.36), we find that the scaling parameter α is canceled and we arrive at the integral equations with mathematical form identical to the equation (2.36). The latter reveals the scale-invariance of λ_k.

The scale-invariance of λ_k implies that resonance values ε_k of the permittivity of nanoparticles and the corresponding resonance frequencies ω_k are scale-invariant. This is true provided that the nanoparticle dimensions remain appreciably smaller than the free-space wavelength of the incident radiation. The latter is one of the conditions for the occurrence of plasmon resonances. The scale-invariance of plasmon resonances is also apparent from formulas (2.54) and (2.65) derived for the examples of spherical and ellipsoidal nanoparticles, respectively. The scale-invariance suggests, among other things, that in the case of ensembles of almost self-similar nanoparticles they will resonate at practically the same free-space wavelength of incoming radiation. This means that plasmon resonances can be simultaneously excited in all nanoparticles of such ensembles.

Next, we discuss

Property 3. All eigenvalues λ_k of integral equation (2.36) are real.

The proof of this property proceeds as follows. Let us first assume that eigenvalues λ_k and corresponding eigenfunctions are complex-valued and then we demonstrate that this leads to the contradiction. To arrive at this contradiction, consider the potential

$$\varphi_k(Q) = \frac{1}{4\pi\varepsilon_0} \oint_S \frac{\sigma_k(M)}{r_{MQ}} ds_M, \tag{2.192}$$

where charge density $\sigma_k(M)$ is an eigenfunction of equation (2.36). According to formulas (2.32), we find

$$\sigma_k(Q) = \varepsilon_0 \left[\frac{\partial \varphi_k^+}{\partial n}(Q) - \frac{\partial \varphi_k^-}{\partial n}(Q) \right], \tag{2.193}$$

$$\frac{1}{2\pi} \oint_S \sigma_k(M) \frac{\mathbf{r}_{MQ} \cdot \mathbf{n}_Q}{r_{MQ}^3} ds_M = -\varepsilon_0 \left[\frac{\partial \varphi_k^+}{\partial n}(Q) + \frac{\partial \varphi_k^-}{\partial n}(Q) \right]. \tag{2.194}$$

By substituting formulas (2.193) and (2.194) into the integral equation (2.36), we obtain

$$\frac{\partial \varphi_k^+}{\partial n}(Q) - \frac{\partial \varphi_k^-}{\partial n}(Q) = -\lambda_k \left[\frac{\partial \varphi_k^+}{\partial n}(Q) + \frac{\partial \varphi_k^-}{\partial n}(Q) \right]. \tag{2.195}$$

Next, we shall multiply both sides of equality (2.195) by the complex conjugate $\varphi_k^*(Q)$ of potential $\varphi_k(Q)$ and integrate over S. The result is

$$\oint_S \varphi_k^*(Q) \left[\frac{\partial \varphi_k^+}{\partial n}(Q) - \frac{\partial \varphi_k^-}{\partial n}(Q) \right] ds_Q$$

$$= -\lambda_k \oint_S \varphi_k^*(Q) \left[\frac{\partial \varphi_k^+}{\partial n}(Q) + \frac{\partial \varphi_k^-}{\partial n}(Q) \right] ds_Q. \tag{2.196}$$

By setting

$$\mathbf{a}_k^\pm = \left(\varphi^\pm \right)^* \operatorname{grad} \varphi_k^\pm \tag{2.197}$$

and by using the "divergence theorem"

$$\int_{V^\pm} \operatorname{div} \mathbf{a}_k^\pm dv = \pm \oint_S \mathbf{n} \cdot \mathbf{a}_k^\pm ds, \tag{2.198}$$

we find

$$\int_{V^\pm} \left[|\text{grad } \varphi_k^\pm|^2 + \left(\varphi_k^\pm\right)^* \nabla^2 \varphi_k^\pm \right] dv = \pm \oint_S \varphi_k^* \frac{\partial \varphi_k^\pm}{\partial n} ds, \tag{2.199}$$

where sign "$-$" in front of the surface integral is due to the fact that $\mathbf{n}$ is the inward normal for V^-.

By taking into account that the potential $\varphi_k(Q)$ defined by formula (2.192) satisfies in V^+ and V^- the Laplace equation

$$\nabla^2 \varphi_k^\pm = 0 \tag{2.200}$$

and by introducing the notation

$$W_k^\pm = \int_{V^\pm} |\text{grad } \varphi_k^\pm|^2 dv > 0, \tag{2.201}$$

from formula (2.199) we find

$$W_k^\pm = \pm \oint_S \varphi_k^*(Q) \frac{\partial \varphi_k^\pm}{\partial n}(Q) ds_Q. \tag{2.202}$$

From equations (2.196) and (2.202), we derive

$$\lambda_k = \frac{W_k^- + W_k^+}{W_k^- - W_k^+}. \tag{2.203}$$

From formulas (2.201) and (2.203) it follows that λ_k are real numbers, which is in contradiction with our original assumption that λ_k are complex numbers. It is easy to see that the eigenfunctions of equation (2.36) can be chosen to be real-valued as well. Indeed, if some $\sigma_k(M)$ is a complex-valued function

$$\sigma_k(M) = \nu_k(M) + j\chi_k(M), \quad (j = \sqrt{-1}), \tag{2.204}$$

then from the integral equation (2.36) we find

$$\nu_k(Q) = \frac{\lambda_k}{2\pi} \oint_S \nu_k(M) \frac{\mathbf{r}_{MQ} \cdot \mathbf{n}_Q}{r_{MQ}^3} ds_M \tag{2.205}$$

and

$$\chi_k(Q) = \frac{\lambda_k}{2\pi} \oint_S \chi_k(M) \frac{\mathbf{r}_{MQ} \cdot \mathbf{n}_Q}{r_{MQ}^3} ds_M, \tag{2.206}$$

which means that there are two real-valued eigenfunctions corresponding to the eigenvalue λ_k. Thus, Property 3 is established.

It follows from this property and the Fredholm theory that the integral equation (2.36) and its adjoint equation (2.138) have the same set of eigenvalues. Consequently, all results concerning the spectrum of integral equation (2.36) are valid for the spectrum of adjoint integral equation (2.138).

It is easy to deduce from the formula (2.203) the following

Property 4. Eigenvalue $\lambda_0 = 1$ and the corresponding eigenfunction $\sigma_0(M)$ correspond to the classical Robin problem of distribution of electric charges on the surface S of a charged conductor, while for all other eigenvalues the following inequality holds:

$$|\lambda_k| > 1, \quad (k = 1, 2, ...). \tag{2.207}$$

Indeed, from the formulas (2.203) and (2.201) we find that the equality $\lambda_0 = 1$ holds if and only if

$$W_0^+ = \int_{V^+} |\operatorname{grad} \varphi_0^+|^2 dv = 0. \tag{2.208}$$

On the other hand, the integral equation (2.36) can be written for $\lambda_0 = 1$ as follows:

$$\sigma_0(Q) - \frac{1}{2\pi} \oint_S \sigma_0(M) \frac{\mathbf{r}_{MQ} \cdot \mathbf{n}_Q}{r_{MQ}^3} ds_M = 0. \tag{2.209}$$

According to formula (2.31), the last equation means that on the interior side of S the normal component of electric field $\mathbf{n}_Q \cdot \mathbf{E}_0^+(Q)$ created by charge distribution $\sigma_0(M)$ is equal to zero,

$$\mathbf{n}_Q \cdot \mathbf{E}_0^+(Q) = 0. \tag{2.210}$$

Formulas (2.208) and (2.210) are only consistent with the interpretation of $\sigma_0(M)$ as the distribution of surface electric charges on the boundary S of the electrically charged conductor V^+ (see Figure 2.14). For this reason,

$$\oint_S \sigma_0(M) ds_M = q \neq 0, \tag{2.211}$$

where q is the total charge of the conductor V^+. The last formula is in contrast with the charge neutrality condition (2.42) for plasmon modes.

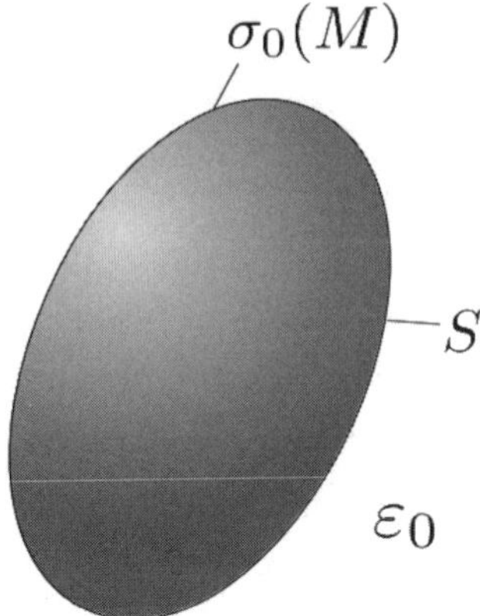

Figure 2.14

It is interesting to note that formulas (2.209) and (2.211) can be combined into one inhomogeneous integral equation for the Robin problem

$$\sigma_0(Q) - \frac{1}{2\pi} \oint_S \sigma_0(M) \left[\frac{\mathbf{r}_{MQ} \cdot \mathbf{n}_Q}{r_{MQ}^3} - \frac{2\pi}{A} \right] ds_M = \frac{q}{A}, \qquad (2.212)$$

where A is the area of S. The last equation can be useful for the calculation of electrostatic fields when the charge q of the conductor is given.

In the case when the potential U of the conductor is specified, then we have

$$\frac{1}{4\pi\varepsilon_0} \oint_S \frac{\sigma_0(M)}{r_{MQ}} ds_M = U. \qquad (2.213)$$

By combining equations (2.209) and (2.213), we obtain the following integral equation

$$\sigma_0(Q) - \frac{1}{2\pi} \oint_S \sigma_0(M) \left[\frac{\mathbf{r}_{MQ} \cdot \mathbf{n}_Q}{r_{MQ}^3} - \frac{\beta}{r_{MQ}} \right] ds_M = 2\beta\varepsilon_0 U, \qquad (2.214)$$

where $\beta > 0$ is some constant coordinating the dimensions of the two terms of the kernel of integral equation (2.214). It can be shown [6] that for any $\beta > 0$, integral equation (2.214) is uniquely solvable. This equation is convenient for the calculation of electric charge distribution $\sigma_0(M)$ when the potential

U of the conductor is given. Integral equations (2.211) and (2.213) can be easily generalized to the case of multiple conductors.

The presented discussion of the Robin problem is of interest in its own right. However, this discussion is a digression as far as the analysis of plasmon modes is concerned. Returning to this analysis, we point out that $W_k^+ \neq 0$ for $\lambda_k \neq 1$ and, according to formula (2.203), the inequality (2.207) is valid. This in turn implies that, as expected, all ε_k are negative.

As discussed, the eigenvalue $\lambda_0 = 1$ of integral equation (2.36) has nothing to do with plasmon resonances. In this sense, this eigenvalue is spurious. It turns out that the integral equation (2.36) can be easily modified to get rid of this eigenvalue. Indeed, by combining formulas (2.36) and (2.42) we arrive at the following eigenvalue problem for the boundary integral equation:

$$\sigma_k(Q) = \frac{\lambda_k}{2\pi} \oint_S \sigma_k(M) \left[\frac{\mathbf{r}_{MQ} \cdot \mathbf{n}_Q}{r_{MQ}^3} - \frac{2\pi}{A} \right] ds_M. \qquad (2.215)$$

Next, we state

Property 5. The spectrum of integral equation (2.215) coincides with the spectrum of integral equation (2.36) except for the eigenvalue $\lambda_0 = 1$.

Indeed, for all $\lambda_k \neq \lambda_0$ the conditions (2.42) hold and equation (2.36) can be equivalently transformed into equation (2.215). On the other hand, integrating both sides of equation (2.215) over S we find that

$$\oint_S \sigma_k(Q)ds_Q = \frac{\lambda_k}{2\pi} \oint_S \sigma_k(M) \left[\oint_S \frac{\mathbf{r}_{MQ} \cdot \mathbf{n}_Q}{r_{MQ}^3} ds_Q - 2\pi \right] ds_M. \qquad (2.216)$$

From the last equation and formula (2.44) it follows that

$$\oint_S \sigma_k(Q)ds_Q = 0. \qquad (2.217)$$

By using formula (2.217), the integral equation (2.215) can be transformed into the integral equation (2.36). This means that for all eigenvalues such that the conditions (2.217) hold the integral equations (2.36) and (2.215) are equivalent. The condition (2.217) does not hold for the eigenfunction $\sigma_0(M)$ corresponding to the eigenvalue $\lambda_0 = 1$. That is why $\lambda_0 = 1$ is not the eigenvalue of integral equation (2.215). Indeed, it is clear from the physical meaning of $\sigma_0(M)$ and it can be easily justified mathematically that if the condition (2.217) holds for $\sigma_0(M)$ then $\sigma_0(M) = 0$. Thus, Property 5 is established.

It is worthwhile to remark that the integral equation (2.215) may present some advantages in numerical computations by using iterative techniques because the conditions (2.217) are strictly imposed by the nature of the kernel of this equation.

In the case when the boundary S of the nanoparticle is convex, the inequality (2.207) can be sharpened. This sharpening is formulated as

Property 6. For a nanoparticle with convex boundary S, the following estimate for the eigenvalues λ_k holds:

$$|\lambda_k| > \frac{1}{1 - \dfrac{A}{4\pi R d}}, \tag{2.218}$$

where as before A is the area of S, R is the maximum radius of curvature of S and d is the diameter of S, i.e., the maximum distance between any two points on S.

The derivation of the inequality (2.218) is based on the fact that for convex boundary S the kernel of integral equation (2.36) is positive and the following estimate is valid:

$$\frac{\mathbf{r}_{MQ} \cdot \mathbf{n}_Q}{r_{MQ}^3} > \frac{1}{2Rd}. \tag{2.219}$$

Indeed, according to the Blaschke theorem [17] for any point Q on the convex surface S, this surface is contained inside the sphere of radius R tangent to S at Q (see Figure 2.15). Consequently,

$$\frac{\mathbf{r}_{MQ} \cdot \mathbf{n}_Q}{r_{MQ}^3} = \frac{\cos\gamma}{r_{MQ}^2} > \frac{\cos\gamma}{r_{M'Q} r_{MQ}} > \frac{1}{2Rd}. \tag{2.220}$$

Now, the proof of (2.218) proceeds as follows. According to the property (2.42) of eigenfunctions, the surface S can be partitioned into two sets S_k^+ where $\sigma_k(M) \geq 0$ and S_k^- where $\sigma_k(M) < 0$. Then the property (2.42) implies that

$$\int_{S_k^+} \sigma_k(M) ds_M = \int_{S_k^-} |\sigma_k(M)| ds_M = \frac{1}{2} \oint_S |\sigma_k(M)| ds_M. \tag{2.221}$$

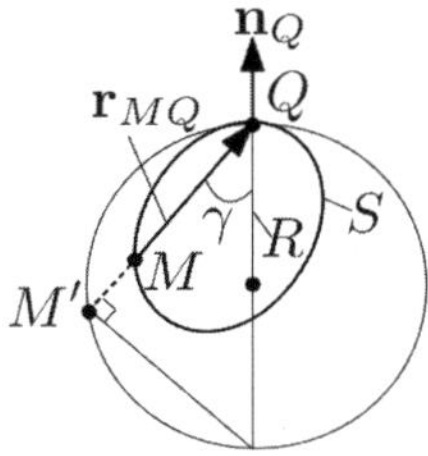

Figure 2.15

Consider first the case when $\lambda_k > 0$. From formula (2.221) and integral equation (2.36) we find

$$\oint_S |\sigma_k(Q)| ds_Q = 2 \int_{S_k^+} \sigma_k(Q) ds_Q = \frac{\lambda_k}{\pi} \oint_S \sigma_k(M) \left[\int_{S_k^+} \frac{\mathbf{r}_{MQ} \cdot \mathbf{n}_Q}{r_{MQ}^3} ds_Q \right] ds_M.$$

$$(2.222)$$

It is clear that the last double integral is strictly positive and the last formula can be transformed as follows:

$$\oint_S |\sigma_k(Q)| ds_Q = \frac{\lambda_k}{\pi} \left[\int_{S_k^+} \sigma_k(M) \left(\int_{S_k^+} \frac{\mathbf{r}_{MQ} \cdot \mathbf{n}_Q}{r_{MQ}^3} ds_Q \right) ds_M \right.$$
$$\left. - \int_{S_k^-} |\sigma_k(M)| \left(\int_{S_k^+} \frac{\mathbf{r}_{MQ} \cdot \mathbf{n}_Q}{r_{MQ}^3} ds_Q \right) ds_M \right]. \qquad (2.223)$$

From formula (2.44) follows that

$$\int_{S_k^+} \frac{\mathbf{r}_{MQ} \cdot \mathbf{n}_Q}{r_{MQ}^3} ds_Q = 2\pi - \int_{S_k^-} \frac{\mathbf{r}_{MQ} \cdot \mathbf{n}_Q}{r_{MQ}^3} ds_Q. \qquad (2.224)$$

By using the last formula in equation (2.223), we find

$$\oint_S |\sigma_k(Q)| ds_Q = \frac{\lambda_k}{\pi} \left[2\pi \int_{S_k^+} \sigma_k(M) ds_M \right.$$
$$- \int_{S_k^+} \sigma_k(M) \left(\int_{S_k^-} \frac{\mathbf{r}_{MQ} \cdot \mathbf{n}_Q}{r_{MQ}^3} ds_Q \right) ds_M$$
$$\left. - \int_{S_k^-} |\sigma_k(M)| \left(\int_{S_k^+} \frac{\mathbf{r}_{MQ} \cdot \mathbf{n}_Q}{r_{MQ}^3} ds_Q \right) ds_M \right]. \qquad (2.225)$$

Now, by using the inequality (2.219) and formula (2.221), we obtain

$$
\left[\int_{S_k^+} \sigma_k(M) \left(\int_{S_k^-} \frac{\mathbf{r}_{MQ} \cdot \mathbf{n}_Q}{r_{MQ}^3} ds_Q \right) ds_M \right.
$$
$$
\left. + \int_{S_k^-} |\sigma_k(M)| \left(\int_{S_k^+} \frac{\mathbf{r}_{MQ} \cdot \mathbf{n}_Q}{r_{MQ}^3} ds_Q \right) ds_M \right] > \frac{A}{4Rd} \oint_S |\sigma_k(M)| ds_M.
$$

$$(2.226)$$

From formulas (2.225) and (2.226), we derive

$$
\oint_S |\sigma_k(Q)| ds_Q < \lambda_k \left(1 - \frac{A}{4\pi Rd} \right) \oint_S |\sigma_k(M)| ds_M, \tag{2.227}
$$

which results in inequality

$$
\lambda_k > \frac{1}{1 - \dfrac{A}{4\pi Rd}}. \tag{2.228}
$$

In the case of $\lambda_k < 0$, from integral equation (2.36) and formula (2.221) we find

$$
\oint_S |\sigma_k(Q)| ds_Q = -2 \oint_{S_k^-} \sigma_k(Q) ds_M
$$
$$
= -\frac{\lambda_k}{\pi} \oint_S \sigma_k(M) \left[\int_{S_k^-} \frac{\mathbf{r}_{MQ} \cdot \mathbf{n}_Q}{r_{MQ}^3} ds_Q \right] ds_M. \tag{2.229}
$$

Now, by using the same line of reasoning as in the derivation of inequality (2.228), we arrive at

$$
-\lambda_k > \frac{1}{1 - \dfrac{A}{4\pi Rd}}, \tag{2.230}
$$

which together with formula (2.228) is tantamount to inequality (2.218), and Property 6 is established.

It is clear that inequality (2.218) suggests that

$$
\frac{A}{4\pi Rd} < 1. \tag{2.231}
$$

Actually, the last inequality can be directly established by using formula (2.219) as follows:

$$2\pi = \oint_S \frac{\mathbf{r}_{MQ} \cdot \mathbf{n}_Q}{r_{MQ}^3} ds_Q > \frac{A}{2Rd}, \tag{2.232}$$

which is equivalent to (2.231).

From formulas (2.37) and (2.218) we find

$$\left| \frac{\varepsilon_k + \varepsilon_0}{\varepsilon_k - \varepsilon_0} \right| < 1 - \frac{A}{4\pi Rd}. \tag{2.233}$$

By using the last inequality, the following upper and lower bounds for resonance values ε_k of dielectric permittivity can be established:

$$-\frac{8\pi Rd - A}{A} < \frac{\varepsilon_k}{\varepsilon_0} < \frac{A}{A - 8\pi Rd}. \tag{2.234}$$

For the dispersion relation (1.8), the previous formula leads to the following upper and lower bounds for the resonance frequencies:

$$\frac{A}{8\pi Rd} < \frac{\omega_k^2}{\omega_p^2} < 1 - \frac{A}{8\pi Rd}, \tag{2.235}$$

which indicate that the bandwidth for resonance frequencies is smaller than $\omega_p \sqrt{1 - A/4\pi Rd}$.

In the case of nanowires with convex boundaries L, it can be shown by using integral equation (2.91) and the same line of reasoning as before that the following inequality is valid:

$$|\lambda_k| > \frac{1}{1 - \dfrac{L}{2\pi R}}, \tag{2.236}$$

where L is the length of the cross-sectional boundary, while R is the maximum radius of its curvature. The last inequality is isoperimetric in the sense that it is exact for nanowires of circular cross sections. It is shown in the next chapter (see section 3.1) that for such nanowires plasmon resonances occur only for $\varepsilon = -\varepsilon_0$ which corresponds to $\lambda_k = \infty$; the latter is consistent with inequality (2.236).

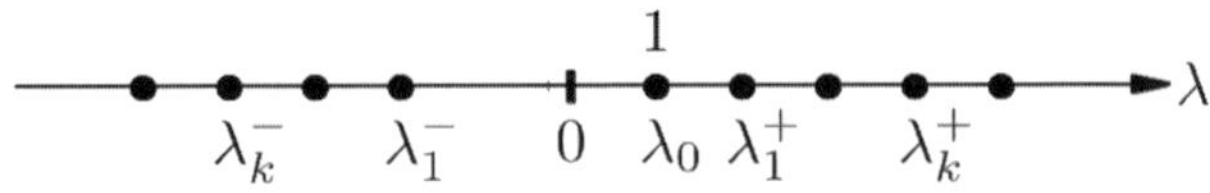

Figure 2.16

It turns out that the two-dimensional integral equation (2.91) exhibits the interesting phenomenon of mirror symmetry (twin) spectra. This fact is formulated as

Property 7. For any shape of nanowire cross-section (for any L) the set of eigenvalues of integral equation (2.91) consists of $\lambda_0 = 1$ and of pairs of positive $\lambda_k^+ > 0$ and negative $\lambda_k^- < 0$ eigenvalues of the same absolute values, i.e.,

$$\lambda_k^+ = -\lambda_k^-, \qquad (k = 1, 2, ...). \tag{2.237}$$

This property is illustrated by Figure 2.16.

This phenomenon of mirror symmetry spectrum can be traced to the symmetry of the mathematical formulation of the plasmon resonance problem in two dimensions. This symmetry appears due to the existence of the unique stream function for each two-dimensional plasmon mode. To reveal this "internal" symmetry, consider a plasmon mode characterized by λ_k and $\sigma_k(M)$ which satisfy the boundary integral equation (2.91) and charge neutrality condition (2.42), where S is replaced by L. The potential of this plasmon mode created by surface charges $\sigma_k(M)$ is the solution of the following boundary value problem:

$$\nabla^2 \varphi_k^+ = 0 \quad \text{in } V^+, \tag{2.238}$$

$$\nabla^2 \varphi_k^- = 0 \quad \text{in } V^-, \tag{2.239}$$

$$\varphi_k^+ = \varphi_k^- \quad \text{on } L, \tag{2.240}$$

$$\varepsilon_k \frac{\partial \varphi_k^+}{\partial n} = \varepsilon_0 \frac{\partial \varphi_k^-}{\partial n} \quad \text{on } L, \tag{2.241}$$

$$\varphi_k^-(\infty) = 0, \tag{2.242}$$

where ε_k is related to λ_k by formulas (2.41).

Now, we introduce the stream function ψ_k for the electric displacement field $\mathbf{D}_k$ of this plasmon mode,

$$\mathbf{D}_k^\pm = \mathbf{e}_z \times \operatorname{grad} \psi_k^\pm, \tag{2.243}$$

where $\mathbf{e}_z$ is the unit vector normal to a cross-sectional plane. It is apparent that due to the charge neutrality condition, the stream function ψ_k defined by (2.243) is single-valued. It is also clear that for any $\psi_k^{\pm}$,

$$\operatorname{div} \mathbf{D}_k^{\pm} = 0. \tag{2.244}$$

By using the equation

$$\operatorname{curl} \mathbf{D}_k^{\pm} = 0 \tag{2.245}$$

and the boundary conditions for $\mathbf{D}_k^{\pm}$ on L, we arrive at the following boundary value problem:

$$\nabla^2 \psi_k^+ = 0 \quad \text{in } V^+, \tag{2.246}$$

$$\nabla^2 \psi_k^- = 0 \quad \text{in } V^-, \tag{2.247}$$

$$\psi_k^+ = \psi_k^- \quad \text{on } L, \tag{2.248}$$

$$\frac{\varepsilon_0^2}{\varepsilon_k} \frac{\partial \psi_k^+}{\partial n} = \varepsilon_0 \frac{\partial \psi_k^-}{\partial n} \quad \text{on } L, \tag{2.249}$$

$$\psi_k^-(\infty) = 0, \tag{2.250}$$

where the last equation implies that the reference point for ψ_k is at infinity.

It is clear now that the existence of nonzero solution for the boundary value problems (2.238)-(2.242) implies the existence of nonzero solution for the boundary value problem (2.246)-(2.250). It is also clear that the boundary value problem (2.246)-(2.250) is mathematically transformed into the boundary value problem (2.238)-(2.242) by replacing $\psi_k^{\pm}$ by $\varphi_k^{\pm}$ and ε_k by $\varepsilon_0^2/\varepsilon_k$. This implies that if there exists a plasmon mode for $\varepsilon = \varepsilon_k$, then there also exists a plasmon mode for $\varepsilon = \varepsilon_0^2/\varepsilon_k$. In other words, plasmon resonances may occur for reciprocal values of nanowire relative dielectric permittivity. According to formula (2.37), this means that the "mirror" symmetry property (2.237) holds and Property 7 is established.

According to Property 5, the spectrum of integral equation

$$\sigma_k(Q) = \frac{\lambda_k}{\pi} \oint_L \sigma_k(M) \left[\frac{\mathbf{r}_{MQ} \cdot \mathbf{n}_Q}{r_{MQ}^2} - \frac{\pi}{L} \right] dl_M \tag{2.251}$$

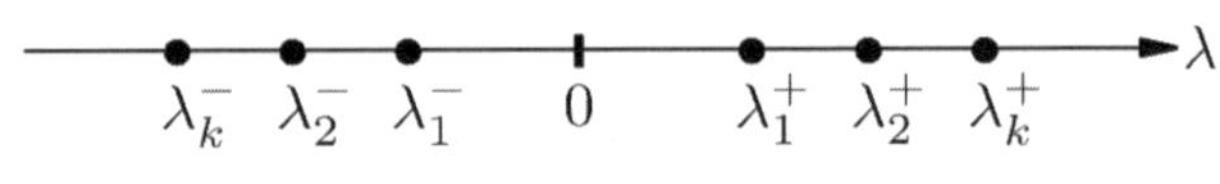

Figure 2.17

coincides with the spectrum of integral equation (2.91) except for the eigenvalue $\lambda_0 = 1$. This means that the spectrum of integral equation (2.251) is purely mirror symmetric (see Figure 2.17), i.e., it consists only of eigenvalue pairs which satisfy the condition (2.237). This mirror symmetry property of the spectrum of integral equation (2.251) will be used in section 2.5 to establish the connection between plasmon mode analysis and the Riemann hypothesis.

The mirror symmetry property (2.237) of the spectrum of integral equation (2.91) has been established by using the mathematical (internal) symmetry of boundary value problems (2.238)-(2.242) and (2.246)-(2.250). There are other properties of the spectrum which are related to geometric symmetries of boundaries S and L of nanoparticles and nanowires, respectively. Namely, geometric multiplicities of eigenvalues λ_k of boundary integral equation (2.36) (or (2.91)) can be predicted by using irreducible representations of the symmetry groups of nanoparticle boundaries S, i.e., the group of geometric transformations with respect to which S (or L) is invariant. This can be stated as

Property 8. The numbers N_i of linearly independent eigenfunctions $\sigma_k^{(i)}(M)$, $(i = 1, 2, ..., N_i)$ corresponding to eigenvalues λ_k of integral equation (2.36) are equal to the dimensions of inequivalent irreducible representations of the symmetry group of the boundary S.

This property can be established in the same way as it is done in quantum mechanical analysis [18, 19] of geometric multiplicities of quantized physical quantities (energy, for instance).

As an example illustrating Property 8, consider a spherical nanoparticle. As shown in section 2.1 (see formulas (2.51) and (2.53)), the spherical harmonics $Y_{km}(\theta, \phi)$, $(m = -k, ..., 0, ..., k)$ are $2k + 1$ linearly independent eigenfunctions of integral equation (2.36) corresponding to eigenvalues $\lambda_k = 2k+1$ (and resonance values $\varepsilon_k = -\varepsilon_0 \frac{k+1}{k}$) of the spherical nanoparticle. Thus, eigenvalues λ_k have $2k + 1$ geometric multiplicity which coincides with the dimensions of irreducible representations of the rotation group, which is the symmetry group of the sphere [15]. It is clear that the practical utility of Property 8 is in the prediction of the number of different plasmon modes that can be excited at the same resonance frequency (wavelength) of incident radiation.

Now, we proceed to the discussion of another issue. It is clear that the kernel of integral equation (2.36) is not symmetric. For this reason, the corresponding integral operator is not Hermitian (i.e., self-adjoint) in the space $\mathcal{L}_2(S)$. This implies that the eigenfunctions $\sigma_k(M)$ and $\sigma_i(M)$ corresponding to different eigenvalues λ_k and λ_i, respectively, are not orthogonal on S,

$$\langle \sigma_k, \sigma_i \rangle = \oint_S \sigma_k(M)\sigma_i(M)ds_M \neq 0. \tag{2.252}$$

It turns out that the electric fields $\mathbf{E}_k$ and $\mathbf{E}_i$ of plasmon modes corresponding to eigenfunctions $\sigma_k(M)$ and $\sigma_i(M)$, respectively, are orthogonal in the whole space $\mathbb{R}^3$. Moreover, the strong orthogonality property for $\mathbf{E}_k$ and $\mathbf{E}_i$ is valid, which states that $\mathbf{E}_k$ and $\mathbf{E}_i$ are orthogonal separately in V^+ and V^-. This is formulated as

Property 9. The following orthogonality conditions are valid:

$$\int_{V^+} \mathbf{E}_k^+ \cdot \mathbf{E}_i^+ dv = 0, \quad (k \neq i), \tag{2.253}$$

$$\int_{V^-} \mathbf{E}_k^- \cdot \mathbf{E}_i^- dv = 0, \quad (k \neq i). \tag{2.254}$$

To establish this property, we turn to formulas (2.30) and (2.31) and shall rewrite integral equation (2.36) as follows:

$$\mathbf{n}_Q \cdot \left[\mathbf{E}_k^-(Q) - \mathbf{E}_k^+(Q)\right] = \lambda_k \mathbf{n}_Q \cdot \left[\mathbf{E}_k^-(Q) + \mathbf{E}_k^+(Q)\right]. \tag{2.255}$$

The last formula can be transformed as

$$(1 - \lambda_k)\, \mathbf{n}_Q \cdot \mathbf{E}_k^-(Q) = (1 + \lambda_k)\, \mathbf{n}_Q \cdot \mathbf{E}_k^+(Q). \tag{2.256}$$

Next, we shall multiply both sides of equation (2.256) by electric potential $\varphi_i(Q)$ of the plasmon mode corresponding to λ_i and integrate over S. This results in

$$(1 - \lambda_k) \oint_S \varphi_i(Q)\mathbf{n}_Q \cdot \mathbf{E}_k^-(Q)ds_Q = (1 + \lambda_k) \oint_S \varphi_i(Q)\mathbf{n}_Q \cdot \mathbf{E}_k^+(Q)ds_Q. \tag{2.257}$$

According to the divergence theorem and equation div $\mathbf{E}_k^+ = 0$, we have

$$\oint_S \varphi_i(Q)\mathbf{n}_Q \cdot \mathbf{E}_k^+(Q)ds_Q = \int_{V+} \text{div}\left(\varphi_i^+ \mathbf{E}_k^+\right) dv$$

$$= \int_{V+} \left(\mathbf{E}_k^+ \cdot \text{grad } \phi_i^+ + \phi_i^+ \text{div } \mathbf{E}_k^+\right) dv$$

$$= -\int_{V+} \mathbf{E}_k^+ \cdot \mathbf{E}_i^+ dv. \tag{2.258}$$

Similarly, we establish that

$$\oint_S \varphi_i(Q)\mathbf{n}_Q \cdot \mathbf{E}_k^- ds_Q = \int_{V-} \mathbf{E}_k^- \cdot \mathbf{E}_i^- dv. \tag{2.259}$$

By substituting the last two formulas into equation (2.257), we find

$$(\lambda_k - 1)\int_{V-} \mathbf{E}_k^- \cdot \mathbf{E}_i^- dv = (1 + \lambda_k)\int_{V+} \mathbf{E}_k^+ \cdot \mathbf{E}_i^+ dv. \tag{2.260}$$

Next, we consider equation (2.36) written for λ_i and $\sigma_i(M)$. By literally repeating the same line of reasoning as before, we derive

$$(\lambda_i - 1)\int_{V-} \mathbf{E}_i^- \cdot \mathbf{E}_k^- dv = (1 + \lambda_i)\int_{V+} \mathbf{E}_i^+ \cdot \mathbf{E}_k^+ dv. \tag{2.261}$$

Since $\lambda_k \neq \lambda_i$ and $|\lambda_k| > 1$, $(k = 1, 2, ...)$, we conclude that equalities (2.260) and (2.261) can be simultaneously valid only if the orthogonality conditions (2.253) and (2.254) hold. Thus, Property 9 is established.

It has been pointed out before that the integral operator in equation (2.36) is not self-adjoint (Hermitian) in $\mathcal{L}_2(S)$. Such an operator may have eigenfunctions along with "generalized" eigenfunctions (or null-functions) which are defined by the following chain of equations:

$$\sigma_k^{(1)}(Q) = \frac{\lambda_k}{2\pi}\oint_S \sigma_k^{(1)}(M)\frac{\mathbf{r}_{MQ} \cdot \mathbf{n}_Q}{r_{MQ}^3}ds_M, \tag{2.262}$$

$$\sigma_k^{(1)}(Q) + \sigma_k^{(2)}(Q) = \frac{\lambda_k}{2\pi}\oint_S \sigma_k^{(2)}(M)\frac{\mathbf{r}_{MQ} \cdot \mathbf{n}_Q}{r_{MQ}^3}ds_M, \tag{2.263}$$

$$\vdots$$

$$\sigma_k^{(n-1)}(Q) + \sigma_k^{(n)}(Q) = \frac{\lambda_k}{2\pi}\oint_S \sigma_k^{(n)}(M)\frac{\mathbf{r}_{MQ} \cdot \mathbf{n}_Q}{r_{MQ}^3}ds_M. \tag{2.264}$$

In operator form, these equations can be written as follows:

$$\left(\hat{I} - \lambda_k \hat{K}\right)\sigma_k^{(1)} = 0, \tag{2.265}$$

$$\sigma_k^{(1)} = -\left(\hat{I} - \lambda_k \hat{K}\right)\sigma_k^{(2)}, \tag{2.266}$$

$$\vdots$$

$$\sigma_k^{(n-1)} = -\left(\hat{I} - \lambda_k \hat{K}\right)\sigma_k^{(n)}, \tag{2.267}$$

where $\hat{I}$ is the identity operator. By applying operation $\left(\hat{I} - \lambda_k \hat{K}\right)$ to both sides of (2.266) and (2.267), we find

$$\left(\hat{I} - \lambda_k \hat{K}\right)^2 \sigma_k^{(2)} = 0, \quad \cdots, \quad \left(\hat{I} - \lambda_k \hat{K}\right)^n \sigma_k^{(n)} = 0. \tag{2.268}$$

This explains why $\sigma_k^{(i)}(M)$, $(i > 1)$ are called "generalized" eigenfunctions. For self-adjoint operators, "generalized" (in the sense of (2.268)) eigenfunctions do not exist. It turns out that the same is true for the integral operator in equation (2.36). This is stated as

Property 10. If $\sigma_k^{(1)}(M)$ is nonzero solution of equation (2.262), then $\sigma_k^{(i)}(M) = 0$ for $i > 1$.

To establish this property, we assume that $\sigma_k^{(1)}(M) \neq 0$ and $\sigma_k^{(2)}(M) \neq 0$ and show that this assumption will lead to the contradiction. To arrive at the contradiction, we introduce potentials

$$\varphi_k^{(1)}(Q) = \frac{1}{4\pi\varepsilon_0} \oint_S \frac{\sigma_k^{(1)}(M)}{r_{MQ}} ds_M, \tag{2.269}$$

$$\varphi_k^{(2)}(Q) = \frac{1}{4\pi\varepsilon_0} \oint_S \frac{\sigma_k^{(2)}(M)}{r_{MQ}} ds_M. \tag{2.270}$$

By using formula (2.32), equations (2.262) and (2.263) can be respectively rewritten as follows:

$$\frac{\partial \varphi_k^{(1)-}}{\partial n}(Q) - \frac{\partial \varphi_k^{(1)+}}{\partial n}(Q) = \lambda_k \left[\frac{\partial \varphi_k^{(1)-}}{\partial n}(Q) + \frac{\partial \varphi_k^{(1)+}}{\partial n}(Q) \right], \tag{2.271}$$

$$\left[\frac{\partial\varphi_k^{(1)-}}{\partial n}(Q) - \frac{\partial\varphi_k^{(1)+}}{\partial n}(Q)\right] + \left[\frac{\partial\varphi_k^{(2)-}}{\partial n}(Q) - \frac{\partial\varphi_k^{(2)+}}{\partial n}(Q)\right]$$
$$= \lambda_k\left[\frac{\partial\varphi_k^{(2)-}}{\partial n}(Q) + \frac{\partial\varphi_k^{(2)+}}{\partial n}(Q)\right]. \tag{2.272}$$

By multiplying both sides of formulas (2.271) and (2.272) by $\varphi_k^{(2)}(Q)$ and $\varphi_k^{(1)}(Q)$, respectively, and integrating over S, we find

$$\oint_S \varphi_k^{(2)}(Q)\left[\frac{\partial\varphi_k^{(1)-}}{\partial n}(Q) - \frac{\partial\varphi_k^{(1)+}}{\partial n}(Q)\right] ds_Q$$
$$= \lambda_k\oint_S \varphi_k^{(2)}(Q)\left[\frac{\partial\varphi_k^{(1)-}}{\partial n}(Q) + \frac{\partial\varphi_k^{(1)+}}{\partial n}(Q)\right] ds_Q, \tag{2.273}$$

$$\oint_S \varphi_k^{(1)}(Q)\left[\frac{\partial\varphi_k^{(1)-}}{\partial n}(Q) - \frac{\partial\varphi_k^{(1)+}}{\partial n}(Q)\right] ds_Q$$
$$+ \oint_S \varphi_k^{(1)}(Q)\left[\frac{\partial\varphi_k^{(2)-}}{\partial n}(Q) - \frac{\partial\varphi_k^{(2)+}}{\partial n}(Q)\right] ds_Q$$
$$= \lambda_k\oint_S \varphi_k^{(1)}(Q)\left[\frac{\partial\varphi_k^{(2)-}}{\partial n}(Q) + \frac{\partial\varphi_k^{(2)+}}{\partial n}(Q)\right] ds_Q. \tag{2.274}$$

Now, by using the well-known Green formula

$$\int_{V^\pm}\left(\varphi_k^{(1)\pm}\nabla^2\varphi_k^{(2)\pm} - \varphi_k^{(2)\pm}\nabla^2\varphi_k^{(1)\pm}\right) dv$$
$$= \pm\oint_S\left[\varphi_k^{(1)}(Q)\frac{\partial\varphi_k^{(2)\pm}}{\partial n}(Q) - \varphi_k^{(2)}(Q)\frac{\partial\varphi_k^{(1)\pm}}{\partial n}(Q)\right] ds_Q \tag{2.275}$$

and the fact that potentials $\varphi_k^{(1)}(Q)$ and $\varphi_k^{(2)}(Q)$ satisfy the Laplace equation in V^+ and V^-, we obtain

$$\oint_S \varphi_k^{(1)}(Q)\frac{\partial\varphi_k^{(2)\pm}}{\partial n}(Q)ds_Q = \oint_S \varphi_k^{(2)}(Q)\frac{\partial\varphi_k^{(1)\pm}}{\partial n}(Q)ds_Q. \tag{2.276}$$

By using formula (2.276), we conclude that the first integral in formula (2.273) is equal to the second integral in formula (2.274) and the second integral in formula (2.273) is equal to the third integral in formula (2.274). This implies that

$$\oint_S \varphi_k^{(1)}(Q) \left[\frac{\partial \varphi_k^{(1)-}}{\partial n}(Q) - \frac{\partial \varphi_k^{(1)+}}{\partial n}(Q) \right] ds_Q = 0. \tag{2.277}$$

Now, by using formulas (2.199) and (2.277), we conclude that

$$\int_{V^+} \left| \operatorname{grad} \varphi_k^{(1)\,|} \right|^2 dv + \int_{V^-} \left| \operatorname{grad} \varphi_k^{(1)-} \right|^2 dv = 0, \tag{2.278}$$

which implies

$$\operatorname{grad} \varphi_k^{(1)\pm} \equiv 0. \tag{2.279}$$

This means according to formula (2.32) that

$$\sigma_k^{(1)}(Q) \equiv 0 \quad \text{on } S. \tag{2.280}$$

The latter contradicts the assumption that $\sigma_k^{(1)}(Q)$ is nonzero. Thus, Property 10 is established.

The established property has important consequences concerning the solution of the following inhomogeneous integral equation:

$$\sigma(Q) - \frac{\lambda}{2\pi} \oint_S \sigma(M) \frac{\mathbf{r}_{MQ} \cdot \mathbf{n}_Q}{r_{MQ}^3} ds_M = f(Q). \tag{2.281}$$

The solution of this inhomogeneous equation can be written in the form

$$\sigma(Q) = f(Q) + \lambda \oint_S f(M)\Gamma(Q, M, \lambda) \, ds_M, \tag{2.282}$$

where $\Gamma(Q, M, \lambda)$ is called the resolvent and can be construed as the kernel of the inverse integral operator. This resolvent is a meromorphic function of complex variable λ and can be represented as the ratio of two entire functions,

$$\Gamma(Q, M, \lambda) = \frac{D(Q, M, \lambda)}{D(\lambda)}, \tag{2.283}$$

where the entire function $D(\lambda)$ is called the Fredholm determinant, while the entire function $D(Q, M, \lambda)$ is called the first Fredholm minor. It is apparent that the eigenvalues λ_k of integral equation (2.36) are poles of the resolvent $\Gamma(Q, M, \lambda)$ and zeros of the Fredholm determinant,

$$D(\lambda_k) = 0, \tag{2.284}$$

otherwise integral equation (2.281) will have a unique solution given by formula (2.282). It turns out (see [20]) that Property 10 implies

Property 11. Eigenvalues λ_k of integral equation (2.36) are simple poles of the resolvent of integral equation (2.281).

In the generic case of nonsymmetric boundary S (and when accidental degeneration of eigenvalues λ_k is not present), the geometric multiplicity of eigenvalues λ_k is equal to one. It can be shown (see [17]) that this combined with the simplicity of resolvent poles leads to the conclusion that all zeros of the Fredholm determinant $D(\lambda)$ of integral equation (2.36) (or (2.91)) are simple.

It is clear from the presented discussion that the integral operator in equation (2.36) has such properties as real eigenvalues, orthogonality of plasmon electric fields created by eigenfunctions corresponding to different eigenvalues and simplicity of poles of the resolvent. These properties are typical for self-adjoint (Hermitian) operators. This observation suggests that the integral operator in equation (2.36) will be self-adjoint in a properly chosen Hilbert space. Orthogonality conditions (2.253)-(2.254) imply that this must be the "energy" space $H(S)$ in which the inner product is defined as follows:

$$\langle \sigma, \nu \rangle_{H(S)} = \oint_S \sigma(Q) \left(\oint_S \frac{\nu(P)}{r_{PQ}} ds_P \right) ds_Q. \tag{2.285}$$

It is easy to see that

$$\langle \sigma, \nu \rangle_{H(S)} = \langle \nu, \sigma \rangle_{H(S)} \tag{2.286}$$

and

$$\langle \sigma, \sigma \rangle_{H(S)} = \oint_S \sigma(Q) \left(\oint_S \frac{\sigma(P)}{r_{PQ}} ds_P \right) ds_Q > 0. \tag{2.287}$$

The latter inequality is transparent from the physical point of view because $\langle \sigma, \sigma \rangle$ is proportional to the energy of the electric field created in free space by surface charge distribution $\sigma(P)$. Now, it can be demonstrated that the operator $\hat{K}$ in integral equation (2.36) is formally self-adjoint in $H(S)$. Indeed, we have

$$\left\langle \hat{K}\sigma, \nu \right\rangle_{H(S)} = \oint_S \left[\frac{1}{2\pi} \oint_S \sigma(M) \frac{\mathbf{r}_{MQ} \cdot \mathbf{n}_Q}{r_{MQ}^3} ds_M \right] \left[\oint_S \frac{\nu(P)}{r_{PQ}} ds_P \right] ds_Q$$

$$= - \oint_S \left[\frac{1}{2\pi} \oint_S \sigma(M) \left(\oint_S \frac{1}{r_{PQ}} \frac{\partial}{\partial n_Q} \left(\frac{1}{r_{MQ}} \right) ds_Q \right) ds_M \right] \nu(P) ds_P.$$

$$(2.288)$$

By using the well-known Green formula

$$\int_{V+} \left(\psi \nabla^2 \varphi - \varphi \nabla^2 \psi \right) dv = \oint_S \left(\psi \frac{\partial \varphi}{\partial n} - \varphi \frac{\partial \psi}{\partial n} \right) ds \qquad (2.289)$$

and choosing $\varphi = 1/r_{PQ}$ and $\psi = 1/r_{MQ}$ which satisfy the Laplace equation with respect to Q when P and M are located on S, we formally obtain [8]

$$\oint_S \frac{1}{r_{PQ}} \frac{\partial}{\partial n_Q} \left(\frac{1}{r_{MQ}} \right) ds_Q = \oint_S \frac{1}{r_{MQ}} \frac{\partial}{\partial n_Q} \left(\frac{1}{r_{PQ}} \right) ds_Q. \qquad (2.290)$$

The word "formally" is used because the issue of singularities at M and P has been ignored. By leaving this and other mathematical technicalities aside and by using formula (2.290) in equation (2.288), we find

$$\left\langle \hat{K}\sigma, \nu \right\rangle_{H(S)} = - \oint_S \frac{1}{2\pi} \left[\oint_S \sigma(M) \left(\oint_S \frac{1}{r_{MQ}} \frac{\partial}{\partial n_Q} \left(\frac{1}{r_{PQ}} \right) ds_Q \right) ds_M \right] \nu(P) ds_P$$

$$= \oint_S \sigma(M) \left[\oint_S \frac{1}{r_{MQ}} \left(\oint_S \nu(P) \frac{\mathbf{r}_{PQ} \cdot \mathbf{n}_Q}{r_{PQ}^3} ds_P \right) ds_Q \right] ds_M$$

$$= \left\langle \sigma, \hat{K}\nu \right\rangle_{H(S)}, \qquad (2.291)$$

which implies that operator $\hat{K}$ is self-adjoint in $H(S)$.

2.4 Plasmon Resonances in Nanoshells

It is apparent that the tunability of plasmon resonances to a desired wavelength of incident radiation may be quite valuable in various applications. It has been found that a wide range of tunability can be achieved by using metallic nanoshells and controlling the plasmon resonance wavelength by

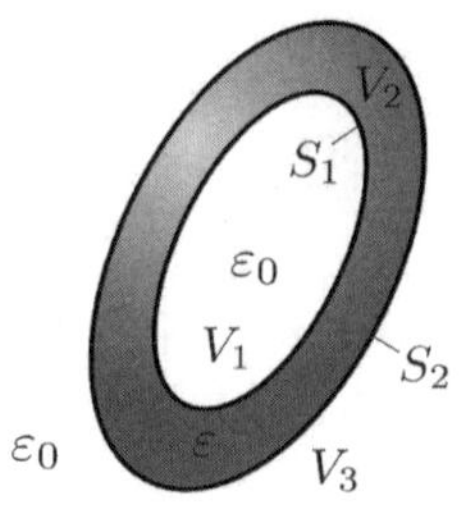

Figure 2.18

adjusting the shell thickness (see, for instance, [21]-[23]). For this reason, plasmon resonances in metallic nanoshells are of great interest. It turns out that the analysis of plasmon modes in nanoshells has unique features which are due to the presence of the two (interior and exterior) boundaries of the nanoshells. In this section, these features are discussed in detail.

We first consider a nanoshell (see Figure 2.18) whose core (i.e., region V_1) has permittivity ε_0. We are interested in such values of dielectric permittivity ε of a nanoshell for which nonzero source-free electric fields may exist. These fields are plasmon modes of a nanoshell. By introducing the potential φ for these electric fields

$$\mathbf{E} = -\mathrm{grad}\ \varphi, \qquad (2.292)$$

we conclude that plasmon modes are nonzero solutions of the following homogeneous boundary value problem:

$$\nabla^2 \varphi = 0 \quad \text{in } V_k \ (k = 1, 2, 3), \qquad (2.293)$$

$$\varphi^+ = \varphi^- \quad \text{on } S_1, \qquad (2.294)$$

$$\varphi^+ = \varphi^- \quad \text{on } S_2, \qquad (2.295)$$

$$\varepsilon_0 \frac{\partial \varphi^+}{\partial n} = \varepsilon \frac{\partial \varphi^-}{\partial n} \quad \text{on } S_1, \qquad (2.296)$$

$$\varepsilon \frac{\partial \varphi^+}{\partial n} = \varepsilon_0 \frac{\partial \varphi^-}{\partial n} \quad \text{on } S_2, \qquad (2.297)$$

$$\varphi(\infty) = 0. \qquad (2.298)$$

Here, superscripts "+" and "−" are used for the notations of limiting values of φ and $\partial \varphi / \partial n$ on S_1 and S_2 when those surfaces are approached from inside and outside, respectively.

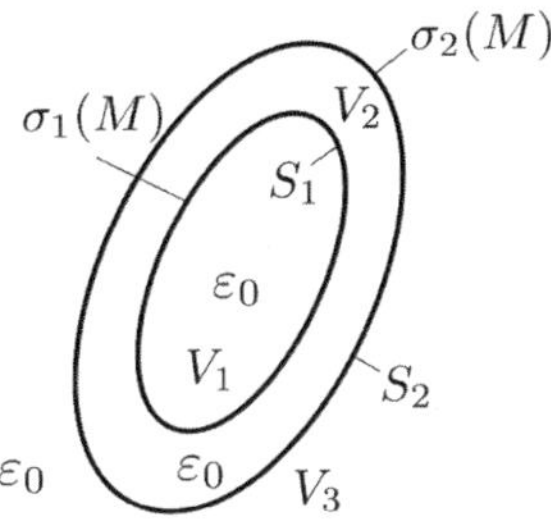

Figure 2.19

To find the specific negative values of ε for which the nonzero source-free plasmon electric fields may exist, we replace the permittivity of the nanoshell by ε_0 (see Figure 2.19) and introduce on its boundaries S_1 and S_2 virtual surface electric charges with densities $\sigma^{(1)}(M)$ and $\sigma^{(2)}(M)$, respectively. The electric potential created by these charges is given by the formula

$$\varphi(Q) = \frac{1}{4\pi\varepsilon_0} \left(\oint_{S_1} \frac{\sigma^{(1)}(M)}{r_{MQ}} ds_M + \oint_{S_2} \frac{\sigma^{(2)}(M)}{r_{MQ}} ds_M \right). \tag{2.299}$$

It is apparent that for any distribution of electric charges $\sigma^{(1)}(M)$ and $\sigma^{(2)}(M)$ the potential given by the formula (2.299) satisfies the Laplace equation (2.293) in the regions V_k, $(k = 1, 2, 3)$ as well as the boundary conditions (2.294)-(2.295) along with the condition at infinity (2.298). Next, we shall find under what restrictions on $\sigma^{(1)}(M)$ and $\sigma^{(2)}(M)$ the boundary conditions (2.296)-(2.297) will also be satisfied. In other words, we shall find under what conditions on $\sigma^{(1)}(M)$ and $\sigma^{(2)}(M)$ the electric fields created by these surface charges in homogeneous space with permittivity ε_0 replicate the plasmon electric fields of the nanoshell. To this end, we shall recall formula (2.32) and find that the boundary conditions (2.296)-(2.297) will be enforced if the following relations are valid on S_1 and S_2, respectively:

$$\frac{\sigma^{(1)}(Q)}{2} - \frac{1}{4\pi} \left(\oint_{S_1} \sigma^{(1)}(M) \frac{\mathbf{r}_{MQ} \cdot \mathbf{n}_Q}{r_{MQ}^3} ds_M + \oint_{S_2} \sigma^{(2)}(M) \frac{\mathbf{r}_{MQ} \cdot \mathbf{n}_Q}{r_{MQ}^3} ds_M \right)$$

$$= -\frac{\varepsilon}{\varepsilon_0} \frac{\sigma^{(1)}(Q)}{2} - \frac{\varepsilon}{4\pi\varepsilon_0} \left(\oint_{S_1} \sigma^{(1)}(M) \frac{\mathbf{r}_{MQ} \cdot \mathbf{n}_Q}{r_{MQ}^3} ds_M \right.$$

$$\left. + \oint_{S_2} \sigma^{(2)}(M) \frac{\mathbf{r}_{MQ} \cdot \mathbf{n}_Q}{r_{MQ}^3} ds_M \right), \tag{2.300}$$

$$\frac{\varepsilon}{\varepsilon_0}\frac{\sigma^{(2)}(Q)}{2} - \frac{\varepsilon}{4\pi\varepsilon_0}\left(\oint_{S_1}\sigma^{(1)}(M)\frac{\mathbf{r}_{MQ}\cdot\mathbf{n}_Q}{r_{MQ}^3}ds_M + \oint_{S_2}\sigma^{(2)}(M)\frac{\mathbf{r}_{MQ}\cdot\mathbf{n}_Q}{r_{MQ}^3}ds_M\right)$$

$$= -\frac{\sigma^{(2)}(Q)}{2} - \frac{1}{4\pi}\left(\oint_{S_1}\sigma^{(1)}(M)\frac{\mathbf{r}_{MQ}\cdot\mathbf{n}_Q}{r_{MQ}^3}ds_M\right.$$

$$\left. + \oint_{S_2}\sigma^{(2)}(M)\frac{\mathbf{r}_{MQ}\cdot\mathbf{n}_Q}{r_{MQ}^3}ds_M\right). \tag{2.301}$$

Now, after simple transformation we arrive at the following coupled integral equations for $\sigma^{(1)}(M)$ and $\sigma^{(2)}(M)$:

$$\sigma^{(1)}(Q) = -\frac{\lambda}{2\pi}\left(\oint_{S_1}\sigma^{(1)}(M)\frac{\mathbf{r}_{MQ}\cdot\mathbf{n}_Q}{r_{MQ}^3}ds_M + \oint_{S_2}\sigma^{(2)}(M)\frac{\mathbf{r}_{MQ}\cdot\mathbf{n}_Q}{r_{MQ}^3}ds_M\right), \tag{2.302}$$

$$\sigma^{(2)}(Q) = \frac{\lambda}{2\pi}\left(\oint_{S_1}\sigma^{(1)}(M)\frac{\mathbf{r}_{MQ}\cdot\mathbf{n}_Q}{r_{MQ}^3}ds_M + \oint_{S_2}\sigma^{(2)}(M)\frac{\mathbf{r}_{MQ}\cdot\mathbf{n}_Q}{r_{MQ}^3}ds_M\right), \tag{2.303}$$

where, as before,

$$\lambda = \frac{\varepsilon - \varepsilon_0}{\varepsilon + \varepsilon_0}. \tag{2.304}$$

Thus, the electric fields created by the distributions of virtual charges $\sigma^{(1)}(M)$ and $\sigma^{(2)}(M)$ over S_1 and S_2, respectively, will mimic actual source-free plasmon electric fields of the nanoshell only for such λ_k (and the corresponding values of ε_k) that nonzero solutions $\sigma_k^{(1)}(M)$ and $\sigma_k^{(2)}(M)$ of coupled homogeneous integral equations (2.302)-(2.303) exist, i.e.,

$$\boxed{\sigma_k^{(1)}(Q) = -\frac{\lambda_k}{2\pi}\left(\oint_{S_1}\sigma_k^{(1)}(M)\frac{\mathbf{r}_{MQ}\cdot\mathbf{n}_Q}{r_{MQ}^3}ds_M + \oint_{S_2}\sigma_k^{(2)}(M)\frac{\mathbf{r}_{MQ}\cdot\mathbf{n}_Q}{r_{MQ}^3}ds_M\right),}$$

$$\tag{2.305}$$

$$\boxed{\sigma_k^{(2)}(Q) = \frac{\lambda_k}{2\pi}\left(\oint_{S_1}\sigma_k^{(1)}(M)\frac{\mathbf{r}_{MQ}\cdot\mathbf{n}_Q}{r_{MQ}^3}ds_M + \oint_{S_2}\sigma_k^{(2)}(M)\frac{\mathbf{r}_{MQ}\cdot\mathbf{n}_Q}{r_{MQ}^3}ds_M\right),}$$

$$\tag{2.306}$$

$$\lambda_k = \frac{\varepsilon_k - \varepsilon_0}{\varepsilon_k + \varepsilon_0}. \tag{2.307}$$

As before, by solving the eigenvalue problem (2.305)-(2.306), we find the resonance values ε_k of dielectric permittivity and the resonance frequencies (see (2.38)) of the nanoshell as well as the electric fields of plasmon modes which are created by charges $\sigma_k^{(1)}(M)$ and $\sigma_k^{(2)}(M)$.

It will be shown below that eigenvalues λ_k corresponding to plasmon modes satisfy the inequality

$$|\lambda_k| > 1. \tag{2.308}$$

By using this inequality, it can be established that boundaries S_1 and S_2 are "charge-neutral." Indeed, by integrating equation (2.305) over S_1 with respect to Q and taking into account that

$$\oint_{S_1} \frac{\mathbf{r}_{MQ} \cdot \mathbf{n}_Q}{r_{MQ}^3} ds_Q = \begin{cases} 2\pi, & \text{if } M \in S_1, \\ 0, & \text{if } M \in S_2, \end{cases} \tag{2.309}$$

we find that

$$\oint_{S_1} \sigma_k^{(1)}(Q)ds_Q = -\lambda_k \oint_{S_1} \sigma_k^{(1)}(M)ds_M. \tag{2.310}$$

Similarly, by integrating equation (2.306) over S_2 with respect to Q and taking into account that

$$\oint_{S_2} \frac{\mathbf{r}_{MQ} \cdot \mathbf{n}_Q}{r_{MQ}^3} ds_Q = \begin{cases} 4\pi, & \text{if } M \in S_1, \\ 2\pi, & \text{if } M \in S_2, \end{cases} \tag{2.311}$$

we obtain

$$\oint_{S_2} \sigma_k^{(2)}(Q)ds_Q = \lambda_k \left(2 \oint_{S_1} \sigma_k^{(1)}(M)ds_M + \oint_{S_2} \sigma_k^{(2)}(M)ds_M \right). \tag{2.312}$$

It is easy to conclude from formulas (2.308), (2.310) and (2.312) that for any plasmon mode the following "charge-neutrality" conditions are valid:

$$\oint_{S_1} \sigma_k^{(1)}(M)ds_M = 0, \tag{2.313}$$

$$\oint_{S_2} \sigma_k^{(2)}(M)ds_M = 0. \tag{2.314}$$

Next, it will be proved that all eigenvalues λ_k of coupled integral equations (2.305)-(2.306) are real for any shape of interior S_1 and exterior S_2 boundaries of the nanoshell. To do this, we introduce the plasmon mode potential

$$\varphi_k(Q) = \frac{1}{4\pi\varepsilon_0} \left(\oint_{S_1} \frac{\sigma_k^{(1)}(M)}{r_{MQ}} ds_M + \oint_{S_2} \frac{\sigma_k^{(2)}(M)}{r_{MQ}} ds_M \right), \tag{2.315}$$

and from formulas (2.32) and (2.315) we find that for $Q \in S_1$ the following relations are valid:

$$\sigma_k^{(1)}(Q) = \varepsilon_0 \left[\frac{\partial \varphi_k^+}{\partial n}(Q) - \frac{\partial \varphi_k^-}{\partial n}(Q) \right], \tag{2.316}$$

$$\frac{1}{2\pi} \left[\oint_{S_1} \sigma_k^{(1)}(M) \frac{\mathbf{r}_{MQ} \cdot \mathbf{n}_Q}{r_{MQ}^3} ds_M + \oint_{S_2} \sigma_k^{(2)}(M) \frac{\mathbf{r}_{MQ} \cdot \mathbf{n}_Q}{r_{MQ}^3} ds_M \right]$$
$$= -\varepsilon_0 \left[\frac{\partial \varphi_k^+}{\partial n}(Q) + \frac{\partial \varphi_k^-}{\partial n}(Q) \right]. \tag{2.317}$$

By substituting the last two formulas into integral equation (2.305) we obtain

$$\frac{\partial \varphi_k^+}{\partial n}(Q) - \frac{\partial \varphi_k^-}{\partial n}(Q) = \lambda_k \left[\frac{\partial \varphi_k^+}{\partial n}(Q) + \frac{\partial \varphi_k^-}{\partial n}(Q) \right] \quad \text{on } S_1. \tag{2.318}$$

By using the same line of reasoning from integral equation (2.306) we find

$$\frac{\partial \varphi_k^+}{\partial n}(Q) - \frac{\partial \varphi_k^-}{\partial n}(Q) = -\lambda_k \left[\frac{\partial \varphi_k^+}{\partial n}(Q) + \frac{\partial \varphi_k^-}{\partial n}(Q) \right] \quad \text{on } S_2. \tag{2.319}$$

If the eigenfunctions $\sigma_k^{(1)}(M)$ and $\sigma_k^{(2)}(M)$ are complex-valued, then we multiply both sides of equations (2.318) and (2.319) by a complex conjugate $\varphi_k^*(Q)$ of $\varphi_k(Q)$ and integrate over S_1 and S_2, respectively. Next, we add "integrated" formulas (2.318) and (2.319) together and obtain

$$\oint_{S_1} \varphi_k^* \frac{\partial \varphi_k^+}{\partial n} ds - \oint_{S_1} \varphi_k^* \frac{\partial \varphi_k^-}{\partial n} ds + \oint_{S_2} \varphi_k^* \frac{\partial \varphi_k^+}{\partial n} ds - \oint_{S_2} \varphi_k^* \frac{\partial \varphi_k^-}{\partial n} ds$$
$$= \lambda_k \left[\oint_{S_1} \varphi_k^* \frac{\partial \varphi_k^+}{\partial n} ds + \oint_{S_1} \varphi_k^* \frac{\partial \varphi_k^-}{\partial n} ds - \oint_{S_2} \varphi_k^* \frac{\partial \varphi_k^+}{\partial n} ds - \oint_{S_2} \varphi_k^* \frac{\partial \varphi_k^-}{\partial n} ds \right]. \tag{2.320}$$

Then, by using the relation similar to (2.199) and taking into account that $\nabla^2 \varphi_k = 0$ in V_k, $(k = 1, 2, 3)$, we find the following expressions for the above surface integrals:

$$\oint_{S_1} \varphi_k^*(Q) \frac{\partial \varphi_k^+}{\partial n}(Q) ds_Q = \int_{V_1} |\text{grad } \varphi_k|^2 \, dv = W_k^{(1)}, \tag{2.321}$$

$$-\oint_{S_2} \varphi_k^*(Q) \frac{\partial \varphi_k^-}{\partial n}(Q) ds_Q = \int_{V_3} |\text{grad } \varphi_k|^2 \, dv = W_k^{(3)}, \tag{2.322}$$

$$-\oint_{S_1} \varphi_k^*(Q) \frac{\partial \varphi_k^-}{\partial n}(Q) ds_Q + \oint_{S_2} \varphi_k^*(Q) \frac{\partial \varphi_k^+}{\partial n}(Q) ds_Q$$
$$= \int_{V_2} |\text{grad } \varphi_k|^2 \, dv = W_k^{(2)}, \tag{2.323}$$

where negative signs in (2.322) and (2.323) account for the fact that the unit normals on S_1 and S_2 have inward directions for the regions V_2 and V_3, respectively.

By substituting formulas (2.321)-(2.323) into equation (2.320), we arrive at the equality

$$W_k^{(1)} + W_k^{(2)} + W_k^{(3)} = \lambda_k \left(W_k^{(1)} - W_k^{(2)} + W_k^{(3)} \right), \tag{2.324}$$

which leads to the expression

$$\lambda_k = \frac{W_k^{(1)} + W_k^{(2)} + W_k^{(3)}}{W_k^{(1)} - W_k^{(2)} + W_k^{(3)}}. \tag{2.325}$$

It is apparent from formulas (2.321)-(2.323) and the last formula that all eigenvalues of coupled integral equations (2.305)-(2.306) are real for any shape of nanoshell boundaries S_1 and S_2. This fact implies that all eigenfunctions of integral equations (2.305)-(2.306) are real as well. It is easy to see that

$$\lambda_0^+ = 1 \quad \text{and} \quad \lambda_0^- = -1 \tag{2.326}$$

are eigenvalues of coupled integral equations (2.305)-(2.306). The first eigenvalue $\lambda_0^+ = 1$ corresponds to the electrostatic Robin problem when the region V_2 is occupied by a charged ideal conductor (see Figure 2.20) and the eigenfunction $\sigma_0^{(1)}(M) \equiv 0$, while $\sigma_0^{(2)}(M)$ has the physical meaning of surface electric charges distributed over the exterior boundary S_2 of this conductor.

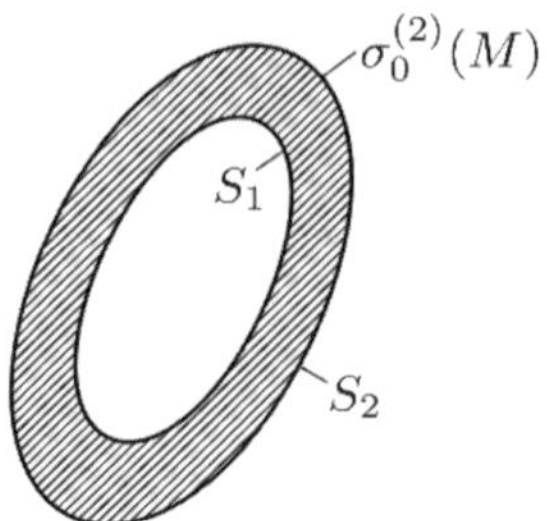

Figure 2.20

It is clear that in this case $W_0^{(2)} = W_0^{(1)} = 0$ and formula (2.325) leads to the first equality in (2.326). On the other hand, the eigenvalue $\lambda_0^- = -1$ corresponds to the electrostatic Robin problem when the regions V_1 and V_3 are occupied by ideal and oppositely charged conductors (see Figure 2.21) and the eigenfunctions $\sigma_0^{(1)}(M)$ and $\sigma_0^{(2)}(M)$ have the physical meaning of surface electric charges distributed over S_1 and S_2, respectively. It is clear that in this case $W_0^{(1)} = W_0^{(3)} = 0$ and formula (2.325) leads to the second equality in (2.326). It follows from the presented discussion that eigenvalues $\lambda_0^+ = 1$ and $\lambda_0^- = -1$ have nothing to do with plasmon modes and, in this sense, they are spurious. It is apparent that all other eigenvalues and eigenfunctions describe plasmon modes in the nanoshell. It is clear from formula (2.325) that for all other eigenvalues λ_k the inequality (2.308) is valid. This inequality and formula (2.307) imply that all ε_k are negative. Actually, by using formulas (2.307) and (2.325), the following expression can be derived:

$$\varepsilon_k = -\varepsilon_0 \frac{W_k^{(1)} + W_k^{(3)}}{W_k^{(2)}}. \tag{2.327}$$

In the same way as has been discussed in the previous section, the spurious eigenvalues $\lambda_0^+ = 1$ and $\lambda_0^- = -1$ can be removed from the spectrum. This is achieved by modifying the coupled integral equations (2.305)-(2.306) as follows:

$$\sigma_k^{(1)}(Q) = -\frac{\lambda_k}{2\pi} \left[\oint_{S_1} \sigma_k^{(1)}(M) \left(\frac{\mathbf{r}_{MQ} \cdot \mathbf{n}_Q}{r_{MQ}^3} - \frac{2\pi}{A_1} \right) ds_M \right.$$
$$\left. + \oint_{S_2} \sigma_k^{(2)}(M) \frac{\mathbf{r}_{MQ} \cdot \mathbf{n}_Q}{r_{MQ}^3} ds_M \right], \tag{2.328}$$

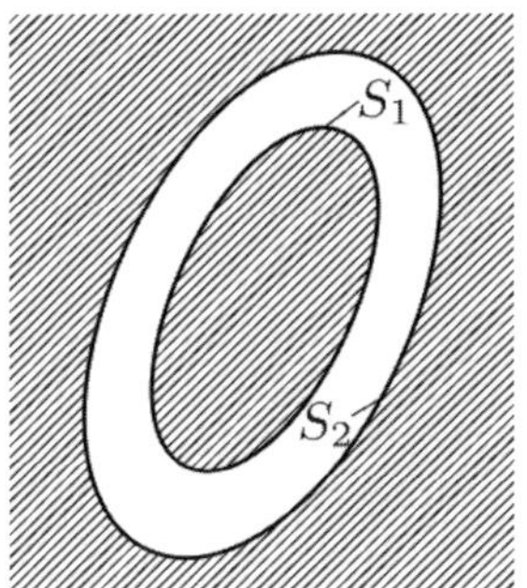

Figure 2.21

$$\sigma_k^{(2)}(Q) = \frac{\lambda_k}{2\pi}\left[\oint_{S_1} \sigma_k^{(1)}(M)\frac{\mathbf{r}_{MQ}\cdot\mathbf{n}_Q}{r_{MQ}^3}ds_M\right.$$

$$\left. + \oint_{S_2} \sigma_k^{(2)}(M)\left(\frac{\mathbf{r}_{MQ}\cdot\mathbf{n}_Q}{r_{MQ}^3} - \frac{2\pi}{A_2}\right)ds_M\right], \tag{2.329}$$

where A_1 and A_2 are the areas of S_1 and S_2, respectively. By using formulas (2.309) and (2.311), it is easy to demonstrate that charge neutrality conditions (2.313) and (2.314) are valid for all λ_k. This implies that equations (2.328)-(2.329) are reduced to (are equivalent to) equations (2.305)-(2.306) when $|\lambda_k| > 1$. On the other hand, $\lambda_0^+ = 1$ and $\lambda_0^- = -1$ are not the eigenvalues for coupled equations (2.328)-(2.329) because for the Robin problem charge-neutrality conditions (2.313) and (2.314) are not valid.

The eigenfunctions $\sigma_k(M)$ and $\sigma_i(M)$ of integral equations (2.305)-(2.306) corresponding to different eigenvalues λ_k and λ_i are not orthogonal. However, the plasmon electric fields $\mathbf{E}_k$ and $\mathbf{E}_i$ created by $\sigma_k(M)$ and $\sigma_i(M)$ are orthogonal in the whole space. Moreover, these fields are separately orthogonal inside and outside the nanoshell. To establish this, we shall use formulas (2.30) and (2.31) and rewrite the integral equations (2.305)-(2.306) as follows:

$$(1 + \lambda_k)\mathbf{n}_Q\cdot\mathbf{E}_k^-(Q) = (1 - \lambda_k)\mathbf{n}_Q\cdot\mathbf{E}_k^+(Q) \quad \text{on } S_1, \tag{2.330}$$

$$-(1 + \lambda_k)\mathbf{n}_Q\cdot\mathbf{E}_k^+(Q) = -(1 - \lambda_k)\mathbf{n}_Q\cdot\mathbf{E}_k^-(Q) \quad \text{on } S_2. \tag{2.331}$$

Next, we shall multiply both sides of formulas (2.330) and (2.331) by the electric potential $\varphi_i(Q)$ created by $\sigma_i^{(1)}(M)$ and $\sigma_i^{(2)}(M)$ and integrate over

S_1 and S_2, respectively. This leads to the following two formulas:

$$(1+\lambda_k) \oint_{S_1} \varphi_i(Q)\mathbf{n}_Q \cdot \mathbf{E}_k^-(Q)ds_Q = (1-\lambda_k) \oint_{S_1} \varphi_i(Q)\mathbf{n}_Q \cdot \mathbf{E}_k^+(Q)ds_Q, \quad (2.332)$$

$$-(1+\lambda_k) \oint_{S_2} \varphi_i(Q)\mathbf{n}_Q \cdot \mathbf{E}_k^+(Q)ds_Q = -(1-\lambda_k) \oint_{S_2} \varphi_i(Q)\mathbf{n}_Q \cdot \mathbf{E}_k^-(Q)ds_Q. \quad (2.333)$$

By adding the last two equations, we find

$$(1+\lambda_k) \left[\oint_{S_1} \varphi_i(Q)\mathbf{n}_Q \cdot \mathbf{E}_k^-(Q)ds_Q - \oint_{S_2} \varphi_i(Q)\mathbf{n}_Q \cdot \mathbf{E}_k^+(Q)ds_Q \right]$$
$$= (1-\lambda_k) \left[\oint_{S_1} \varphi_i(Q)\mathbf{n}_Q \cdot \mathbf{E}_k^+(Q)ds_Q - \oint_{S_2} \varphi_i(Q)\mathbf{n}_Q \cdot \mathbf{E}_k^-(Q)ds_Q \right]. \quad (2.334)$$

By using the divergence theorem and the same line of reasoning as in the discussion of "Property 9" in the previous section, we derive that

$$\oint_{S_1} \varphi_i(Q)\mathbf{n}_Q \cdot \mathbf{E}_k^+(Q)ds_Q = - \int_{V_1} \mathbf{E}_i \cdot \mathbf{E}_k dv, \quad (2.335)$$

$$\oint_{S_2} \varphi_i(Q)\mathbf{n}_Q \cdot \mathbf{E}_k^-(Q)ds_Q = \int_{V_3} \mathbf{E}_i \cdot \mathbf{E}_k dv, \quad (2.336)$$

$$\oint_{S_1} \varphi_i(Q)\mathbf{n}_Q \cdot \mathbf{E}_k^-(Q)ds_Q - \oint_{S_2} \varphi_i(Q)\mathbf{n}_Q \cdot \mathbf{E}_k^+(Q)ds_Q = \int_{V_2} \mathbf{E}_i \cdot \mathbf{E}_k dv. \quad (2.337)$$

By substituting the last three formulas in equation (2.334), we find

$$(\lambda_k + 1) \left[\int_{V_1} \mathbf{E}_i \cdot \mathbf{E}_k dv + \int_{V_3} \mathbf{E}_i \cdot \mathbf{E}_k dv \right] = (\lambda_k - 1) \int_{V_2} \mathbf{E}_i \cdot \mathbf{E}_k dv. \quad (2.338)$$

By writing expressions (2.330) and (2.331) for the plasmon electric field $\mathbf{E}_i(Q)$ created by $\sigma_i^{(1)}(M)$ and $\sigma_i^{(2)}(M)$, by multiplying them by $\varphi_k(Q)$ and by using the same reasoning as above, we obtain

$$(\lambda_i + 1) \left[\int_{V_1} \mathbf{E}_k \cdot \mathbf{E}_i dv + \int_{V_3} \mathbf{E}_k \cdot \mathbf{E}_i dv \right] = (\lambda_i - 1) \int_{V_2} \mathbf{E}_k \cdot \mathbf{E}_i dv. \quad (2.339)$$

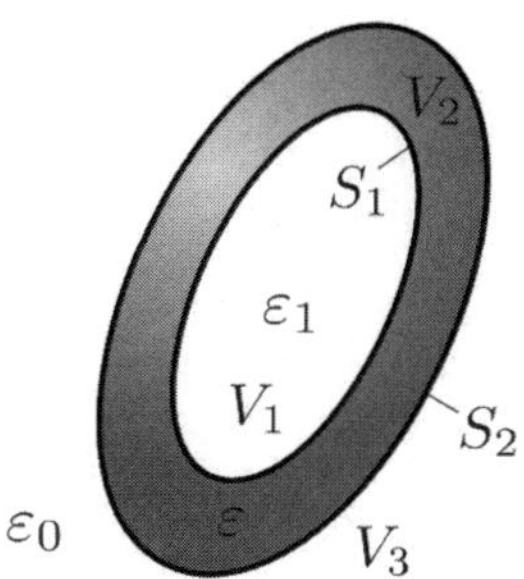

Figure 2.22

Since $\lambda_k \neq \lambda_i$ and $|\lambda_k| > 1$, $(k = 1, 2, ...)$, we conclude that equalities (2.338) and (2.339) can be simultaneously valid only if the following orthogonality conditions hold:

$$\int_{V_2} \mathbf{E}_i \cdot \mathbf{E}_k dv = 0, \tag{2.340}$$

$$\int_{V_1} \mathbf{E}_i \cdot \mathbf{E}_k dv + \int_{V_3} \mathbf{E}_i \cdot \mathbf{E}_k dv = 0. \tag{2.341}$$

It is apparent that from orthogonality conditions (2.340) and (2.341) follow the orthogonality conditions of the plasmon electric fields $\mathbf{E}_k$ and $\mathbf{E}_i$ in the whole space, but not the other way around. In this sense, formulas (2.340) and (2.341) can be called "strong orthogonality conditions."

Up to this point, our discussion has been concerned with nanoshells whose core permittivity is ε_0. For some core materials (silicon, for instance), the core permittivity ε_1 can be appreciably larger than ε_0. This justifies the special consideration of nanoshells shown in Figure 2.22. It is easy to see that the electric potential φ_k for electric field $\mathbf{E}_k = -\text{grad } \varphi_k$ of a specific plasmon mode is the nonzero solution of the following boundary value problem:

$$\nabla^2 \varphi_k = 0 \quad \text{in } V_j \ (j = 1, 2, 3), \tag{2.342}$$

$$\varphi_k^+ = \varphi_k^- \quad \text{on } S_1, \tag{2.343}$$

$$\varphi_k^+ = \varphi_k^- \quad \text{on } S_2, \tag{2.344}$$

$$\varepsilon_1 \frac{\partial \varphi_k^+}{\partial n} = \varepsilon_k \frac{\partial \varphi_k^-}{\partial n} \quad \text{on } S_1, \tag{2.345}$$

$$\varepsilon_k \frac{\partial \varphi_k^+}{\partial n} = \varepsilon_0 \frac{\partial \varphi_k^-}{\partial n} \quad \text{on } S_2, \tag{2.346}$$

$$\varphi_k(\infty) = 0. \tag{2.347}$$

As has been done before, the permittivity of the nanoshell and the permittivity of the core are replaced by ε_0 and virtual surface charges $\sigma_k^{(1)}(M)$ and $\sigma_k^{(2)}(M)$ are distributed over S_1 and S_2, respectively. The plasmon mode potential φ_k created by these charges is given by the formula

$$\varphi_k(Q) = \frac{1}{4\pi\varepsilon_0} \left(\oint_{S_1} \frac{\sigma_k^{(1)}(M)}{r_{MQ}} ds_M + \oint_{S_2} \frac{\sigma_k^{(2)}(M)}{r_{MQ}} ds_M \right). \tag{2.348}$$

This potential satisfies all equations and conditions of the above boundary value problem except for the boundary conditions (2.345) and (2.346). By using formula (2.32), we find that the latter boundary conditions will be satisfied if the surface charges $\sigma_k^{(1)}(M)$ and $\sigma_k^{(2)}(M)$ are the solutions of the following integral equations posed on S_1 and S_2, respectively,

$$\varepsilon_1 \left[2\pi\sigma_k^{(1)}(Q) \right.$$
$$\left. - \oint_{S_1} \sigma_k^{(1)}(M) \frac{\mathbf{r}_{MQ} \cdot \mathbf{n}_Q}{r_{MQ}^3} ds_M - \oint_{S_2} \sigma_k^{(2)}(M) \frac{\mathbf{r}_{MQ} \cdot \mathbf{n}_Q}{r_{MQ}^3} ds_M \right]$$
$$= -\varepsilon_k \left[2\pi\sigma_k^{(1)}(Q) \right.$$
$$\left. + \oint_{S_1} \sigma_k^{(1)}(M) \frac{\mathbf{r}_{MQ} \cdot \mathbf{n}_Q}{r_{MQ}^3} ds_M + \oint_{S_2} \sigma_k^{(2)}(M) \frac{\mathbf{r}_{MQ} \cdot \mathbf{n}_Q}{r_{MQ}^3} ds_M \right], \tag{2.349}$$

$$\varepsilon_0 \left[2\pi\sigma_k^{(2)}(Q) \right.$$
$$\left. + \oint_{S_1} \sigma_k^{(1)}(M) \frac{\mathbf{r}_{MQ} \cdot \mathbf{n}_Q}{r_{MQ}^3} ds_M + \oint_{S_2} \sigma_k^{(2)}(M) \frac{\mathbf{r}_{MQ} \cdot \mathbf{n}_Q}{r_{MQ}^3} ds_M \right]$$
$$= -\varepsilon_k \left[2\pi\sigma_k^{(2)}(Q) \right.$$
$$\left. - \oint_{S_1} \sigma_k^{(1)}(M) \frac{\mathbf{r}_{MQ} \cdot \mathbf{n}_Q}{r_{MQ}^3} ds_M - \oint_{S_2} \sigma_k^{(2)}(M) \frac{\mathbf{r}_{MQ} \cdot \mathbf{n}_Q}{r_{MQ}^3} ds_M \right]. \tag{2.350}$$

It is apparent that plasmon modes exist for such values of σ_k that homogeneous equations (2.349) and (2.350) have nonzero solutions. It is also clear that these equations have the mathematical structure of a generalized eigenvalue problem. This structure becomes more transparent if the above coupled equations are written in the following operator form:

$$\hat{K}_{11}\sigma_k^{(1)} + \hat{K}_{12}\sigma_k^{(2)} = \varepsilon_k \left(\hat{T}_{11}\sigma_k^{(1)} + \hat{T}_{12}\sigma_k^{(2)} \right), \qquad (2.351)$$

$$\hat{K}_{21}\sigma_k^{(1)} + \hat{K}_{22}\sigma_k^{(2)} = \varepsilon_k \left(\hat{T}_{12}\sigma_k^{(1)} + \hat{T}_{22}\sigma_k^{(2)} \right), \qquad (2.352)$$

where the following operators have been introduced:

$$\hat{K}_{11}\sigma_k^{(1)} = \varepsilon_1 \left[2\pi\sigma_k^{(1)}(Q) - \oint_{S_1} \sigma_k^{(1)}(M)\frac{\mathbf{r}_{MQ} \cdot \mathbf{n}_Q}{r_{MQ}^3} ds_M \right], \quad (Q \in S_1), \quad (2.353)$$

$$\hat{K}_{12}\sigma_k^{(2)} = -\varepsilon_1 \oint_{S_2} \sigma_k^{(2)}(M)\frac{\mathbf{r}_{MQ} \cdot \mathbf{n}_Q}{r_{MQ}^3} ds_M, \quad (Q \in S_1), \qquad (2.354)$$

$$\hat{T}_{11}\sigma_k^{(1)} = -2\pi\sigma_k^{(1)}(Q) - \oint_{S_1} \sigma_k^{(1)}(M)\frac{\mathbf{r}_{MQ} \cdot \mathbf{n}_Q}{r_{MQ}^3} ds_M, \quad (Q \in S_1), \qquad (2.355)$$

$$\hat{T}_{12}\sigma_k^{(2)} = - \oint_{S_2} \sigma_k^{(2)}(M)\frac{\mathbf{r}_{MQ} \cdot \mathbf{n}_Q}{r_{MQ}^3} ds_M, \quad (Q \in S_1), \qquad (2.356)$$

$$\hat{K}_{21}\sigma_k^{(1)} = \varepsilon_0 \oint_{S_1} \sigma_k^{(1)}(M)\frac{\mathbf{r}_{MQ} \cdot \mathbf{n}_Q}{r_{MQ}^3} ds_M, \quad (Q \in S_2), \qquad (2.357)$$

$$\hat{K}_{22}\sigma_k^{(2)} = \varepsilon_0 \left[2\pi\sigma_k^{(2)}(Q) + \oint_{S_2} \sigma_k^{(2)}(M)\frac{\mathbf{r}_{MQ} \cdot \mathbf{n}_Q}{r_{MQ}^3} ds_M \right], \quad (Q \in S_2), \quad (2.358)$$

$$\hat{T}_{21}\sigma_k^{(1)} = \oint_{S_1} \sigma_k^{(1)}(M)\frac{\mathbf{r}_{MQ} \cdot \mathbf{n}_Q}{r_{MQ}^3} ds_M, \quad (Q \in S_2), \qquad (2.359)$$

$$\hat{T}_{22}\sigma_k^{(2)} = -2\pi\sigma_k^{(2)}(Q) + \oint_{S_2} \sigma_k^{(2)}(M)\frac{\mathbf{r}_{MQ} \cdot \mathbf{n}_Q}{r_{MQ}^3} ds_M, \quad (Q \in S_2). \qquad (2.360)$$

In numerical computations, boundary integral equations (2.349)-(2.350) will be discretized and reduced to the following generalized eigenvalue problem in linear algebra:

$$\hat{B}\mathbf{x}_k = \varepsilon_k \hat{A}\mathbf{x}_k, \qquad (2.361)$$

where $\hat{B}$ and $\hat{A}$ are finite-dimensional matrix approximations of $\hat{K}$- and $\hat{T}$-operators in (2.351)-(2.352). This fact is another justification for using the term "generalized eigenvalue problem" for integral equations (2.349)-(2.350).

It can be proved that generalized eigenvalues ε_k are real and negative for any shape of S_1 and S_2 and any positive value ε_1 of the nanoshell permittivity. The proof proceeds as follows. For any nonzero solutions $\sigma_k^{(1)}(M)$ and $\sigma_k^{(2)}(M)$ of integral equations (2.349)-(2.350), the potential φ_k defined by formula (2.348) satisfies the boundary conditions (2.345) and (2.346) on S_1 and S_2, respectively. We shall multiply these boundary conditions by $\varphi_k^*(Q)$ and integrate over S_1 and S_2. As a result, we obtain

$$\varepsilon_1 \oint_{S_1} \varphi_k^*(Q)\frac{\partial \varphi_k^+}{\partial n}(Q)ds_Q = \varepsilon_k \oint_{S_1} \varphi_k^*(Q)\frac{\partial \varphi_k^-}{\partial n}(Q)ds_Q, \qquad (2.362)$$

$$\varepsilon_0 \oint_{S_2} \varphi_k^*(Q)\frac{\partial \varphi_k^-}{\partial n}(Q)ds_Q = \varepsilon_k \oint_{S_2} \varphi_k^*(Q)\frac{\partial \varphi_k^+}{\partial n}(Q)ds_Q. \qquad (2.363)$$

By subtracting formula (2.363) from formula (2.362), we find

$$\varepsilon_1 \oint_{S_1} \varphi_k^*(Q)\frac{\partial \varphi_k^+}{\partial n}(Q)ds_Q - \varepsilon_0 \oint_{S_2} \varphi_k^*(Q)\frac{\partial \varphi_k^-}{\partial n}(Q)ds_Q$$

$$= -\varepsilon_k \left[\oint_{S_2} \varphi_k^*(Q)\frac{\partial \varphi_k^+}{\partial n}(Q)ds_Q - \oint_{S_1} \varphi_k^*(Q)\frac{\partial \varphi_k^-}{\partial n}(Q)ds_Q \right]. \qquad (2.364)$$

By recalling formulas (2.321), (2.322) and (2.323), we obtain

$$\varepsilon_1 W_k^{(1)} + \varepsilon_0 W_k^{(3)} = -\varepsilon_k W_k^{(2)}, \qquad (2.365)$$

which leads to

$$\varepsilon_k = -\frac{\varepsilon_1 W_k^{(1)} + \varepsilon_0 W_k^{(3)}}{W_k^{(2)}}. \qquad (2.366)$$

Since $W_k^{(1)}$, $W_k^{(2)}$ and $W_k^{(3)}$ are positive, from the last formula we conclude that ε_k are real and negative. It is also apparent that formula (2.366) is the natural generalization of formula (2.327).

Finally, it can be shown that the following orthogonality conditions are valid for plasmon electric fields $\mathbf{E}_k$ and $\mathbf{E}_i$ corresponding to different generalized eigenvalues ε_k and ε_i, respectively:

$$\int_{V_2} \mathbf{E}_k \cdot \mathbf{E}_i dv = 0, \qquad (2.367)$$

$$\varepsilon_1 \int_{V_1} \mathbf{E}_k \cdot \mathbf{E}_i \, dv + \varepsilon_0 \int_{V_3} \mathbf{E}_k \cdot \mathbf{E}_i \, dv = 0. \tag{2.368}$$

These orthogonality conditions are natural generalizations of orthogonality conditions (2.340)-(2.341).

In all our previous discussion, the plasmon mode analysis was reduced to eigenvalue problems for boundary integral equations by introducing virtual single layer or double layer charges and exploiting their properties. There exists, however, an alternative approach for the derivation of the integral equations, which is based on the Green formula and which leads to equivalent results. This approach is briefly discussed below. The essence of this approach is to use the following Green formulas for the displacement potential Φ_k of the plasmon modes:

$$\frac{1}{4\pi} \oint_S \left[\Phi_k^+(M) \frac{\mathbf{r}_{QM} \cdot \mathbf{n}_M}{r_{QM}^3} + \frac{\partial \Phi_k^+}{\partial n}(M) \frac{1}{r_{MQ}} \right] ds_M = \begin{cases} \Phi_k(Q), & \text{if } Q \in V^+ \\ \Phi_k^+(Q)/2, & \text{if } Q \in S \\ 0, & \text{if } Q \in V^-, \end{cases} \tag{2.369}$$

$$-\frac{1}{4\pi} \oint_S \left[\Phi_k^-(M) \frac{\mathbf{r}_{QM} \cdot \mathbf{n}_M}{r_{QM}^3} + \frac{\partial \Phi_k^-}{\partial n}(M) \frac{1}{r_{QM}} \right] ds_M = \begin{cases} \Phi_k(Q), & \text{if } Q \in V^- \\ \Phi_k^-(Q)/2, & \text{if } Q \in S \\ 0, & \text{if } Q \in V^+, \end{cases} \tag{2.370}$$

where, as before, the minus sign in front of the left-hand side of formula (2.370) reflects the fact that the normal to S, which is outward with respect to V^+, has inward direction with respect to V^-.

By adding equations (2.369) and (2.370) in the case when $Q \in S$ and taking into account the equality of $\partial \Phi_k^+/\partial n$ and $\partial \Phi_k^-/\partial n$ (see formula (2.118)), we obtain

$$\frac{\Phi_k^+(Q) + \Phi_k^-(Q)}{2} = \frac{1}{4\pi} \oint_S \left[\Phi_k^+(Q) - \Phi_k^-(Q) \right] \frac{\mathbf{r}_{QM} \cdot \mathbf{n}_M}{r_{QM}^3} \, ds_M. \tag{2.371}$$

Now by using the relation

$$\Phi_k^-(Q) = \frac{\varepsilon_0}{\varepsilon_k} \Phi_k^+(Q), \tag{2.372}$$

which follows from the boundary condition (2.117), we derive

$$\Phi_k^+(Q) = \frac{\lambda_k}{2\pi} \oint_S \Phi_k^+(M) \frac{\mathbf{r}_{QM} \cdot \mathbf{n}_M}{r_{QM}^3} \, ds_M, \tag{2.373}$$

$$\lambda_k = \frac{\varepsilon_k - \varepsilon_0}{\varepsilon_k + \varepsilon_0}. \tag{2.374}$$

It is apparent that the eigenvalue problem for the boundary integral equation (2.373) is identical to the eigenvalue problem for the integral equation (2.138). This is not surprising because $\Phi_k^+(Q)$ and $\tau_k(Q)$ are related by the formula (2.166).

If we use in the Green formulas the electric field potential φ_k of the plasmon modes and follow almost the same line of reasoning as before, we arrive at the following eigenvalue problem:

$$\frac{\partial \varphi_k^+}{\partial n}(Q) = \frac{\lambda_k}{2\pi} \oint_S \frac{\partial \varphi_k^-}{\partial n}(Q) \frac{\mathbf{r}_{MQ} \cdot \mathbf{n}_Q}{r_{MQ}^3} ds_M, \tag{2.375}$$

where λ_k is given by the formula (2.374). It is clear that this eigenvalue problem is identical to one described by equation (2.36). This is not surprising again because $\frac{\partial \varphi_k^+}{\partial n}(Q)$ and $\sigma_k(Q)$ are related by the formula (2.146).

The Green formulas have been used for the derivation of integral equations (2.373) and (2.375) in the case of a single nanoparticle. The derivation is more involved in the case of multiple nanoparticles or nanoshells. We consider below only the derivation for nanoshells whose core permittivity is ε_0. Other cases can be analyzed in a similar way. In the case of a nanoshell, the Green formula can be written for the displacement field potential Φ_k of plasmon modes in the regions V_1, V_2 and V_3, respectively,

$$\frac{1}{4\pi} \oint_{S_1} \left[\Phi_k^+(M) \frac{\mathbf{r}_{QM} \cdot \mathbf{n}_M}{r_{QM}^3} + \frac{\partial \Phi_k^+}{\partial n}(M) \frac{1}{r_{MQ}} \right] ds_M = \begin{cases} \Phi_k(Q), & \text{if } Q \in V_1 \\ \Phi_k^+(Q)/2, & \text{if } Q \in S_1 \\ 0, & \text{if } Q \notin V_1, \end{cases} \tag{2.376}$$

$$-\frac{1}{4\pi} \oint_{S_1} \left[\Phi_k^-(M) \frac{\mathbf{r}_{QM} \cdot \mathbf{n}_M}{r_{QM}^3} + \frac{\partial \Phi_k^-}{\partial n}(M) \frac{1}{r_{MQ}} \right] ds_M$$

$$+\frac{1}{4\pi} \oint_{S_2} \left[\Phi_k^+(M) \frac{\mathbf{r}_{QM} \cdot \mathbf{n}_M}{r_{QM}^3} + \frac{\partial \Phi_k^+}{\partial n}(M) \frac{1}{r_{MQ}} \right] ds_M = \begin{cases} \Phi_k(Q), & \text{if } Q \in V_2 \\ \Phi_k^-(Q)/2, & \text{if } Q \in S_1 \\ \Phi_k^+(Q)/2, & \text{if } Q \in S_2 \\ 0, & \text{if } Q \notin V_2, \end{cases} \tag{2.377}$$

$$-\frac{1}{4\pi} \oint_{S_2} \left[\Phi_k^-(M) \frac{\mathbf{r}_{QM} \cdot \mathbf{n}_M}{r_{QM}^3} + \frac{\partial \Phi_k^-}{\partial n}(M) \frac{1}{r_{MQ}} \right] ds_M = \begin{cases} \Phi_k(Q), & \text{if } Q \in V_3 \\ \Phi_k^-(Q)/2, & \text{if } Q \in S_2 \\ 0, & \text{if } Q \notin V_3. \end{cases}$$

$$(2.378)$$

We shall next add formulas (2.376) and (2.377) for the case $Q \in S_1$ and take into account the continuity of $\partial \Phi_k^+/\partial n$ and $\partial \Phi_k^-/\partial n$ on S_1. This leads to

$$\frac{\Phi_k^+(Q) + \Phi_k^-(Q)}{2} = \frac{1}{4\pi} \oint_{S_1} \left[\Phi_k^+(M) - \Phi_k^-(M) \right] \frac{\mathbf{r}_{QM} \cdot \mathbf{n}_M}{r_{QM}^3} ds_M$$

$$+ \frac{1}{4\pi} \oint_{S_2} \left[\Phi_k^+(M) \frac{\mathbf{r}_{QM} \cdot \mathbf{n}_M}{r_{QM}^3} + \frac{\partial \Phi_k^+}{\partial n}(M) \frac{1}{r_{MQ}} \right] ds_M.$$

$$(2.379)$$

From the boundary condition

$$\frac{\Phi_k^+(Q)}{\varepsilon_0} = \frac{\Phi_k^-(Q)}{\varepsilon_k}, \quad (Q \in S_1), \tag{2.380}$$

follows that

$$\Phi_k^-(Q) = \frac{\varepsilon_k}{\varepsilon_0} \Phi_k^+(Q), \quad (Q \in S_1). \tag{2.381}$$

By using the continuity of $\partial \Phi_k^+/\partial n$ and $\partial \Phi_k^-/\partial n$ on S_2, we find

$$\frac{1}{4\pi} \oint_{S_2} \frac{\partial \Phi_k^+}{\partial n}(M) \frac{1}{r_{MQ}} ds_M = \frac{1}{4\pi} \oint_{S_2} \frac{\partial \Phi_k^-}{\partial n}(M) \frac{1}{r_{MQ}} ds_M. \tag{2.382}$$

Since $Q \in S_1$ and, consequently, $Q \notin V_3$, from formulas (2.382) and (2.378) we obtain

$$\frac{1}{4\pi} \oint_{S_2} \frac{\partial \Phi_k^+}{\partial n}(M) \frac{1}{r_{MQ}} ds_M = -\frac{1}{4\pi} \oint_{S_2} \Phi_k^-(M) \frac{\mathbf{r}_{QM} \cdot \mathbf{n}_M}{r_{QM}^3} ds_M. \tag{2.383}$$

From the boundary condition

$$\frac{\Phi_k^+(Q)}{\varepsilon_k} = \frac{\Phi_k^-(Q)}{\varepsilon_0}, \quad (Q \in S_2), \tag{2.384}$$

follows that

$$\Phi_k^+(Q) = \frac{\varepsilon_k}{\varepsilon_0} \Phi_k^-(Q), \quad (Q \in S_2). \tag{2.385}$$

By substituting formulas (2.381), (2.383) and (2.385) into equation (2.379), after simple transformations we find for $Q \in S_1$ that

$$\Phi_k^+(Q) = -\frac{\lambda_k}{2\pi} \left[\oint_{S_1} \Phi_k^+(M) \frac{\mathbf{r}_{QM} \cdot \mathbf{n}_M}{r_{QM}^3} ds_M - \oint_{S_2} \Phi_k^-(M) \frac{\mathbf{r}_{QM} \cdot \mathbf{n}_M}{r_{QM}^3} ds_M \right],$$

$$(2.386)$$

where λ_k is given by formula (2.374). By summing the equations (2.377) and (2.378) in the case of $Q \in S_2$ and by using the same line of reasoning, we derive

$$\Phi_k^-(Q) = -\frac{\lambda_k}{2\pi} \left[\oint_{S_1} \Phi_k^+(M) \frac{\mathbf{r}_{QM} \cdot \mathbf{n}_M}{r_{QM}^3} ds_M - \oint_{S_2} \Phi_k^-(M) \frac{\mathbf{r}_{QM} \cdot \mathbf{n}_M}{r_{QM}^3} ds_M \right].$$

$$(2.387)$$

Equations (2.386) and (2.387) define the eigenvalue problem for coupled integral equations. In conclusion, we point out that other cases (i.e., nanoshells whose core permittivity is not equal to ε_0 and multi-nanoparticle systems) can be treated in a similar way. The discussion of this section can also be easily extended to the case when nanoshells are placed on dielectric substrates. This can be done in the same way as in sections 2.1 and 2.2, namely by using the appropriate Green functions in the kernels of the integral equations.

2.5 Relation to the Riemann Hypothesis

In this section, an interesting connection between the Riemann hypothesis and the eigenvalue treatment of plasmon resonances is considered. The discussion begins with the statement of the Riemann hypothesis. This hypothesis deals with the zeta function of complex variables s which for $\mathrm{Re}(s) > 1$ is defined as the following absolutely convergent series:

$$\zeta(s) = \sum_{n=1}^{\infty} \frac{1}{n^s}.$$

$$(2.388)$$

The significance of this zeta function in the number theory stems from its representation by the Euler product

$$\zeta(s) = \sum_{n=1}^{\infty} \frac{1}{n^s} = \prod_i \left(1 - \frac{1}{p_i^s} \right)^{-1},$$

$$(2.389)$$

where p_i are prime numbers and the product is taken over all prime numbers. Thus, the last formula establishes the remarkable connection between natural numbers (positive integers) n and prime numbers p_i. For this reason, the zeta function is very instrumental in the study of distribution of prime numbers. One of the central results of this study is the Prime Number Theorem, which states that

$$\lim_{x \to \infty} \frac{\pi(x)}{Li(x)} = 1, \tag{2.390}$$

where $\pi(x)$ is the number of primes up to x, and $Li(x)$ is the logarithmic integral function defined as follows:

$$Li(x) = \int_2^x \frac{dt}{\ln t}. \tag{2.391}$$

Formula (2.390) implies that $Li(x)$ is a good approximation for $\pi(x)$ for large x.

In the fundamental paper [24] (its English translation can be found in the book [25]), Riemann studied the analytical continuation of the zeta function defined by the infinite series (2.388). Riemann found that $\zeta(s)$ extends to the whole complex plane as a meromorphic function with only one simple pole at $s = 1$ with residue 1 and derived the functional equation for the extended $\zeta(s)$ which relates $\zeta(s)$ to $\zeta(1 - s)$. Riemann then introduced the entire function

$$\xi(\lambda) = \frac{1}{2}s(s-1)\pi^{-\frac{s}{2}}\Gamma\left(\frac{s}{2}\right)\zeta(s),^\dagger \tag{2.392}$$

$$s = \frac{1}{2} + i\lambda, \quad (i = \sqrt{-1}), \tag{2.393}$$

where Γ is the Euler gamma function.

The functional equation for $\zeta(s)$ implies that $\xi(\lambda)$ is an even function of λ whose zeros have imaginary parts between $-i/2$ and $i/2$. Then, Riemann conjectured that all zeros of $\xi(\lambda)$ are real. This conjecture is the Riemann hypothesis.

The zeta function has zeros at negative integers $-2, -4, ...$, which correspond to the simple poles of the gamma function $\Gamma(s/2)$. These zeros are referred to as trivial zeros. The other nontrivial zeros of $\zeta(s)$ are $s_k = \frac{1}{2} + i\tilde{\lambda}_k$, where $\tilde{\lambda}_k$ are zeros of the xi-function, i.e., $\xi(\tilde{\lambda}_k) = 0$. The nontrivial zeros s_k are in the strip $0 \le \operatorname{Re}(s) \le 1$ (see Figure 2.23). In terms of s_k, the Riemann

---

†In literature, notation Ξ is often used instead of ξ in the case when λ is the argument.

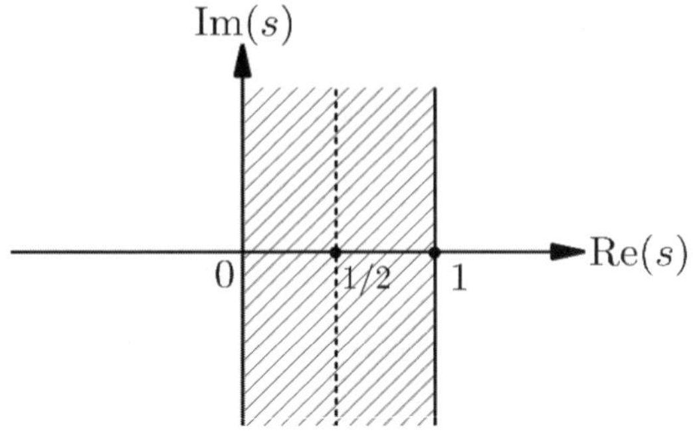

Figure 2.23

hypothesis is stated as follows: all nontrivial zeros of the zeta function have real part equal to 1/2.

In his paper, Riemann derived the integral representation for $\xi(\lambda)$ and stated that $\xi(\lambda)$ "can be developed as a power series ... which converges rapidly." This power series can be written as follows (see the book [25]):

$$\xi(\lambda) = \sum_{n=0}^{\infty} \frac{(-1)^n}{(2n)!} c_{2n} \lambda^{2n}, \tag{2.394}$$

where the expansion coefficients are given by the formulas

$$c_{2n} = \int_1^{\infty} H(x) \left(\ln \sqrt{x} \right)^{2n} dx, \tag{2.395}$$

$$H(x) = 4x^{-\frac{1}{4}} \frac{d}{dx} \left[x^{\frac{3}{2}} \frac{d\psi(x)}{dx} \right] \tag{2.396}$$

and

$$\psi(x) = \sum_{n=1}^{\infty} e^{-n^2 \pi x}. \tag{2.397}$$

It can be shown [25] that for $x \geq 1$,

$$H(x) > 0 \tag{2.398}$$

and, consequently,

$$c_{2n} > 0. \tag{2.399}$$

Indeed, from formulas (2.396) and (2.397) we derive

$$\frac{d}{dx}\left[x^{\frac{3}{2}}\frac{d\psi}{dx}\right] = \frac{d}{dx}\left[-\pi\sum_{n=1}^{\infty}n^2x^{\frac{3}{2}}e^{-n^2\pi x}\right]$$

$$= \sum_{n=1}^{\infty}\left(\pi^2 n^4 x^{\frac{3}{2}}e^{-n^2\pi x} - \frac{3}{2}\pi n^2 x^{\frac{1}{2}}e^{-n^2\pi x}\right)$$

$$= \pi x^{\frac{1}{2}}\sum_{n=1}^{\infty}\left(\pi n^4 x - \frac{3}{2}n^2\right)e^{-n^2\pi x}. \tag{2.400}$$

It is clear that inequality (2.398) follows from formula (2.400) for $x \geq 1$.

Now, the Riemann hypothesis can be stated this way: all zeros of the entire function of complex variable λ defined by the formulas (2.394)-(2.397) are aligned along the real axis.

It is apparent from formulas (2.394)-(2.397) that $\xi(\lambda)$ assumes real values for real λ. Moreover, since $\xi(\lambda)$ is an even function of λ, the set of its zeros consists of pairs $\pm\tilde{\lambda}_k$, $k \in \mathbb{N}$. In the paper, Riemann also suggested the validity of the following product formula:

$$\xi(\lambda) = \xi(0)\prod_{k}\left(1 - \frac{\lambda}{\tilde{\lambda}_k}\right)\left(1 + \frac{\lambda}{\tilde{\lambda}_k}\right), \tag{2.401}$$

which was rigorously proved by Hadamard.

It has been observed in numerical calculations and then conjectured that all zeros of $\xi(\lambda)$ are simple. Numerical calculations also reveal that $\left|\tilde{\lambda}_k\right| > 14$.

It is our intention to discuss below the relation of the Riemann xi-function $\xi(\lambda)$ to the eigenvalue treatment of plasmon resonances. Before proceeding with this discussion, it is worthwhile to remark at this point that the Riemann hypothesis stated for the xi-function defined by formulas (2.394)-(2.397) can be treated as a problem of complex analysis, detached from ξ's number-theoretic origin. It is in this context that the following two questions can be naturally posed:

1. If the Riemann hypothesis is valid, then how is the alignment of all zeros of $\xi(\lambda)$ along the real axis encoded in the mathematical structure of the expansion coefficients defined by formulas (2.395)-(2.397)?

2. What are the necessary and sufficient conditions (in terms of c_{2n}) for the alignment along the real axis of all zeros of an entire function defined by the power series (2.394)?

It is attempted below to discuss these questions at least partially.

It has been shown in section 2.3 that all eigenvalues of integral equation

$$\sigma_k(Q) = \frac{\lambda_k}{\pi} \oint_L \sigma_k(M) \left[\frac{\mathbf{r}_{MQ} \cdot \mathbf{n}_Q}{r_{MQ}^2} - \frac{\pi}{L} \right] dl_M \qquad (2.402)$$

are real, can be grouped in pairs $\pm\lambda_k$, $k \in \mathbb{N}$ and $|\lambda_k| > 1$. These properties of the spectrum of integral equation (2.402) are valid for any sufficiently smooth curve L. These properties of the spectrum imply that the Fredholm determinant $D(\lambda)$ of integral equation (2.402) is the entire function of complex variable λ that assumes real values for real λ and the set of its zeros consists of pairs of real numbers $\pm\lambda_k$, $k \in \mathbb{N}$, with $|\lambda_k| > 1$. It has been discussed in section 2.3 that these zeros are simple poles of the resolvent of the integral equation (2.402). This implies that if a curve L does not have any symmetry and geometric multiplicity of all λ_k is 1 (i.e., there is no accidental degeneration of λ_k), then λ_k are simple zeros of the Fredholm determinant $D(\lambda)$. Furthermore, it has been proved (see [20]) that if a kernel of an integral equation is Hölder-continuous with respect to the variable of integration (which is certainly the case for the kernel of integral equation (2.402) when the curve L is smooth), then the Fredholm determinant is an entire function of genus zero. This implies the validity of the product formula for $D(\lambda)$,

$$D(\lambda) = D(0) \prod_k \left(1 - \frac{\lambda}{\lambda_k} \right) \left(1 + \frac{\lambda}{\lambda_k} \right). \qquad (2.403)$$

Since $D(\lambda)$ is an even (see the last formula) entire function of λ, the following power series can be used for its representation[‡]:

$$D(\lambda) = \sum_{n=0}^{\infty} \frac{(-1)^n}{(2n)!} b_{2n} \lambda^{2n}. \qquad (2.404)$$

It is clear from the presented facts that for various smooth curves L, Fredholm determinants $D(\lambda)$ of integral equation (2.402) form the class of entire functions with properties that have been conjectured or proved for the Riemann xi-function $\xi(\lambda)$. For this reason, the problem can be posed to find such curve L that

$$\boxed{\xi(\lambda) = D(\lambda),} \qquad (2.405)$$

which will prove the Riemann hypothesis.

[‡]By using formulas (2.408)-(2.411) and the representations of iterated kernels (2.413) in terms of eigenvalues, it can be shown that $b_{2n} > 0$.

The last formula is consistent with a spectral interpretation of the Riemann hypothesis which is usually referred to as the Hilbert-Pólya conjecture. This conjecture deals with self-adjoint operator spectral interpretation. It is apparent that the integral operator in equation (2.402) is not self-adjoint in the space of square-summable functions because its kernel is not symmetric. However, this integral operator has real eigenvalues and they are simple poles of its resolvent. These properties are typical for self-adjoint operators. In this sense, it is not surprising that, as discussed at the end of section 2.3, this operator will be self-adjoint in the "energy" function space. In this "energy" function space, the inner product is introduced by the formula

$$\langle \nu, \sigma \rangle = \oint_L \nu(M) \left(\oint_L \sigma(P) \ln \frac{1}{r_{PM}} dl_P \right) dl_M, \qquad (2.406)$$

which is defined on functions with zero mean over L. It is easy to see that $\langle \sigma, \sigma \rangle$ is positive. This can be formally proved, but it is also immediately clear from the fact that, up to a factor, $\langle \sigma, \sigma \rangle$ has the physical meaning of energy stored in the electric field created by the charge distribution σ over L. This explains the term "energy" function space.

By using the relation

$$\oint_L \left(\ln \frac{1}{r_{PQ}} \right) \frac{\partial}{\partial n_Q} \left(\ln \frac{1}{r_{MQ}} \right) dl_Q = \oint_L \left(\ln \frac{1}{r_{MQ}} \right) \frac{\partial}{\partial n_Q} \left(\ln \frac{1}{r_{PQ}} \right) dl_Q, \qquad (2.407)$$

which follows from the Green formula, and by following the same line of reasoning as at the end of section 2.3, it can be shown that the integral operator in the equation (2.402) is self-adjoint in the "energy" function space.

Next, we shall outline a possible approach of attempting to find such curve L that equality (2.405) holds. This approach is based on formulas from the theory of integral equations [20], which relate the expansion coefficients b_{2n} for the Fredholm determinant to the iterated traces of the kernel of the corresponding integral equation. By using these formulas as well as equations (2.394), (2.404) and (2.405), the following recurrent relations can be deduced:

$$(-1)^n c_{2n} = \oint_L B_{2n-1}(Q, Q) dl_Q, \qquad (2.408)$$

$$B_0(Q, M) = K(Q, M), \qquad (2.409)$$

     Plasmon Resonances in Nanoparticles

where

$$B_{2n-1}(Q, M) = -(2n - 1) \oint_S K(Q, P)B_{2n-2}(P, M)ds_P, \qquad (2.410)$$

$$B_{2n}(Q, M) = (-1)^n c_{2n} K(Q, M) - 2n \oint_S K(Q, P)B_{2n-2}(P, M)ds_P \qquad (2.411)$$

and $K(Q, M)$ is the kernel of integral equation (2.402),

$$K(Q, M) = \frac{1}{\pi}\left(\frac{\mathbf{r}_{MQ} \cdot \mathbf{n}_Q}{r_{MQ}^2} - \frac{\pi}{L}\right). \qquad (2.412)$$

It is clear that, by using formulas (2.408)-(2.411), the iterated traces of the kernel,

$$q_{2n} = \oint_L K_{2n}(Q, Q)dl_Q, \qquad (q_{2n-1} = 0), \qquad (2.413)$$

$$K_n(Q, M) = \oint_L K(Q, P)K_{n-1}(P, M)dl_P, \qquad (2.414)$$

can be sequentially found. It is not clear how the kernel (2.412) can be reconstructed from its iterated traces (2.413). However, if the latter is possible, then plane curve L can be reconstructed from this kernel. This is because it can be shown that

$$\lim_{M \to Q} \frac{\mathbf{r}_{MQ} \cdot \mathbf{n}_Q}{r_{MQ}^2} = \frac{k(Q)}{2}, \qquad (2.415)$$

where $k(Q)$ is the curvature of L at Q, and it is known from the differential geometry that a plane curve can be reconstructed from its curvature.

It is worthwhile to mention that there is another way to connect the iterated traces to the Fredholm determinant, which is based on the following formula from the integral equation theory:

$$-\frac{D'(\lambda)}{\lambda D(\lambda)} = \sum_{n=0}^{\infty} q_{2n}\lambda^{2n}, \qquad (2.416)$$

which is valid for sufficiently small λ.

This formula coupled with the equality (2.405) leads to

$$-\frac{\xi'(\lambda)}{\lambda \xi(\lambda)} = \sum_{n=0}^{\infty} q_{2n}\lambda^{2n}. \qquad (2.417)$$

It turns out that the last formula offers another (unrelated to the previous discussion) approach to the pursuit of the proof of the Riemann hypothesis. Indeed, from the product formula (2.401) we find

$$-\frac{\xi'(\lambda)}{\lambda\xi(\lambda)} = \sum_{k=1}^{\infty} \frac{1}{\tilde{\lambda}_k} \left(\frac{1}{\tilde{\lambda}_k - \lambda} + \frac{1}{\tilde{\lambda}_k + \lambda} \right), \tag{2.418}$$

which after substitution into formula (2.417) leads to

$$\sum_{k=1}^{\infty} \frac{1}{\tilde{\lambda}_k} \left(\frac{1}{\tilde{\lambda}_k - \lambda} + \frac{1}{\tilde{\lambda}_k + \lambda} \right) = \sum_{n=0}^{\infty} q_{2n} \lambda^{2n}. \tag{2.419}$$

From formula (2.418) follows that $\xi'(\lambda)/\lambda\xi(\lambda)$ is a meromorphic function whose poles are zeros of the xi-function. It is clear from the power series expansion (2.394) that the meromorphic function $\xi'(\lambda)/\lambda\xi(\lambda)$ is real for real λ. In this sense, this function can be called a real meromorphic function. The right-hand side of (2.419) can be construed as a power series expansion of this real meromorphic function at the origin of the complex λ-plane. It is apparent that the Riemann hypothesis will be proven if it is established that all poles of this real meromorphic function are real. It turns out that the necessary and sufficient conditions that all poles of this real meromorphic function are real can be formulated in terms of expansion coefficients q_{2n}. These necessary and sufficient conditions are given by the Grommer theorem [26, 27], which states that the real meromorphic function in the right-hand side of formula (2.419) has only real poles if and only if the expansion coefficients q_{2n} form the positive sequence, i.e., Hankel quadratic forms are positive,

$$\sum_{i=j=0}^{N} q_{2i+2j} \gamma_{2i} \gamma_{2j} > 0 \tag{2.420}$$

for any N.

The necessity of the condition (2.420) can be easily established. Indeed, by introducing the notation

$$\mathcal{F}(\lambda) = \sum_{k=1}^{\infty} \frac{1}{\tilde{\lambda}_k} \left(\frac{1}{\tilde{\lambda}_k - \lambda} + \frac{1}{\tilde{\lambda}_k + \lambda} \right), \tag{2.421}$$

we find through differentiation that

$$\mathcal{F}^{(2n)}(0) = 2(2n)! \sum_{k=1}^{\infty} \frac{1}{\tilde{\lambda}_k^{2n+2}}, \quad \mathcal{F}^{(2n-1)}(0) = 0. \tag{2.422}$$

From formulas (2.419) and (2.421) follows that

$$\mathcal{F}^{(2n)}(0) = (2n)!q_{2n}. \tag{2.423}$$

The last two formulas imply that

$$q_{2n} = 2\sum_{k=1}^{\infty} \frac{1}{\tilde{\lambda}_k^{2n+2}}. \tag{2.424}$$

Now, suppose that all $\tilde{\lambda}_k$ are real. Then we derive

$$\sum_{i,j=0}^{N} q_{2i+2j}\gamma_{2i}\gamma_{2j} = 2\sum_{k=1}^{\infty} \frac{1}{\tilde{\lambda}_k^2}\left(\sum_{i,j=0}^{N} \frac{\gamma_{2i}\gamma_{2j}}{\tilde{\lambda}_k^{2i+2j}}\right)$$

$$= 2\sum_{k=1}^{\infty} \frac{1}{\tilde{\lambda}_k^2}\left(\sum_{i=0}^{N} \frac{\gamma_{2i}}{\tilde{\lambda}_k^{2i}}\right)^2 > 0, \tag{2.425}$$

which proves the necessity of the condition (2.420). The simple proof of the sufficiency of the condition (2.420) can be carried out by using the theory of canonical solutions of the problem of moments [27]. This theory is beyond the scope of our present discussion.

Let us next consider how the positivity of the Hankel quadratic forms may be verified. To this end, we introduce a linear functional f defined on the set of polynomials

$$p(z) = \sum_{i=0}^{m} p_{2i}z^{2i} \tag{2.426}$$

by the following formula:

$$f(p(z)) = \sum_{i=0}^{m} q_{2i}p_{2i}. \tag{2.427}$$

It is clear that the inequality (2.420) is satisfied if for any "complete square" polynomial

$$\gamma(z) = \left(\sum_{i=0}^{N} \gamma_{2i}z^{2i}\right)^2 = \sum_{i,j=0}^{N} \gamma_{2i}\gamma_{2j}z^{2i+2j} \tag{2.428}$$

with arbitrary real γ_{2i} we have

$$f(\gamma(z)) > 0. \tag{2.429}$$

Indeed, from expressions (2.426), (2.427), (2.428) and (2.429) we find

$$f(\gamma(z)) = \sum_{i,j=0}^{N} q_{2i+2j}\gamma_{2i}\gamma_{2j} > 0. \tag{2.430}$$

Now, we shall return to formula (2.417) and shall rewrite it in the following form:

$$\chi(\lambda) = \frac{\xi'(\lambda)}{\lambda} = -\xi(\lambda)\mathcal{F}(\lambda), \tag{2.431}$$

where, according to the relations (2.419) and (2.421), we have

$$\mathcal{F}(\lambda) = \sum_{n=0}^{\infty} q_{2n}\lambda^{2n}. \tag{2.432}$$

By using the Leibniz rule for higher-order derivatives of a product of two functions as well as the fact that $\chi(\lambda)$, $\xi(\lambda)$ and $\mathcal{F}(\lambda)$ are even functions and, consequently, their odd derivatives are equal to zero at the origin, from formula (2.431) we derive

$$\chi^{(2n)}(0) = -\sum_{k=0}^{n} \frac{(2n)!}{(2n-2k)!(2k)!}\xi^{(2n-2k)}(0)\mathcal{F}^{(2k)}(0). \tag{2.433}$$

Then, by using the power series (2.394), we find

$$\chi^{(2n)}(0) = \frac{(-1)^{n+1}}{2n+1}c_{2n+2}, \tag{2.434}$$

$$\xi^{(2n-2k)}(0) = (-1)^{n-k}c_{2n-2k}. \tag{2.435}$$

By substituting formulas (2.423), (2.434) and (2.435) into formula (2.433), we obtain the following recurrent relations:

$$c_{2n+2} = \sum_{k=0}^{n} \frac{(-1)^{k}(2n+1)!}{(2n-2k)!}c_{2n-2k}q_{2k}, \quad (n = 0, 1, 2, ...). \tag{2.436}$$

Thus, according to the Grommer theorem, the necessary and sufficient condition for the validity of the Riemann hypothesis is that the expansion coefficients c_{2n} in the power series (2.394) are such that the solution q_{2n}, $(n = 0, 1, 2, ...)$ of simultaneous equations (2.436) forms the positive sequence in the sense of (2.420).

Next, we shall interpret the recurrent relations (2.436) in terms of the functional f defined by formulas (2.426) and (2.427). By using inequality (2.399), we find that the recurrent relations (2.436) are equivalent to

$$c_{2n+2} = f\left(R_{2n}(z)\right) > 0, \tag{2.437}$$

where the polynomials $R_{2n}(z)$ are defined as follows:

$$R_{2n}(z) = \sum_{k=0}^{n} r_{2k}^{(2n)} z^{2k}, \tag{2.438}$$

$$r_{2k}^{(2n)} = \frac{(-1)^k (2n+1)!}{(2n-2k)!} c_{2n-2k}. \tag{2.439}$$

The set of polynomials on which the functional f is positive can be appreciably extended by using the following property of the expansion coefficients c_{2n} for $\xi(\lambda)$:

$$\sum_{i,j=0}^{N} c_{2i+2j+2m} a_{2i} a_{2j} > 0, \tag{2.440}$$

which is valid for arbitrary real a's and any N and m.

The proof of this property proceeds as follows. According to formula (2.395) and inequality (2.398), we have

$$\sum_{i,j=0}^{N} c_{2i+2j+2m} a_{2i} a_{2j} = \int_{1}^{\infty} H(x) \left(\ln \sqrt{x}\right)^{2m} \left(\sum_{i,j=0}^{N} \left(\ln \sqrt{x}\right)^{2i+2j} a_{2i} a_{2j}\right) dx$$

$$= \int_{1}^{\infty} H(x) \left(\ln \sqrt{x}\right)^{2m} \left(\sum_{i=0}^{N} a_{2i} \left(\ln \sqrt{x}\right)^{2i}\right)^{2} dx > 0. \tag{2.441}$$

By substituting formula (2.437) into inequality (2.440) and using the linearity of the functional f, we derive

$$f\left(\sum_{i,j=0}^{N} a_{2i}a_{2j}R_{2i+2j+2m-2}(z)\right) > 0. \tag{2.442}$$

It is not clear at this time if the set of polynomials on which functional f is positive (see the last inequality) contains all "complete-square" polynomials (2.428). In other words, it is not clear if the alignment of all zeros of $\xi(\lambda)$ along the real axis is fully encoded in inequalities (2.399) and (2.440). It is conceivable that other properties of expansion coefficients c_{2n} can be instrumental in the proof of inequality (2.429). Among such properties may be the Turán inequalities

$$c_{2n}^2 > \frac{2n-1}{2n+1} c_{2n-2}c_{2n+2}, \tag{2.443}$$

which were conjectured by Pólya and which are necessary conditions for the validity of the Riemann hypothesis. These inequalities were later proved in [28]-[32]. The subtle nature of the Turán inequalities is revealed by the following inequalities of "opposite" sense,

$$c_{2n}^2 \leq c_{2n-2}c_{2n+2}, \tag{2.444}$$

which can be easily proved by using formula (2.395). Indeed, from (2.395) we find

$$\begin{aligned}
c_{2n}^2 &= \left\{\int_1^\infty H(x)\left(\ln\sqrt{x}\right)^{2n} dx\right\}^2 \\
&= \left\{\int_1^\infty [H(x)]^{\frac{1}{2}}\left(\ln\sqrt{x}\right)^{n-1} [H(x)]^{\frac{1}{2}}\left(\ln\sqrt{x}\right)^{n+1} dx\right\}^2 \\
&\leq \int_1^\infty H(x)\left(\ln\sqrt{x}\right)^{2n-2} dx \int_1^\infty H(x)\left(\ln\sqrt{x}\right)^{2n+2} dx, \tag{2.445}
\end{aligned}$$

which is tantamount to inequality (2.444).

It is interesting to illustrate the inequalities (2.443) and (2.444) through the numerical computations based on formulas (2.395)-(2.397)[§]. The results of computations of c_{2n} are (partially) illustrated by Table 2.1. It is worthwhile to note that c_{2n} are first decreasing and achieve the minimum value at $2n = 42$, and then c_{2n} are strictly increasing (see Figures 2.24 and 2.25).

[§]These computations were carried out by Dr. P. McAvoy on a laptop computer.

Table 2.1

$2n$	c_{2n}	$2n$	c_{2n}	$2n$	c_{2n}
0	4.9712077819e-01	70	1.1034972672e-06	140	6.5854400400e+01
2	2.2971944315e-02	72	1.4987338856e-06	142	1.2975774631e+02
4	2.9628484337e-03	74	2.0682197174e-06	144	2.5751657002e+02
6	5.9929594660e-04	76	2.8984715546e-06	146	5.1469481580e+02
8	1.6096657455e-04	78	4.1232099025e-06	148	1.0359003609e+03
10	5.3038634278e-05	80	5.9511535306e-06	150	2.0992397437e+03
12	2.0475115211e-05	82	8.7112806537e-06	152	4.2828773003e+03
14	8.9877558933e-06	84	1.2927173344e-05	154	8.7961729142e+03
16	4.3933042509e-06	86	1.9440186976e-05	156	1.8184136503e+04
18	2.3548833836e-06	88	2.9615258660e-05	158	3.7834664853e+04
20	1.3679861516e-06	90	4.5687724037e-05	160	7.9221678025e+04
22	8.5331439117e-07	92	7.1352680650e-05	162	1.6692250644e+05
24	5.6729724758e-07	94	1.1277489189e-04	164	3.5388595563e+05
26	3.9950482182e-07	96	1.8033363386e-04	166	7.5483359514e+05
28	2.9649456827e-07	98	2.9166204200e-04	168	1.6197256817e+06
30	2.3089199551e-07	100	4.7698430706e-04	170	3.4962132955e+06
32	1.8796716106e-07	102	7.8856323556e-04	172	7.5907615840e+06
34	1.5945436290e-07	104	1.3175565691e-03	174	1.6575606414e+07
36	1.4055599169e-07	106	2.2243283035e-03	176	3.6401280978e+07
38	1.2842339050e-07	108	3.7933784913e-03	178	8.0388592164e+07
40	1.2135730923e-07	110	6.5336509213e-03	180	1.7851340751e+08
42	1.1837577135e-07	112	1.1363096268e-02	182	3.9857938527e+08
44	1.1897898196e-07	114	1.9950810709e-02	184	8.9473836430e+08
46	1.2302648724e-07	116	3.5355986359e-02	186	2.0192258278e+09
48	1.3068418807e-07	118	6.3229904408e-02	188	4.5809167477e+09
50	1.4242125693e-07	120	1.1409348807e-01	190	1.0446488090e+10
52	1.5905025529e-07	122	2.0768369383e-01	192	2.3944789142e+10
54	1.8181316499e-07	124	3.8130656817e-01	194	5.5162954154e+10
56	2.1252471257e-07	126	7.0600244661e-01	196	1.2771838594e+11
58	2.5379455536e-07	128	1.3180451023e+00	198	2.9716860759e+11
60	3.0936393387e-07	130	2.4807380050e+00	200	6.9481758598e+11
62	3.8461333397e-07	132	4.7064635587e+00	202	1.6324216652e+12
64	4.8733013560e-07	134	8.9993339511e+00	204	3.8535714440e+12
66	6.2887645146e-07	136	1.7340778337e+01	206	9.1398770473e+12
68	8.2597927193e-07	138	3.3667582666e+01	208	2.1779104575e+13

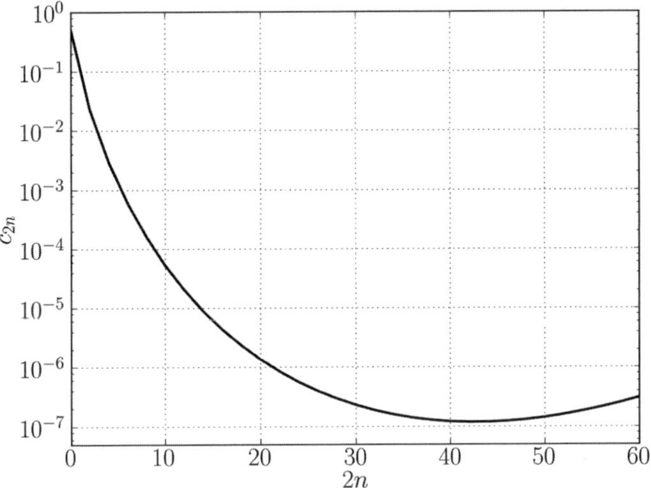

Figure 2.24

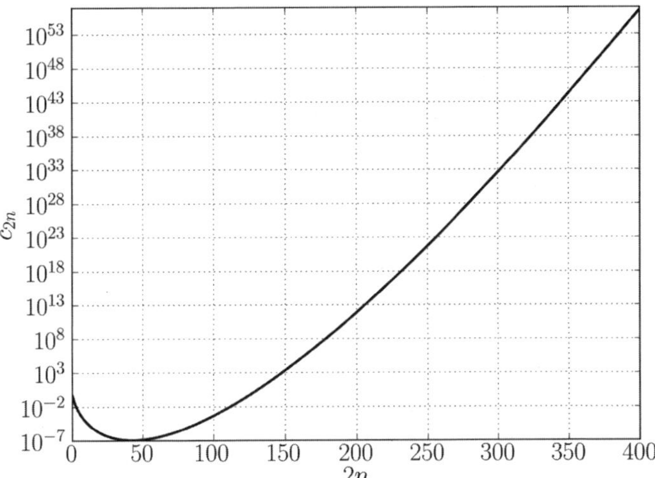

Figure 2.25

Table 2.2

n	$\hat{b}_n$	n	$\hat{b}_n$	n	$\hat{b}_n$
0	6.2140097274e-02	7	6.8571135660e-11	14	1.3806604234e-16
1	7.1787325985e-04	8	8.3795628565e-12	15	2.6879365965e-17
2	2.3147253388e-05	9	1.1228959005e-12	16	5.4705643870e-18
3	1.1704998957e-06	10	1.6307665725e-13	17	1.1601831858e-18
4	7.8596960230e-08	11	2.5430750584e-14	18	2.5566985950e-19
5	6.4744426609e-09	12	4.2266938655e-15	19	5.8400196623e-20
6	6.2485092806e-10	13	7.4413571846e-16	20	1.3796728721e-20

Similar computations were carried out by the authors of [30] and [31] for $\hat{b}_n$ which are the scaled versions of c_{2n} and related to the latter by the formula $c_{2n} = 8 \cdot 2^{2n}\hat{b}_n$. For comparison purposes, the values of $\hat{b}_n$ computed by using the last formula are given for $0 \le n \le 20$ in Table 2.2. These values of $\hat{b}_n$ coincide with (possible) exception of their last digit with the values reported in the Table 3.1 in [31]. It is also remarked in [31] that $\hat{b}_n$ reaches the minimum value for $\hat{b}_{339}$, while our computations show that for c_{2n} the minimum occurs for c_{42}. This difference is related to the scaling by the factor $8 \cdot 2^{2n}$. Tables 2.3a and 2.3b illustrate the validity of inequalities (2.443) and (2.444) for different ranges of n. The closeness of the upper and lower bounds given by the above inequalities for c_{2n}^2 is quite remarkable.

Finally, the computed values of c_{2n} were used in equations (2.436) to calculate q_{2n}. The results of the calculations are presented in Table 2.4. This table reveals that the computed q_{2n} are positive as implied by formula (2.424). The positivity of q_{2n} for small n can be established analytically by using the Turán inequalities (2.443). It would be interesting to prove positivity of q_{2n} for any n. Table 2.4 also suggests that the ratio q_{2n-2}/q_{2n} converges to the number whose first ten digits are 199.7904548. This number is $\tilde{\lambda}_1^2 = (14.134725)^2$. In other words, the results of computations are consistent with the formula

$$\lim_{n \to \infty} \frac{q_{2n-2}}{q_{2n}} = \tilde{\lambda}_1^2, \tag{2.446}$$

which is implied by (2.424).

By introducing

$$q_{2n}^{(1)} = q_{2n} - \frac{1}{\tilde{\lambda}_1^{2n+2}}, \tag{2.447}$$

it is expected from (2.424) that

$$\lim_{n\to\infty} \frac{q_{2n-2}^{(1)}}{q_{2n}^{(1)}} = \tilde{\lambda}_2^2 = (21.02204)^2, \tag{2.448}$$

which is consistent with the last column of Table 2.4. Similarly, the validity of the Riemann hypothesis and formula (2.424) imply that

$$\lim_{n\to\infty} \frac{q_{2n-2}^{(k)}}{q_{2n}^{(k)}} = \tilde{\lambda}_{k+1}^2, \tag{2.449}$$

where

$$q_{2n}^{(k)} = q_{2n} - \sum_{i=0}^{k} \frac{1}{\tilde{\lambda}_i^{2n+2}}. \tag{2.450}$$

The presented discussion clearly suggests the intimate connection between the q-numbers and the zeros of the xi-function.

It is interesting to note that from formulas (2.418) and (2.419) follows the explicit relation between q_{2n} and $\xi(\lambda)$

$$q_{2n} = -\frac{1}{2\pi i} \oint_C \frac{\xi'(\lambda)}{\lambda^{2n+2}\xi(\lambda)} d\lambda, \tag{2.451}$$

where $i = \sqrt{-1}$ and C is a circle of sufficiently small radius in the complex λ-plane. Finally, the necessary and sufficient condition (2.420) can be written in terms of the xi-function as follows:

$$-\frac{1}{2\pi i} \oint_C \frac{\xi'(\lambda)}{\lambda^2\xi(\lambda)} \left[\sum_{i=0}^{N} \frac{\gamma_{2i}}{\lambda^{2i}} \right]^2 d\lambda > 0. \tag{2.452}$$

The mathematical tools used in the presented discussion of the Riemann hypothesis are quite modest by modern standards. Nevertheless, it cannot be precluded that these tools may be helpful in the pursuit of the elusive proof of this hypothesis. One of the reasons is that the interpretation of the Grommer theorem in terms of the functional f reduces the search for the proof of the Riemann hypothesis to the polynomial analysis in which the property of the expansion coefficients c_{2n} for the xi-function may be naturally accounted for.

Table 2.3a

$2n$	$\frac{2n-1}{2n+1}c_{2n-2}c_{2n+2}$	c_{2n}^2	$c_{2n-2}c_{2n+2}$
2	4.90964506336e-04	5.27710225618e-04	1.47289351901e-03
4	8.26019586812e-06	8.77847084101e-06	1.37669931135e-05
6	3.40656830916e-07	3.59155631608e-07	4.76919563282e-07
8	2.47223188614e-08	2.59102381224e-08	3.17858385361e-08
10	2.69657113009e-09	2.81309672611e-09	3.29580915900e-09
12	4.03360098143e-10	4.19230342894e-10	4.76698297806e-10
14	7.79596226008e-11	8.07797559970e-11	8.99534106933e-11
16	1.86751032430e-11	1.93011222410e-11	2.11651170087e-11
18	5.37734996707e-12	5.54547575026e-12	6.00997937496e-12
20	1.81807913019e-12	1.87138611094e-12	2.00945588073e-12
22	7.08571754303e-13	7.28145450176e-13	7.76054778522e-13
24	3.13630956696e-13	3.21826167110e-13	3.40903213800e-13
26	1.55741252314e-13	1.59604102657e-13	1.68200552499e-13
28	8.58809161798e-14	8.79090290113e-14	9.22424655265e-14
30	5.21356782501e-14	5.33111135914e-14	5.57312422674e-14
32	3.45854187066e-14	3.53316536378e-14	3.68167360426e-14
34	2.49102015447e-14	2.54256938472e-14	2.64199107292e-14
36	1.93707688375e-14	1.97559868014e-14	2.04776699140e-14
38	1.61827535151e-14	1.64925672285e-14	1.70574969483e-14
40	1.44606463044e-14	1.47275965036e-14	1.52022179097e-14
42	1.37673891470e-14	1.40128232417e-14	1.44389691054e-14
44	1.39160950866e-14	1.41559981478e-14	1.45633553232e-14
46	1.48870260522e-14	1.51355165629e-14	1.55486716546e-14
48	1.68064201341e-14	1.70783570114e-14	1.75215869483e-14
50	1.99702415737e-14	2.02838144249e-14	2.07853534747e-14
52	2.49169251631e-14	2.52969837075e-14	2.58940594832e-14
54	3.25729421602e-14	3.30560269619e-14	3.38021097889e-14
56	4.45241320190e-14	4.51667534513e-14	4.61431913651e-14
58	6.35187529494e-14	6.44116763280e-14	6.57474811231e-14
60	9.44123513880e-14	9.57060435767e-14	9.76127700792e-14

Table 2.3b

$2n$	$\frac{2n-1}{2n+1}c_{2n-2}c_{2n+2}$	c_{2n}^2	$c_{2n-2}c_{2n+2}$
340	4.78875859414e+83	4.80450347992e+83	4.81701085724e+83
342	4.30481490124e+84	4.31889624812e+84	4.33006308248e+84
344	3.88981411259e+85	3.90247324757e+85	3.91249524444e+85
346	3.53289649067e+86	3.54433590398e+86	3.55337705004e+86
348	3.22511898864e+87	3.23550929340e+87	3.24370757071e+87
350	2.95909278772e+88	2.96857834283e+88	2.97605033951e+88
352	2.72869420664e+89	2.73739767689e+89	2.74424232177e+89
354	2.52883334732e+90	2.53685941582e+90	2.54316101501e+90
356	2.35526834049e+91	2.36270674258e+91	2.36853745790e+91
358	2.20445573817e+92	2.21138375051e+92	2.21680563026e+92
360	2.07342966608e+93	2.07991415685e+93	2.08498080628e+93
362	1.95970394405e+94	1.96580306858e+94	1.97056102962e+94
364	1.86119262013e+95	1.86695722930e+95	1.87144712493e+95
366	1.77614532612e+96	1.78162014160e+96	1.78587762927e+96
368	1.70309461437e+97	1.70831921411e+97	1.71237578393e+97
370	1.64081302484e+98	1.64582267394e+98	1.64970632037e+98
372	1.58827809527e+99	1.59310445084e+99	1.59684024134e+99
374	1.54464389452e+100	1.54931560777e+100	1.55292616742e+100
376	1.50921795029e+101	1.51376118507e+101	1.51726711269e+101
378	1.48144267584e+102	1.48588155939e+102	1.48930178818e+102
380	1.46088058779e+103	1.46523760692e+103	1.46858972018e+103
382	1.44720275970e+104	1.45149911351e+104	1.45479962458e+104
384	1.44018008106e+105	1.44443599979e+105	1.44770060367e+105
386	1.43967699628e+106	1.44391203438e+106	1.44715583781e+106
388	1.44564748643e+107	1.44988079723e+107	1.45311853287e+107
390	1.45813313421e+108	1.46238373274e+108	1.46562996267e+108
392	1.47726318160e+109	1.48155020162e+109	1.48481951501e+109
394	1.50325655418e+110	1.50759950539e+110	1.51090671477e+110
396	1.53642588819e+111	1.54084491921e+111	1.54420525978e+111
398	1.57718365878e+112	1.58169983195e+112	1.58512916840e+112

Table 2.4

$2n$	q_{2n}	q_{2n-2}/q_{2n}	$q_{2n-2}^{(1)}/q_{2n}^{(1)}$
0	4.6209986231e-02		
2	7.4345198571e-05	621.5597929568	1493.362544191
4	2.8834786280e-07	257.8316268694	645.3671650902
6	1.3260633605e-09	217.4465198193	530.4321233275
8	6.4273283012e-12	206.3164192580	490.0966591086
10	3.1753869030e-14	202.4108714184	471.0943309063
12	1.5806651448e-16	200.8892847046	460.6828068729
14	7.8929428000e-19	200.2630938628	454.4309289670
16	3.9465301475e-21	199.9970228282	450.4496105052
18	1.9744330767e-23	199.8816872622	447.8159551478
20	9.8805127046e-26	199.8310346534	446.0307003010
22	4.9449888887e-28	199.8085926386	444.8016078314
24	2.4749868886e-30	199.7985892909	443.9471015699
26	1.2387686847e-32	199.7941116283	443.3493762128
28	6.2002885780e-35	199.7921014594	442.9296742922
30	3.1033842662e-37	199.7911971651	442.6342778671
32	1.5533169821e-39	199.7907897676	442.4260670493
34	7.7747248127e-42	199.7906060410	442.2791770981
36	3.8914382380e-44	199.7905231246	442.1754908920
38	1.9477595365e-46	199.7904856850	442.1022764432
40	9.7490112940e-49	199.7904687736	442.0505679351
42	4.8796179951e-51	199.7904611329	442.0140436926
44	2.4423678947e-53	199.7904576801	441.9882429196
46	1.2224647474e-55	199.7904561196	441.9700164201
48	6.1187344754e-58	199.7904554142	441.9571403050
50	3.0625759737e-60	199.7904550954	441.9480438401
52	1.5328940386e-62	199.7904549513	441.9416174984
54	7.6725088768e-65	199.7904548861	441.9370774898
56	3.8402779964e-67	199.7904548567	441.9338701105
58	1.9221528873e-69	199.7904548434	441.9316041924
60	9.6208444438e-72	199.7904548374	441.9300033887
62	4.8154675116e-74	199.7904548346	441.9288724688
64	2.4102590464e-76	199.7904548334	441.9280735082
66	1.2063934929e-78	199.7904548328	441.9275090670
68	6.0382939409e-81	199.7904548326	441.9271103067
70	3.0223135264e-83	199.7904548325	441.9268285949
72	1.5127417018e-85	199.7904548324	441.9266295744

References

[1] F. Ouyang and M. Isaacson, Philosophical Magazine B 60, 481 (1989).

[2] F. Ouyang and M. Isaacson, Ultramicroscopy 31, 345 (1989).

[3] D.R. Fredkin and I.D. Mayergoyz, Physical Review Letters 91, 253902 (2003).

[4] I.D. Mayergoyz, D.R. Fredkin, and Z. Zhang, Physical Review B 72, 155412 (2005).

[5] M.I. Stockman, S.V. Faleev, and D.J. Bergman, Physical Review Letters 87, 167401 (2001).

[6] K. Li, M.I. Stockman, and D.J. Bergman, Physical Review Letters 91, 227402 (2003).

[7] E.A. Stern and R.A. Ferrell, Physical Review 120, 130 (1960).

[8] O.V. Tozoni and I.D. Mayergoyz, Analysis of 3D Electromagnetic Fields, Technika, Kyiv (1974).

[9] I.D. Mayergoyz, Iterative Techniques for the Analysis of Static Fields in Inhomogeneous, Anisotropic and Nonlinear Media, Naukova Dumka, Kyiv, 1979.

[10] J.D. Jackson, Classical Electrodynamics, John Wiley, New York (1999).

[11] C.F. Bohren and D.R. Huffman, Absorption and Scattering of Light by Small Particles, John Wiley, New York (1983).

[12] A. Ishimaru, Electromagnetic Wave Propagation, Radiation and Scattering, Prentice Hall, New Jersey (1999).

[13] S.G. Mikhlin, Mathematical Physics, an Advanced Course, North-Holland, Amsterdam (1970).

[14] L.V. Kantorovich and G.P. Akilov, Functional Analysis in Normed Spaces, Pergamon Press, New York (1982).

[15] N.M. Günter, Potential Theory, Ungar, New York (1967).

[16] O.D. Kellogg, Foundation of Potential Theory, Dover, New York (1953).

[17] W. Blaschke, Kreis und Kugel, de Gruyter, Berlin (1956).

[18] G.Y. Lyubarskii, Application of Group Theory in Physics, Pergamon Press, New York (1960).

[19] A.W. Joshi, Elements of Group Theory for Physicists, Wiley, New Delhi (1982).

[20] E. Goursat, Course in Mathematical Analysis, Dover, New York (1964).

[21] H. Wang, D.W. Brandt, F. Le, P. Nordlander, and N.J. Halas, Nanoletters 6, 827 (2006).

[22] R.D. Averitt, S.L. Westcott, and N.J. Halas, Journal of Optical Society of America B 16, 1824 (1999).

[23] E. Prodan, C. Radloff, N.J. Halas, and P. Nordlander, Science 302, 419 (2003).

[24] B. Riemann, Monat. der Königl. Preuss. Akad. der Wissen. zu Berlin aus der Jahre 1859, 671 (1860).

[25] H.M. Edwards, Riemann's Zeta Function, Academic Press, New York-London (1974).

[26] J. Grommer, J. Reine Angew. Math. 144, 114 (1914).

[27] N.I. Akhiezer and M.G. Krein, Some Questions in the Theory of Moments, Transl. Math. Monographs, American Mathematical Society (1962).

[28] E. Grosswald, Illinois Journal of Mathematics 10, 9 (1966).

[29] Yu.V. Matiyasevich, Cybernetics 18, 705 (1983).

[30] G. Csordas, T.S. Norfolk, and R.S. Varga, Transactions of the American Mathematical Society 296, 521 (1986).

[31] R.S. Varga, Scientific Computations: Problems and Conjectures, SIAM (1990).

[32] J.B. Conrey and A. Ghosh, Transactions of the American Mathematical Society 342, 407 (1994).

Chapter 3

Analytical and Numerical Analysis of Plasmon Resonances

3.1 Some Analytical Solutions for Plasmon Modes

This chapter deals mostly with numerical issues related to the implementation of the plasmon mode analysis presented in the previous chapter. Before proceeding with the discussion of these issues, analytical solutions for plasmon modes in certain nanostructures are presented in this section. There are two reasons why these analytical solutions are important. First, these analytical solutions are used in the book as test examples for numerical computations. Second, these analytical solutions for plasmon modes are of interest in their own right. This is because these analytical solutions are derived for nanostructures that have appeared (or will appear) in various applications of plasmon resonances.

The analytical results obtained in this section are based on the method of separation of variables in different coordinate systems. In this method, possible solutions of the Laplace equation are expressed as products of functions, each of which depends only on one of the variables of the coordinate system used. These product solutions are plasmon modes, and they are realized for specific negative values of dielectric permittivity of metallic nanoparticles or

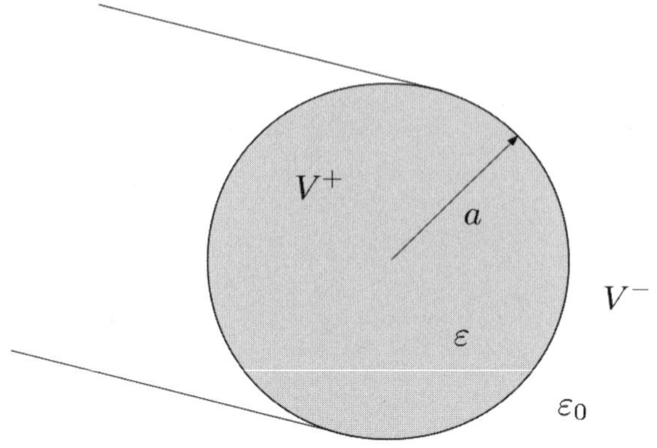

Figure 3.1

nanowires. These specific resonance values of permittivity are found from the interface boundary conditions.

The review of various coordinate systems and the corresponding product solutions of the Laplace equations can be found in the books [1]-[4].

We begin with the simplest example.

1. Plasmon modes in nanowires of circular cross sections

We consider plasmon modes excited in such nanowires by plane waves whose directions of propagation are normal to the axes of the nanowires and whose incident electric fields are in the nanowire cross-sectional planes. Under these conditions, two-dimensional plasmon modes can be excited. The electric potential of such modes satisfies the two-dimensional Laplace equation inside and outside the circular cross section (see Figure 3.1),

$$\nabla^2 \varphi^\pm = 0 \quad \text{in } V^\pm, \tag{3.1}$$

and the following boundary conditions

$$\varphi^+\big|_{r=a} = \varphi^-\big|_{r=a}, \tag{3.2}$$

$$\varepsilon \frac{\partial \varphi^+}{\partial r}\bigg|_{r=a} = \varepsilon_0 \frac{\partial \varphi^-}{\partial r}\bigg|_{r=a}, \tag{3.3}$$

as well as "zero-condition" at infinity

$$\varphi^-(\infty) = 0. \tag{3.4}$$

The last condition is justified because all plasmon modes are "charge-neutral."

It is known that in the polar coordinate system the product solutions of the Laplace equation which are finite in V^+ and satisfy the condition (3.4) in V^- are given by the formulas

$$\varphi_n^+(r, \theta) = A_n r^n \frac{\cos}{\sin}(n\theta), \tag{3.5}$$

$$\varphi_n^-(r, \theta) = B_n r^{-n} \frac{\cos}{\sin}(n\theta). \tag{3.6}$$

Next, we find that these product solutions satisfy the boundary conditions (3.2) and (3.3) only for special value ε_n of nanowire permittivity ε such that the following relations are valid:

$$A_n a^n = B_n a^{-n}, \tag{3.7}$$

$$\varepsilon_n n A_n a^{n-1} = -\varepsilon_0 n B_n a^{-n-1}. \tag{3.8}$$

From the last two formulas follows that

$$\boxed{\begin{aligned} \varepsilon_n &= -\varepsilon_0, \\ B_n &= a^{2n} A_n. \end{aligned}} \tag{3.9} \tag{3.10}$$

Now, the conclusion can be drawn that all possible plasmon modes are of the form

$$\varphi_n^+(r, \theta) = A_n r^n \frac{\cos}{\sin}(n\theta), \tag{3.11}$$

$$\varphi_n^-(r, \theta) = A_n a^{2n} r^{-n} \frac{\cos}{\sin}(n\theta), \tag{3.12}$$

and they all can be excited at the same resonance value of permittivity ε_n specified by formula (3.9) and, consequently, at the same frequency (wavelength) of incident radiation. If the dispersion relation (1.8) is used, then this plasmon resonance frequency is given by the formula

$$\omega_n = \frac{\omega_p}{\sqrt{2}}. \tag{3.13}$$

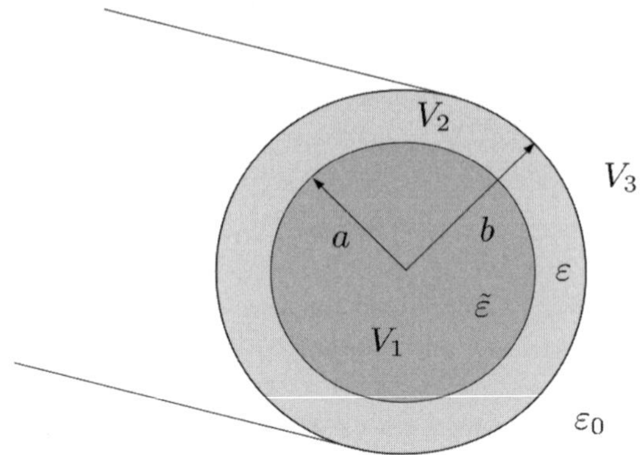

Figure 3.2

2. Plasmon modes in circular cross-section nanotubes

A typical cross section of such nanotubes is shown in Figure 3.2, where the core of the metallic nanotube has permittivity $\tilde{\varepsilon}$.

The electric potential φ of plasmon modes satisfies the Laplace equation

$$\nabla^2 \varphi^{(k)} = 0 \quad \text{in } V_k, \quad (k = 1, 2, 3), \tag{3.14}$$

the interface boundary conditions

$$\varphi^{(1)}\Big|_{r=a} = \varphi^{(2)}\Big|_{r=a}, \tag{3.15}$$

$$\varphi^{(2)}\Big|_{r=b} = \varphi^{(3)}\Big|_{r=b}, \tag{3.16}$$

$$\tilde{\varepsilon}\frac{\partial \varphi^{(1)}}{\partial r}\Big|_{r=a} = \varepsilon\frac{\partial \varphi^{(2)}}{\partial r}\Big|_{r=a}, \tag{3.17}$$

$$\varepsilon\frac{\partial \varphi^{(2)}}{\partial r}\Big|_{r=b} = \varepsilon_0\frac{\partial \varphi^{(3)}}{\partial r}\Big|_{r=b}, \tag{3.18}$$

and the following condition at infinity:

$$\varphi^{(3)}(\infty) = 0. \tag{3.19}$$

As before, we shall use polar coordinates and the appropriate product solutions of the Laplace equation

$$\varphi_n^{(1)}(r,\theta) = A_n r^n \, {\cos \atop \sin} (n\theta) , \tag{3.20}$$

$$\varphi_n^{(2)}(r,\theta) = \left(B_n r^n + C_n r^{-n}\right) {\cos \atop \sin} (n\theta) , \tag{3.21}$$

$$\varphi_n^{(3)}(r,\theta) = D_n r^{-n} \, {\cos \atop \sin} (n\theta) . \tag{3.22}$$

These product solutions satisfy the boundary conditions (3.15)-(3.18) only for special values ε_n of ε such that the following relations are valid:

$$A_n a^n = B_n a^n + C_n a^{-n}, \tag{3.23}$$

$$B_n b^n + C_n b^{-n} = D_n b^{-n}, \tag{3.24}$$

$$\tilde{\varepsilon} A_n n a^{n-1} = \varepsilon_n \left(B_n n a^{n-1} - C_n n a^{-n-1}\right), \tag{3.25}$$

$$\varepsilon_n \left(B_n n b^{n-1} - C_n n b^{-n-1}\right) = -\varepsilon_0 D_n n b^{-n-1}. \tag{3.26}$$

From equations (3.23) and (3.24) we find

$$A_n = B_n + C_n a^{-2n}, \tag{3.27}$$

$$D_n = B_n b^{2n} + C_n. \tag{3.28}$$

By substituting formulas (3.27) and (3.28) into equations (3.25) and (3.26), we end up with the following homogeneous equations for B_n and C_n:

$$(\tilde{\varepsilon} - \varepsilon_n) n a^{n-1} B_n + (\tilde{\varepsilon} + \varepsilon_n) n a^{-n-1} C_n = 0, \tag{3.29}$$

$$- (\varepsilon_0 + \varepsilon_n) n b^{n-1} B_n + (\varepsilon_n - \varepsilon_0) n b^{-n-1} C_n = 0. \tag{3.30}$$

These homogeneous equations have nonzero solutions only if the determinant of the corresponding matrix is equal to zero. This leads to the quadratic equation for ε_n. After some straightforward transformations, this quadratic equation can be written as follows:

$$\varepsilon_n^2 + 2(\tilde{\varepsilon} + \varepsilon_0)\alpha_n \varepsilon_n + \tilde{\varepsilon}\varepsilon_0 = 0, \tag{3.31}$$

where

$$\boxed{\alpha_n = \frac{1}{2} \frac{\left(\frac{b}{a}\right)^{2n} + 1}{\left(\frac{b}{a}\right)^{2n} - 1} > \frac{1}{2}.} \tag{3.32}$$

It follows from equation (3.31) that the plasmon modes whose electric potential is described by formulas (3.20)-(3.22) may exist for two values of dielectric permittivity given by the equation

$$\varepsilon_n^{\pm} = -\left(\tilde{\varepsilon} + \varepsilon_0\right)\alpha_n \pm \sqrt{\left(\tilde{\varepsilon} + \varepsilon_0\right)^2 \alpha_n^2 - \tilde{\varepsilon}\varepsilon_0}. \tag{3.33}$$

It is also apparent that

$$\varepsilon_n^{+}\varepsilon_n^{-} = \tilde{\varepsilon}\varepsilon_0. \tag{3.34}$$

Thus, there are two distinct bands of plasmon resonances corresponding to ε_n^{-} and ε_n^{+}, respectively.

The expressions (3.32) and (3.33) can be appreciably simplified in the case when the thickness δ of the nanotube is sufficiently small. Indeed, in this case we find

$$b = a + \delta, \quad b^{2n} = (a + \delta)^{2n} \approx a^{2n} + 2na^{2n-1}\delta, \tag{3.35}$$

and

$$\left(\frac{b}{a}\right)^{2n} \approx 1 + 2n\left(\frac{\delta}{a}\right), \tag{3.36}$$

which leads to

$$\alpha_n \approx \frac{1 + n\left(\frac{\delta}{a}\right)}{2n\left(\frac{\delta}{a}\right)}. \tag{3.37}$$

For plasmon modes such that

$$n\left(\frac{\delta}{a}\right) \ll 1, \tag{3.38}$$

we derive

$$\alpha_n \approx \frac{a}{2n\delta}, \tag{3.39}$$

$$\varepsilon_n^{-} \approx -(\tilde{\varepsilon} + \varepsilon_0)\frac{a}{n\delta}, \tag{3.40}$$

$$\varepsilon_n^{+} \approx -\frac{\tilde{\varepsilon}\varepsilon_0}{(\tilde{\varepsilon} + \varepsilon_0)}\frac{n\delta}{a}. \tag{3.41}$$

In particular, for dipole modes ($n = 1$), we find

$$\varepsilon_1^{-} \approx -(\tilde{\varepsilon} + \varepsilon_0)\frac{a}{\delta}. \tag{3.42}$$

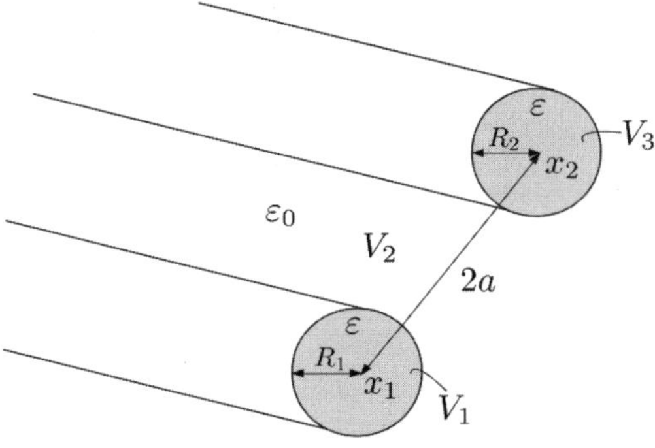

Figure 3.3

The last three formulas reveal that the resonance values of permittivity and, consequently, the resonance wavelength can be effectively controlled by the thickness of the nanotube. This is a typical situation for nanotubes and nanoshells, and it explains why thin-wall nanostructures are attractive for tuning plasmon resonances to a desired wavelength.

To finish the discussion of this example, we point out that by using equations (3.27)-(3.29) the plasmon mode potential can be determined up to one scaling factor B_n. Indeed, from equation (3.29) we determine C_n in terms of B_n. (By the way, equation (3.30) will lead to the same result, because the permittivity ε_n is chosen from the condition that the determinant of the linear equations (3.29) and (3.30) is equal to zero.) Having found C_n in terms of B_n, we can use formulas (3.27) and (3.28) to find A_n and D_n in terms of B_n. In the end, formulas (3.20)-(3.22) will be expressed in terms of B_n.

3. Plasmon modes in two adjacent circular cross-section nanowires

A typical cross-sectional view of such nanostructure is shown in Figure 3.3. For the solution of this problem we shall use bipolar coordinates (μ, η) which are related to the rectangular coordinates (x, y) by the formulas

$$x = \frac{a\sinh\mu}{\cosh\mu - \cos\eta},\tag{3.43}$$

$$y = \frac{a\sin\eta}{\cosh\mu - \cos\eta},\tag{3.44}$$

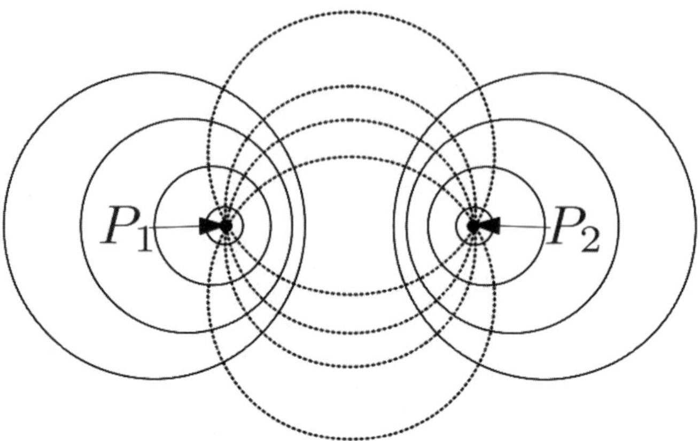

Figure 3.4

where

$$-\infty < \mu < \infty, \tag{3.45}$$

$$0 \leq \eta < 2\pi. \tag{3.46}$$

Figure 3.4 illustrates bipolar coordinates by representing two families of orthogonal circular lines corresponding to lines $\mu = const$ and $\eta = const$, respectively. It is apparent that bipolar coordinates admit the following physical interpretation: lines $\eta = const$ and $\mu = const$ are respectively electric field lines and equipotential lines of the electric field created by two infinite in extent and oppositely charged filamentary lines with locations defined by equations $x = a$, $y = 0$ and $x = -a$, $y = 0$ (see points P_1 and P_2 in Figure 3.4). The utility of the bipolar coordinates for our problem is based on the fact that the circular boundaries of nanowires coincide with lines $\mu = \mu_1$ and $\mu = -\mu_2$. Furthermore, the following formulas are valid:

$$R_1 = \frac{a}{\sinh \mu_1}, \quad R_2 = \frac{a}{\sinh \mu_2}, \tag{3.47}$$

$$x_1 = a \coth \mu_1, \quad x_2 = -a \coth \mu_2, \tag{3.48}$$

where x_1 and x_2 are the x-coordinates of the centers of the circular cross sections of the nanowires.

As before, the electric potential of plasmon modes satisfies the Laplace equation

$$\nabla^2 \varphi^{(k)} = 0 \quad \text{in } V_k, \ (k = 1, 2, 3) \tag{3.49}$$

and the boundary conditions

$$\varphi^{(1)}\big|_{\mu_1} = \varphi^{(2)}\big|_{\mu_1}, \tag{3.50}$$

$$\varphi^{(2)}\big|_{-\mu_2} = \varphi^{(3)}\big|_{-\mu_2}, \tag{3.51}$$

$$\varepsilon \frac{\partial \varphi^{(1)}}{\partial \mu}\bigg|_{\mu_1} = \varepsilon_0 \frac{\partial \varphi^{(2)}}{\partial \mu}\bigg|_{\mu_1}, \tag{3.52}$$

$$\varepsilon_0 \frac{\partial \varphi^{(2)}}{\partial \mu}\bigg|_{-\mu_2} = \varepsilon \frac{\partial \varphi^{(3)}}{\partial \mu}\bigg|_{-\mu_2}, \tag{3.53}$$

where the last two boundary conditions are written after cancellation of metric h_μ-coefficients.

The appropriate product solutions of the Laplace equations in bipolar coordinates can be written as follows:

$$\varphi_n^{(1)}(\mu, \eta) = A_n e^{-n\mu} \frac{\cos}{\sin}(n\eta), \tag{3.54}$$

$$\varphi_n^{(2)}(\mu, \eta) = \left(B_n e^{-n\mu} + C_n e^{n\mu}\right) \frac{\cos}{\sin}(n\eta), \tag{3.55}$$

$$\varphi_n^{(3)}(\mu, \eta) = D_n e^{n\mu} \frac{\cos}{\sin}(n\eta). \tag{3.56}$$

These product solutions satisfy the boundary conditions (3.50)-(3.53) only for special values ε_n of ε such that the following relations are valid:

$$A_n e^{-n\mu_1} = B_n e^{-n\mu_1} + C_n e^{n\mu_1}, \tag{3.57}$$

$$B_n e^{n\mu_2} + C_n e^{-n\mu_2} = D_n e^{-n\mu_2}, \tag{3.58}$$

$$-\varepsilon_n A_n n e^{-n\mu_1} = \varepsilon_0 \left[-B_n n e^{-n\mu_1} + C_n n e^{n\mu_1}\right], \tag{3.59}$$

$$\varepsilon_0 \left[-B_n n e^{n\mu_2} + C_n n e^{n\mu_2}\right] = \varepsilon_n D_n n e^{-n\mu_2}. \tag{3.60}$$

By substituting equations (3.57) and (3.58) into formulas (3.59) and (3.60), we obtain after straightforward transformations the following homogeneous equations for B_n and C_n:

$$(\varepsilon_0 - \varepsilon_n)e^{-n\mu_1}B_n - (\varepsilon_0 + \varepsilon_n)e^{n\mu_1}C_n = 0, \tag{3.61}$$

$$(\varepsilon_0 + \varepsilon_n)e^{n\mu_2}B_n - (\varepsilon_0 - \varepsilon_n)e^{-n\mu_2}C_n = 0. \tag{3.62}$$

These homogeneous equations have nonzero solutions only if their determinant is equal to zero. This leads to the following quadratic equation for ε_n:

$$\varepsilon_n^2 + 2\varepsilon_n\varepsilon_0 \coth n\,(\mu_1 + \mu_2) + \varepsilon_0^2 = 0. \tag{3.63}$$

Thus, the plasmon modes with electric potential described by formulas (3.54)-(3.56) may exist for the following two values of ε_n:

$$\boxed{\varepsilon_n^{\pm} = -\varepsilon_0 \coth n\,(\mu_1 + \mu_2) \pm \varepsilon_0 \sqrt{\coth^2 n\,(\mu_1 + \mu_2) - 1}.} \tag{3.64}$$

It is apparent from formulas (3.63) and (3.64) that

$$\varepsilon_n^{+}\varepsilon_n^{-} = \varepsilon_0^2. \tag{3.65}$$

It is also clear that $\coth n(\mu_1 + \mu_2) \to 1$ as $n \to \infty$. Consequently,

$$\lim_{n \to \infty} \varepsilon_n^{\pm} = -\varepsilon_0. \tag{3.66}$$

Formulas (3.65) and (3.66) are fully consistent with the general properties of plasmon resonances in nanowires discussed in the previous chapter.

It can be observed from formulas (3.47) and (3.48) that it is possible to increase infinitely a, μ_1 and μ_2 in such a way that R_1 and R_2 remain unchanged, while x_1 and x_2 become increasingly large. In this case of increasing separation between two circular cross-section nanowires, from formula (3.64) we find

$$\varepsilon_n \to -\varepsilon_0, \tag{3.67}$$

which is consistent (as it must be) with formula (3.9).

Having found ε_n from formula (3.64), we can then find B_n in terms of C_n from equation (3.61). Next, we can determine A_n and D_n in terms of C_n by using equations (3.57) and (3.58). In this way, the plasmon mode potential φ_n defined by formulas (3.54)-(3.56) can be found up to a scaling constant C_n.

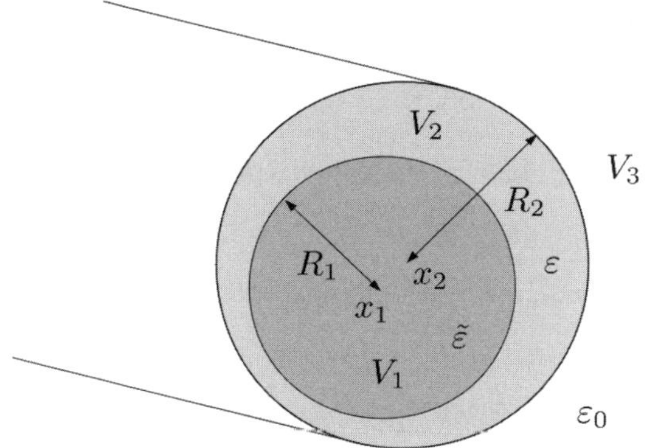

Figure 3.5

4. Plasmon modes in eccentric nanotubes

It turns out that bipolar coordinates can also be used for the analysis of plasmon modes in eccentric nanotubes whose cross section is illustrated by Figure 3.5. In this case, for the inner and outer circular boundaries we have

$$R_1 = \frac{a}{\sinh \mu_1}, \quad R_2 = \frac{a}{\sinh \mu_2}, \tag{3.68}$$

$$x_1 = a \coth \mu_1, \quad x_2 = a \coth \mu_2, \tag{3.69}$$

where x_1 and x_2 are x-coordinates of the centers of the inner and outer boundaries, respectively.

The electric potential of plasmon modes satisfies the Laplace equation

$$\nabla^2 \varphi_k = 0 \quad \text{in } V_k, \quad (k = 1, 2, 3) \tag{3.70}$$

and the boundary conditions

$$\varphi^{(1)}\Big|_{\mu_1} = \varphi^{(2)}\Big|_{\mu_1}, \tag{3.71}$$

$$\varphi^{(2)}\Big|_{\mu_2} = \varphi^{(3)}\Big|_{\mu_2}, \tag{3.72}$$

$$\tilde{\varepsilon}\,\frac{\partial \varphi^{(1)}}{\partial \mu}\bigg|_{\mu_1} = \varepsilon\,\frac{\partial \varphi^{(2)}}{\partial \mu}\bigg|_{\mu_1}, \tag{3.73}$$

$$\varepsilon\,\frac{\partial \varphi^{(2)}}{\partial \mu}\bigg|_{\mu_2} = \varepsilon_0\,\frac{\partial \varphi^{(3)}}{\partial \mu}\bigg|_{\mu_2}. \tag{3.74}$$

The appropriate product solutions are given by formulas (3.54)-(3.56). The product solutions satisfy the boundary conditions (3.71)-(3.74) for such special values ε_n of ε that the following relations are valid:

$$A_n e^{-n\mu_1} = B_n e^{-n\mu_1} + C_n e^{n\mu_1}, \tag{3.75}$$

$$B_n e^{-n\mu_2} + C_n e^{n\mu_2} = D_n e^{n\mu_2}, \tag{3.76}$$

$$-\tilde{\varepsilon} n A_n e^{-n\mu_1} = \varepsilon_n \left[n B_n e^{-n\mu_1} + n C_n e^{n\mu_1} \right], \tag{3.77}$$

$$\varepsilon_n \left[-n B_n e^{-n\mu_2} + n C_n e^{n\mu_2} \right] = \varepsilon_0 n D_n e^{n\mu_2}. \tag{3.78}$$

By substituting formulas (3.75) and (3.76) into equations (3.77) and (3.78), respectively, we obtain the following homogeneous equations for B_n and C_n:

$$(\varepsilon_n - \tilde{\varepsilon}) \, e^{-n\mu_1} B_n - (\tilde{\varepsilon} + \varepsilon_n) \, e^{n\mu_1} C_n = 0, \tag{3.79}$$

$$(\varepsilon_n + \varepsilon_0) \, e^{-n\mu_2} B_n - (\varepsilon_n - \varepsilon_0) \, e^{n\mu_2} C_n = 0. \tag{3.80}$$

These homogeneous equations have nonzero solutions only if their determinant is equal to zero. This leads to the following equation for ε_n:

$$\varepsilon_n^2 + \varepsilon_n \left(\tilde{\varepsilon} + \varepsilon_0 \right) \coth n \left(\mu_1 - \mu_2 \right) + \tilde{\varepsilon} \varepsilon_0 = 0. \tag{3.81}$$

From the last equation we find that the plasmon modes with electric potential described by formulas (3.54)-(3.56) may exist for the following values of ε_n:

$$\varepsilon_n^{\pm} = -\frac{\tilde{\varepsilon} + \varepsilon_0}{2} \coth n \left(\mu_1 - \mu_2 \right) \pm \sqrt{ \frac{(\tilde{\varepsilon} + \varepsilon_0)^2}{4} \coth^2 n \left(\mu_1 - \mu_2 \right) - \tilde{\varepsilon} \varepsilon_0 }. \tag{3.82}$$

In the particular case when the permittivity $\tilde{\varepsilon}$ of the nanotube core is equal to ε_0, we obtain

$$\boxed{\varepsilon_n^{\pm} = -\varepsilon_0 \coth n \left(\mu_1 - \mu_2 \right) \pm \varepsilon_0 \sqrt{\coth^2 n \left(\mu_1 - \mu_2 \right) - 1}.} \tag{3.83}$$

It is obvious from the last formula that

$$\varepsilon_n^{+} \varepsilon_n^{-} = \varepsilon_0^2, \tag{3.84}$$

as expected according to the general theory developed in the previous chapter.

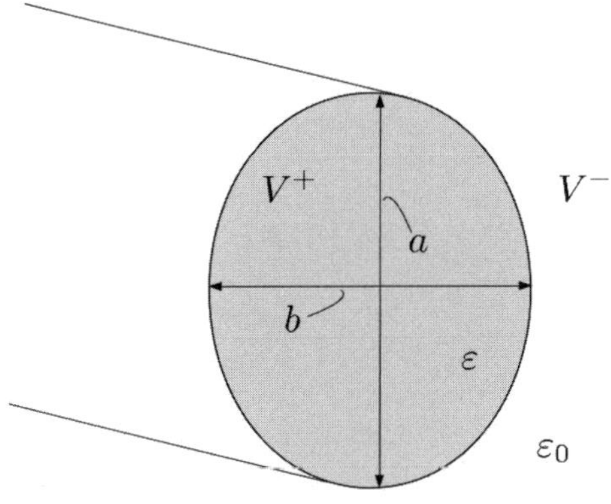

Figure 3.6

It is also apparent that by using bipolar coordinates, the plasmon modes can be studied in two adjacent eccentric nanotubes. In this case, quartic equations for ε_n can be derived.

5. Plasmon modes in nanowires of elliptical cross sections

A typical cross-sectional view of such nanowires is shown in Figure 3.6. The analysis of plasmon modes in such nanowires can be handled by using elliptic coordinates. These coordinates are related to rectangular (Cartesian) coordinates by the formulas

$$x = c \cosh \mu \cos \nu, \quad y = c \sinh \mu \sin \nu, \tag{3.85}$$

where

$$0 \leq \mu < \infty, \quad 0 < \nu < 2\pi, \tag{3.86}$$

and $2c$ is the distance between foci of ellipses which are coordinate lines $\mu = const.$

The usefulness of elliptic coordinates for our problem stems from the fact that the elliptic coordinates can be chosen in such a way that the cross-sectional boundary will coincide with the line $\mu = \mu_0$. Indeed, from formulas (3.85) we find that

$$\tanh \mu_0 = \frac{b}{a}, \tag{3.87}$$

where b and a are minor and major semi-axes of the elliptical cross-sectional boundary.

The electric potential of plasmon modes satisfies the Laplace equation

$$\frac{\partial^2 \varphi^\pm}{\partial \mu^2} + \frac{\partial^2 \varphi^\pm}{\partial \nu^2} = 0 \quad \text{in } V^\pm \tag{3.88}$$

and the following boundary conditions:

$$\varphi^+\big|_{\mu_0} = \varphi^-\big|_{\mu_0}, \tag{3.89}$$

$$\varepsilon \frac{\partial \varphi^+}{\partial \mu}\bigg|_{\mu_0} = \varepsilon_0 \frac{\partial \varphi^-}{\partial \mu}\bigg|_{\mu_0}, \tag{3.90}$$

where the last boundary condition is written after cancellation of the metric coefficient $h_\mu = h_\nu = c\sqrt{\cosh^2 \mu - \cos^2 \nu}$.

The product solutions of the Laplace equation (3.88) must be chosen in such a way that electric field components

$$E_\mu^+ = -\frac{1}{c\sqrt{\cosh^2 \mu - \cos^2 \nu}} \frac{\partial \varphi^+}{\partial \mu}, \tag{3.91}$$

$$E_\nu^+ = -\frac{1}{c\sqrt{\cosh^2 \mu - \cos^2 \nu}} \frac{\partial \varphi^+}{\partial \nu}, \tag{3.92}$$

will be finite at foci $(\mu = 0, \nu = 0)$ and $(\mu = 0, \nu = \pi)$ where the metric coefficients $h_\mu = h_\nu$ are equal to zero. This requires that the following conditions

$$\frac{\partial \varphi^+}{\partial \mu}\bigg|_{\substack{\mu = 0 \\ \nu = 0}} = \frac{\partial \varphi^+}{\partial \mu}\bigg|_{\substack{\mu = 0 \\ \nu = \pi}} = 0, \tag{3.93}$$

$$\frac{\partial \varphi^+}{\partial \nu}\bigg|_{\substack{\mu = 0 \\ \nu = 0}} = \frac{\partial \varphi^+}{\partial \nu}\bigg|_{\substack{\mu = 0 \\ \nu = \pi}} = 0 \tag{3.94}$$

are satisfied for product solutions. These conditions will be satisfied for the following two kinds of product solutions:
the first kind,

$$\varphi_n^+(\mu, \nu) = A_n^+ \sinh n\mu \sin n\nu, \tag{3.95}$$

$$\varphi_n^-(\mu, \nu) = A_n^- e^{-n\mu} \sin n\nu, \tag{3.96}$$

and the second kind,

$$\varphi_n^+(\mu, \nu) = B_n^+ \cosh n\mu \cos n\nu, \tag{3.97}$$

$$\varphi_n^-(\mu, \nu) = B_n^- e^{-n\mu} \cos n\nu. \tag{3.98}$$

We first consider the product solutions of the first kind. It is apparent that these product solutions satisfy the boundary conditions (3.89)-(3.90) only for such special values ε_n^- of ε that the following equalities are valid:

$$A_n^+ \sinh n\mu_0 = A_n^- e^{-n\mu_0}, \tag{3.99}$$

$$\varepsilon_n^- A_n^+ \cosh n\mu_0 = -\varepsilon_0 A_n^- e^{-n\mu_0}. \tag{3.100}$$

From the last two formulas we easily find

$$\varepsilon_n^- = -\varepsilon_0 \frac{\sinh n\mu_0}{\cosh n\mu_0}. \tag{3.101}$$

Next, we shall express ε_n^- in terms of a and b. To this end, from formula (3.87) we derive

$$\frac{e^{2\mu_0} - 1}{e^{2\mu_0} + 1} = \frac{b}{a}, \tag{3.102}$$

which leads to

$$e^{2\mu_0} = \frac{a + b}{a - b} \tag{3.103}$$

and

$$e^{2n\mu_0} = \left(\frac{a + b}{a - b}\right)^n. \tag{3.104}$$

By using the last equality, the formula (3.101) can be written in the form

$$\boxed{\varepsilon_n^- = -\varepsilon_0 \frac{\left(\frac{a+b}{a-b}\right)^n - 1}{\left(\frac{a+b}{a-b}\right)^n + 1}.} \tag{3.105}$$

Recalling that (see Chapter 2)

$$\lambda_n^- = \frac{\varepsilon_n^- - \varepsilon_0}{\varepsilon_n^- + \varepsilon_0}, \tag{3.106}$$

from formula (3.105) we find

$$\lambda_n^- = -\left(\frac{a+b}{a-b}\right)^n.$$

$$(3.107)$$

It is apparent that λ_n^- is negative, and this justifies the use of superscript "$-$" for ε_n and λ_n in the previous formulas.

Now, we consider the product solution of the second kind (see formulas (3.97) and (3.98)). These product solutions satisfy the boundary conditions (3.89)-(3.90) only for such special values ε_n^+ of ε that the following equalities are valid:

$$B_n^+ \cosh n\mu_0 = B_n^- e^{-n\mu_0},$$

$$(3.108)$$

$$\varepsilon_n^+ B_n^+ \sinh n\mu_0 = -\varepsilon_0 B_n^- e^{-n\mu_0}.$$

$$(3.109)$$

From the last two equations we easily derive

$$\varepsilon_n^+ = -\varepsilon_0 \frac{\cosh n\mu_0}{\sinh n\mu_0}.$$

$$(3.110)$$

By using formula (3.104), the last equation can be written in the form

$$\varepsilon_n^+ = -\varepsilon_0 \frac{\left(\frac{a+b}{a-b}\right)^n + 1}{\left(\frac{a+b}{a-b}\right)^n - 1},$$

$$(3.111)$$

which leads to the following expression:

$$\lambda_n^+ = \left(\frac{a+b}{a-b}\right)^n.$$

$$(3.112)$$

It is apparent that λ_n^+ is positive, and this justifies the use of superscript "$+$" for ε_n and λ_n in the previous formulas. It is also clear that

$$\lambda_n^+ = -\lambda_n^-,$$

$$(3.113)$$

and

$$\varepsilon_n^+ \varepsilon_n^- = \varepsilon_0^2,$$

$$(3.114)$$

which is consistent with the general properties of plasmon spectrum discussed in Chapter 2. Furthermore, we find that $\varepsilon^{\pm} \to -\varepsilon_0$ as $a \to b$, which is in agreement with formula (3.9).

It can be shown by using formulas (3.91)-(3.92) and (3.95), (3.97) that for $n = 1$ electric fields of the corresponding plasmon modes are spatially uniform inside the nanowires. These are dipole plasmon modes. By recalling formulas (3.105) and (3.111) we find that these dipole plasmon modes occur for the following values of dielectric permittivity:

$$\varepsilon_1^- = -\varepsilon_0 \frac{b}{a}, \tag{3.115}$$

$$\varepsilon_1^+ = -\varepsilon_0 \frac{a}{b}. \tag{3.116}$$

By using the dispersion relation (1.8) we derive the following expressions for the resonance frequencies of these plasmon modes:

$$\omega_- = \omega_p \left(\frac{a}{a+b} \right)^{\frac{1}{2}}, \tag{3.117}$$

$$\omega_+ = \omega_p \left(\frac{b}{a+b} \right)^{\frac{1}{2}}. \tag{3.118}$$

The analysis of plasmon resonances in nanoparticles and nanowires carried out in this book is based on classical electrodynamics and presumes the validity of the macroscopic constitutive relations. It is of interest, of course, to evaluate the limits of this approach by using quantum mechanical treatment of plasmon resonances. Such treatment has been undertaken in [5] for nanowires with elliptical cross sections, where it is demonstrated that formulas (3.117) and (3.118) are fully accurate provided that cross-sectional dimensions exceed 8 nm. Another (empirical) approach to test the validity of classical electrodynamics for the plasmon mode analysis is to compare the numerical (and analytical) results of such analysis with experimental data. This approach is practiced throughout the book.

6. Plasmon modes in spherical nanoparticles

The analysis of plasmon modes in a spherical nanoparticle (see Figure 3.7) requires the finding of special values of ε for which nonzero solutions of the

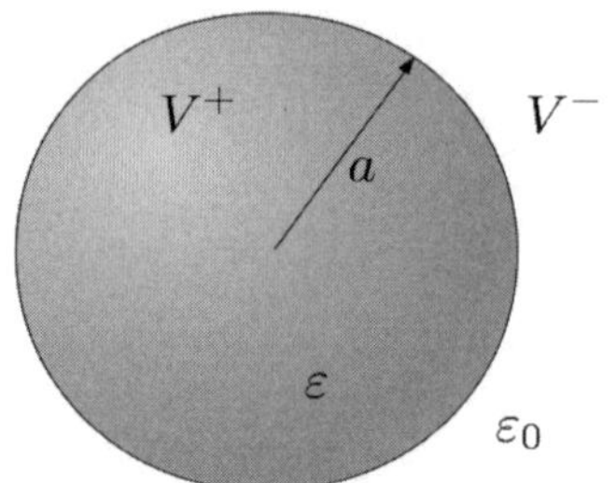

Figure 3.7

following boundary value problem exist:

$$\nabla^2 \varphi^\pm = 0 \quad \text{in } V^\pm, \tag{3.119}$$

$$\varphi^+\big|_{r=a} = \varphi^-\big|_{r=a}, \tag{3.120}$$

$$\varepsilon \left.\frac{\partial \varphi^+}{\partial r}\right|_{r=a} = \varepsilon_0 \left.\frac{\partial \phi^-}{\partial r}\right|_{r=a}, \tag{3.121}$$

$$\varphi^-(\infty) = 0. \tag{3.122}$$

It is well known that the appropriate product solutions of the Laplace equation in the spherical coordinates can be written in the form

$$\varphi_{nm}^+(r, \theta, \phi) = A_{nm}^+ r^n Y_n^m(\theta, \phi), \tag{3.123}$$

$$\varphi_{nm}^-(r, \theta, \phi) = A_{nm}^- r^{-n-1} Y_n^m(\theta, \phi). \tag{3.124}$$

It is apparent that the above product solutions will satisfy the boundary conditions (3.120)-(3.121) only for such special values ε_n of ε that the following relations are valid:

$$A_{nm}^+ a^n = A_{nm}^- a^{-n-1}, \tag{3.125}$$

$$\varepsilon_n A_{nm}^+ n a^{n-1} = -\varepsilon_0 A_{nm}^- (n + 1) a^{-n-2}. \tag{3.126}$$

From the last two equations we find

$$\boxed{A_{nm}^- = A_{nm}^+ a^{2n+1},} \tag{3.127}$$

$$\boxed{\varepsilon_n = -\varepsilon_0 \frac{n + 1}{n}.} \tag{3.128}$$

It is clear that the last formula coincides with formula (2.54) which has been derived by using integral equation eigenvalue analysis of plasmon modes. It is also clear that $\varepsilon_n \to -\varepsilon_0$ as $n \to \infty$. Furthermore, for each ε_n there are $2n+1$ plasmon modes. By using formulas (3.123), (3.124) and (3.127), the electric potential of these plasmon modes can be written in the form

$$\varphi_{nm}^+(r,\theta,\phi) = C_{nm}\left(\frac{r}{a}\right)^n Y_n^m(\theta,\phi), \quad (m = 0, \pm 1, ..., \pm n), \tag{3.129}$$

$$\varphi_{nm}^-(r,\theta,\phi) = C_{nm}\left(\frac{a}{r}\right)^{n+1} Y_n^m(\theta,\phi), \quad (m = 0, \pm 1, ..., \pm n), \tag{3.130}$$

where

$$C_{nm} = A_{nm}^+ a^n. \tag{3.131}$$

It is apparent from formulas (3.129)-(3.130) that with increase in n plasmon modes exhibit increased oscillations with respect to θ and ϕ (i.e., along the spherical boundary) and decrease faster in the radial direction. In other words, with increase in n plasmon modes are increasingly concentrated near the nanoparticle boundary. For this reason, they are often termed "surface" plasmon modes. These two features (highly oscillatory nature and concentration near the boundary) of high-order plasmon modes are generic and valid for nanoparticles of arbitrary (smooth) shapes. These two generic features are the physical reason behind the universality of the limiting behavior of ε_n, i.e., $\lim_{n \to \infty} \varepsilon_n = -\varepsilon_0$. Indeed, high-order plasmon modes are not appreciably affected by local curvature of the nanoparticle boundary. These modes "feel" as if this boundary is flat, and (see the end of this section) for the flat boundary all plasmon modes exist only for one value of ε equal to $-\varepsilon_0$.

7. Plasmon modes in spherical nanoshells

For spherical nanoshells (see Figure 3.8), the analysis of plasmon modes requires the finding of special values of ε such that nonzero solutions of the following boundary value problem exist:

$$\nabla^2 \varphi_k = 0 \quad \text{in } V_k, \quad (k = 1, 2, 3), \tag{3.132}$$

$$\varphi^{(1)}\big|_{r=a} = \varphi^{(2)}\big|_{r=a}, \tag{3.133}$$

$$\varphi^{(2)}\big|_{r=b} = \varphi^{(3)}\big|_{r=b}, \tag{3.134}$$

$$\tilde{\varepsilon}\,\frac{\partial \varphi^{(1)}}{\partial r}\bigg|_{r=a} = \varepsilon\,\frac{\partial \varphi^{(2)}}{\partial r}\bigg|_{r=a}, \tag{3.135}$$

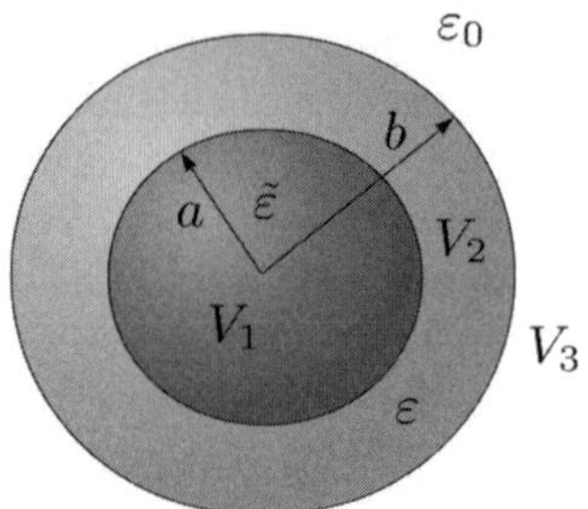

Figure 3.8

$$\varepsilon \left.\frac{\partial\varphi^{(2)}}{\partial r}\right|_{r=b} = \varepsilon_0 \left.\frac{\partial\varphi^{(3)}}{\partial r}\right|_{r=b}, \tag{3.136}$$

$$\varphi^{(3)}(\infty) = 0. \tag{3.137}$$

The appropriate product solutions of the Laplace equation in the spherical coordinates can be written in the following form:

$$\varphi_{nm}^{(1)}(r,\theta,\phi) = A_{nm}r^n Y_n^m(\theta,\phi), \tag{3.138}$$

$$\varphi_{nm}^{(2)}(r,\theta,\phi) = \left(B_{nm}r^n + C_{nm}r^{-n-1}\right) Y_n^m(\theta,\phi), \tag{3.139}$$

$$\varphi_{nm}^{(3)}(r,\theta,\phi) = D_{nm}r^{-n-1} Y_n^m(\theta,\phi). \tag{3.140}$$

These product solutions will satisfy the boundary conditions (3.133)-(3.136) only for special values ε_n of ε such that the following relations are valid:

$$A_{nm}a^n = B_{nm}a^n + C_{nm}a^{-n-1}, \tag{3.141}$$

$$B_{nm}b^n + C_{nm}b^{-n-1} = D_{nm}b^{-n-1}, \tag{3.142}$$

$$\tilde{\varepsilon}nA_{nm}a^{n-1} = \varepsilon_n\left[nB_{nm}a^{n-1} - (n+1)C_{nm}a^{-n-2}\right], \tag{3.143}$$

$$\varepsilon_n\left[nB_{nm}b^{n-1} - (n+1)C_{nm}b^{-n-2}\right] = -\varepsilon_0(n+1)D_{nm}b^{-n-2}. \tag{3.144}$$

Equations (3.141) and (3.142) can be used to find the expressions for A_{nm} and D_{nm} in terms of B_{nm} and C_{nm}. By substituting then these expressions in formulas (3.143) and (3.144) we end up with the following homogeneous equations for B_{nm} and C_{nm}:

$$(\varepsilon_n - \tilde{\varepsilon})na^{n-1}B_{nm} - \left[(n+1)\varepsilon_n + n\tilde{\varepsilon}\right]a^{-n-2}C_{nm} = 0, \tag{3.145}$$

$$[n\varepsilon_n + (n+1)\varepsilon_0]\, b^{n-1} B_{nm} - (\varepsilon_n - \varepsilon_0)(n+1)b^{-n-2}C_{nm} = 0. \qquad (3.146)$$

The above homogeneous equations have nonzero solutions only for such values of ε_n that the determinant of these equations is equal to zero. This leads to the following quadratic equation for ε_n:

$$\varepsilon_n^2 + 2\beta_n\varepsilon_n + \tilde{\varepsilon}\varepsilon_0 = 0, \qquad (3.147)$$

where

$$\beta_n = \frac{(\varepsilon_0 + \tilde{\varepsilon}) + \left(\frac{n+1}{n}\varepsilon_0 + \frac{n}{n+1}\tilde{\varepsilon}\right)\left(\frac{b}{a}\right)^{2n+1}}{2\left[\left(\frac{b}{a}\right)^{2n+1} - 1\right]}. \qquad (3.148)$$

By solving quadratic equation (3.147), we find

$$\varepsilon_n^{\pm} = -\beta_n \pm \sqrt{\beta_n^2 - \tilde{\varepsilon}\varepsilon_0}. \qquad (3.149)$$

In the case of spherical nanoshells, there exist two distinct bands of plasmon resonances corresponding to ε_n^+ and ε_n^-, respectively. It is clear from formula (3.147) (as well as formula (3.149)) that

$$\varepsilon_n^+\varepsilon_n^- = \tilde{\varepsilon}\varepsilon_0. \qquad (3.150)$$

For dipole ($n = 1$) plasmon modes, we have

$$\varepsilon_1^{\pm} = -\beta_1 \pm \sqrt{\beta_1^2 - \tilde{\varepsilon}\varepsilon_0}, \qquad (3.151)$$

$$\beta_1 = \frac{2(\varepsilon_0 + \tilde{\varepsilon}) + (4\varepsilon_0 + \tilde{\varepsilon})\left(\frac{b}{a}\right)^3}{4\left[\left(\frac{b}{a}\right)^3 - 1\right]}. \qquad (3.152)$$

The last expressions can be appreciably simplified in the practically important case when the thickness δ of the nanoshell is sufficiently small (i.e., $\delta \ll a$). In this case, we find

$$\left(\frac{b}{a}\right)^3 \approx 1 + 3\frac{\delta}{a}. \qquad (3.153)$$

By substituting formula (3.153) into equation (3.152) and then by using relations (3.150) and (3.151), we derive

$$\beta_1 \approx \frac{(2\varepsilon_0 + \tilde{\varepsilon})a}{4\delta}, \qquad (3.154)$$

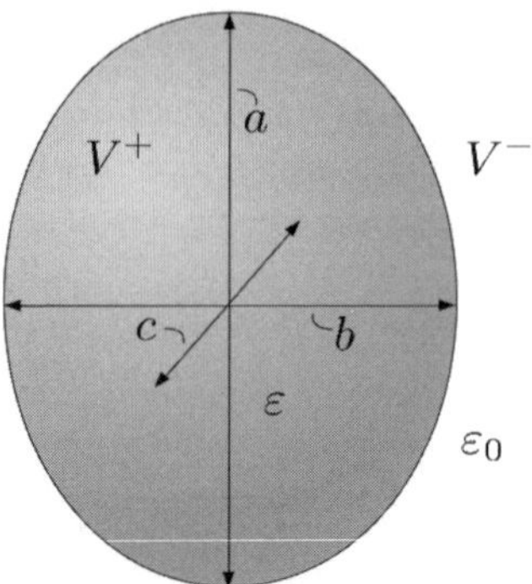

Figure 3.9

$$\varepsilon_1^- \approx -\frac{(2\varepsilon_0 + \tilde{\varepsilon})a}{2\delta},$$

(3.155)

$$\varepsilon_1^+ \approx -\frac{2\tilde{\varepsilon}\varepsilon_0\delta}{(2\varepsilon_0 + \tilde{\varepsilon})a}.$$

(3.156)

The last formulas reveal that the resonance values of permittivity and, consequently, the resonance wavelengths can be effectively controlled by the thickness of the spherical nanoshells [6]. This is a generic feature of thin-wall nanostructures which makes them attractive for tuning of plasmon resonances to a desired wavelength.

8. Plasmon modes in ellipsoidal nanoparticles

Next, we shall analyze plasmon modes in an ellipsoidal nanoparticle with semi-axes a, b and c (see Figure 3.9) which satisfy the inequalities

$$a > b > c.$$

(3.157)

To solve this problem, we shall use ellipsoidal coordinates defined by numbers a, b and c. By using these three numbers, we can introduce

$$k = \sqrt{a^2 - c^2}, \qquad h = \sqrt{a^2 - b^2},$$

(3.158)

which according to (3.157) satisfy the inequalities

$$-\infty < h^2 < k^2 < \infty.$$

(3.159)

Now, we consider the second-order equation

$$\frac{x^2}{\lambda^2} + \frac{y^2}{\lambda^2 - h^2} + \frac{z^2}{\lambda^2 - k^2} = 1. \tag{3.160}$$

We define the ellipsoidal coordinates ξ, η and ζ as follows:

$$\lambda = \xi, \quad \text{if } k^2 < \lambda^2 < \infty, \tag{3.161}$$

$$\lambda = \eta, \quad \text{if } h^2 < \lambda^2 < k^2, \tag{3.162}$$

$$\lambda = \zeta, \quad \text{if } -\infty < \lambda^2 < h^2. \tag{3.163}$$

It is clear that equations

$$\xi = const, \tag{3.164}$$

$$\eta = const, \tag{3.165}$$

$$\zeta = const, \tag{3.166}$$

define the family of confocal ellipsoids, one-sheet hyperboloids and two-sheet hyperboloids, respectively. It is also clear from formulas (3.158), (3.160) and (3.161) that the surface $\xi = a$ coincides with the boundary of the ellipsoidal nanoparticle.

Analysis of plasmon modes in an ellipsoidal nanoparticle requires the finding of such ε that nonzero solutions of the following boundary value problem exist:

$$\nabla^2 \varphi^{\pm} = 0 \quad \text{in } V^{\pm}, \tag{3.167}$$

$$\varphi^{+}\big|_{\xi=a} = \varphi^{-}\big|_{\xi=a}, \tag{3.168}$$

$$\varepsilon \left.\frac{\partial \varphi^{+}}{\partial \xi}\right|_{\xi=a} = \varepsilon_0 \left.\frac{\partial \varphi^{-}}{\partial \xi}\right|_{\xi=a}, \tag{3.169}$$

$$\varphi^{-}(\infty) = 0. \tag{3.170}$$

The appropriate product solutions of the Laplace equation in the ellipsoidal coordinates can be written as follows:

$$\varphi_{nm}^{+}(\xi, \eta, \zeta) = A_{nm}^{+} E_n^m(\xi) E_n^m(\eta) E_n^m(\zeta), \tag{3.171}$$

$$\varphi_{nm}^{-}(\xi, \eta, \zeta) = A_{nm}^{-} F_n^m(\xi) E_n^m(\eta) E_n^m(\zeta). \tag{3.172}$$

Here, $E_n^m(\xi)$, $E_n^m(\eta)$ and $E_n^m(\zeta)$ are Lamé polynomials called ellipsoidal harmonics of the first kind, while $F_n^m(\xi)$ are ellipsoidal harmonics of the second kind which are related to $E_n^m(\xi)$ by the formula

$$F_n^m(\xi) = (2n + 1)E_n^m(\xi) \int_\xi^\infty \frac{d\lambda}{[E_n^m(\lambda)]^2 \sqrt{(\lambda^2 - h^2)(\lambda^2 - k^2)}}. \qquad (3.173)$$

The product solutions (3.171) and (3.172) will satisfy the boundary conditions (3.168) and (3.169) only for special values ε_{nm} of ε such that the following relations are valid:

$$A_{nm}^+ E_n^m(a) = A_{nm}^- F_n^m(a), \qquad (3.174)$$

$$\varepsilon_{nm} A_{nm}^+ \dot{E}_n^m(a) = \varepsilon_0 A_{nm}^- \dot{F}_n^m(a), \qquad (3.175)$$

where $\dot{E}_n^m$ and $\dot{F}_n^m$ are the derivatives of E_n^m and F_n^m, respectively.

From the last two equations, we derive

$$\varepsilon_{nm} = \varepsilon_0 \frac{E_n^m(a)\dot{F}_n^m(a)}{\dot{E}_n^m(a)F_n^m(a)}. \qquad (3.176)$$

As an example, we consider the dipole plasmon mode corresponding to $m = n = 1$. According to formula (3.176), we have

$$\varepsilon_{11} = \varepsilon_0 \frac{E_1^1(a)\dot{F}_1^1(a)}{\dot{E}_1^1(a)F_1^1(a)}. \qquad (3.177)$$

It is known that $E_1^1(\xi) = \xi$. Consequently,

$$E_1^1(a) = a, \qquad \dot{E}_1^1(a) = 1. \qquad (3.178)$$

Furthermore, according to formula (3.173) we have

$$F_1^1(a) = 3a \int_a^\infty \frac{d\lambda}{\lambda^2 \sqrt{(\lambda^2 - h^2)(\lambda^2 - k^2)}}, \qquad (3.179)$$

$$\dot{F}_1^1(a) = 3 \int_a^\infty \frac{d\lambda}{\lambda^2 \sqrt{(\lambda^2 - h^2)(\lambda^2 - k^2)}} - \frac{3}{a\sqrt{(a^2 - h^2)(a^2 - k^2)}}. \qquad (3.180)$$

By substituting formulas (3.178), (3.179) and (3.180) into equation (3.177) and taking into account formula (3.158), we derive

$$\varepsilon_{11} = \varepsilon_0 \left(1 - \frac{1}{abc \displaystyle\int_a^\infty \frac{d\lambda}{\lambda^2 \sqrt{(\lambda^2 - h^2)(\lambda^2 - k^2)}}} \right). \tag{3.181}$$

Finally, by using the change of variables

$$\lambda^2 - a^2 = \nu, \tag{3.182}$$

we find

$$\int_a^\infty \frac{d\lambda}{\lambda^2 \sqrt{(\lambda^2 - h^2)(\lambda^2 - k^2)}} = \frac{1}{2} \int_0^\infty \frac{d\nu}{(\nu + a^2)\sqrt{(\nu + a^2)(\nu + b^2)(\nu + c^2)}}. \tag{3.183}$$

Thus, formula (3.181) can be written as follows:

$$\boxed{\varepsilon_{11} = \varepsilon_0 \left(1 - \frac{1}{N_1} \right),} \tag{3.184}$$

where N_1 is the depolarizing coefficient given by the expression

$$\boxed{N_1 = \frac{abc}{2} \int_0^\infty \frac{d\nu}{(\nu + a^2)\sqrt{(\nu + a^2)(\nu + b^2)(\nu + c^2)}}.} \tag{3.185}$$

Similar results are valid for ε_{12} and ε_{13} with the replacement of N_1 in formula (3.185) by N_2 and N_3, respectively. The expressions for N_2 and N_3 are obtained from formula (3.185) by permutation of a and b and a and c, respectively. By the way, in the case of a spheroidal nanoparticle $b = c$, and the formulas for depolarizing coefficients are appreciably simplified and can be written as follows:

$$\boxed{\begin{aligned} N_1 &= \frac{1 - e^2}{2e^3} \left[\ln \frac{1 + e}{1 - e} - 2e \right], &\qquad (3.186) \\[2mm] N_2 &= N_3 = \frac{1 - N_1}{2}, &\qquad (3.187) \end{aligned}}$$

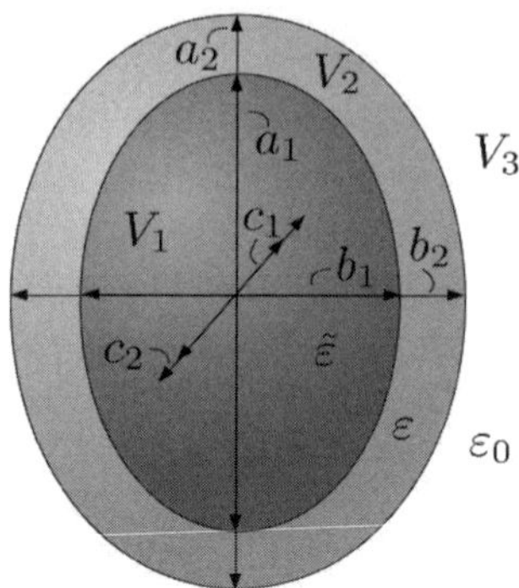

Figure 3.10

where

$$e^2 = 1 - \frac{b^2}{a^2}.$$ (3.188)

It is worthwhile to point out that formula (3.184) coincides with formula (2.65) which has been derived in the previous chapter by using much simpler and elegant reasoning. However, this reasoning cannot be applied to the derivation of the general formula (3.176), which is valid for all (not only dipole) plasmon modes.

9. Plasmon modes in ellipsoidal nanoshells

Ellipsoidal coordinates can be used for the analysis of plasmon modes in ellipsoidal nanoshells (see Figure 3.10) provided that inner and outer ellipsoidal boundaries have the equations

$$\xi = a_1, \qquad \xi = a_2,$$ (3.189)

where a_1 and a_2 are semi-axes of these boundaries along the x-direction. From formulas (3.158) and (3.160) we conclude that equations (3.189) imply that

$$a_2^2 - a_1^2 = b_2^2 - b_1^2 = c_2^2 - c_1^2.$$ (3.190)

Thus, only ellipsoidal nanoshells which satisfy the above condition can be analyzed using ellipsoidal coordinates. It is appropriate to remark here that the thickness of such ellipsoidal nanoshells is not constant.

By using the ellipsoidal coordinates the analysis of plasmon modes in the ellipsoidal nanoshell can be reduced as before to the following homogeneous boundary value problem:

$$\nabla^2 \varphi^{(i)} = 0 \quad \text{in } V_i, \quad (i = 1, 2, 3), \tag{3.191}$$

$$\left. \varphi^{(1)} \right|_{\xi = a_1} = \left. \varphi^{(2)} \right|_{\xi = a_1}, \tag{3.192}$$

$$\left. \varphi^{(2)} \right|_{\xi = a_2} = \left. \varphi^{(3)} \right|_{\xi = a_2}, \tag{3.193}$$

$$\tilde{\varepsilon} \left. \frac{\partial \varphi^{(1)}}{\partial \xi} \right|_{\xi = a_1} = \varepsilon \left. \frac{\partial \varphi^{(2)}}{\partial \xi} \right|_{\xi = a_1}, \tag{3.194}$$

$$\varepsilon \left. \frac{\partial \varphi^{(2)}}{\partial \xi} \right|_{\xi = a_2} = \varepsilon_0 \left. \frac{\partial \varphi^{(3)}}{\partial \xi} \right|_{\xi = a_2}, \tag{3.195}$$

$$\varphi^{(3)}(\infty) = 0. \tag{3.196}$$

The appropriate product solutions of the Laplace equation which are regular in V_i, $(i = 1, 2, 3)$ can be written as follows:

$$\varphi_{nm}^{(1)}(\xi, \eta, \zeta) = A_{nm} E_n^m(\xi) E_n^m(\eta) E_n^m(\zeta), \tag{3.197}$$

$$\varphi_{nm}^{(2)}(\xi, \eta, \zeta) = \left[B_{nm} E_n^m(\xi) + C_{nm} F_n^m(\xi) \right] E_n^m(\eta) E_n^m(\zeta), \tag{3.198}$$

$$\varphi_{nm}^{(3)}(\xi, \eta, \zeta) = D_{nm} F_n^m(\xi) E_n^m(\eta) E_n^m(\zeta). \tag{3.199}$$

The above product solutions will satisfy the boundary conditions (3.192)-(3.195) only for special values ε_{nm} of ε such that the following relations are valid:

$$A_{nm} E_n^m(a_1) = B_{nm} E_n^m(a_1) + C_{nm} F_n^m(a_1), \tag{3.200}$$

$$B_{nm} E_n^m(a_2) + C_{nm} F_n^m(a_2) = D_{nm} F_n^m(a_2), \tag{3.201}$$

$$\tilde{\varepsilon} A_{nm} \dot{E}_n^m(a_1) = \varepsilon_{nm} \left[B_{nm} \dot{E}_n^m(a_1) + C_{nm} \dot{F}_n^m(a_1) \right], \tag{3.202}$$

$$\varepsilon_0 D_{nm} \dot{F}_n^m(a_2) = \varepsilon_{nm} \left[B_{nm} \dot{E}_n^m(a_2) + C_{nm} \dot{F}_n^m(a_2) \right]. \tag{3.203}$$

Equations (3.200) and (3.201) can be used to find the expressions for A_{nm} and D_{nm} in terms of B_{nm} and C_{nm}. By substituting these expressions into formulas (3.202) and (3.203), we end up with the following homogeneous equations for B_{nm} and C_{nm}:

$$\tilde{\varepsilon}\left[B_{nm}E_n^m(a_1)\dot{E}_n^m(a_1) + C_{nm}F_n^m(a_1)\dot{E}_n^m(a_1)\right]$$
$$= \varepsilon_{nm}\left[B_{nm}E_n^m(a_1)\dot{E}_n^m(a_1) + C_{nm}E_n^m(a_1)\dot{F}_n^m(a_1)\right], \qquad (3.204)$$

$$\varepsilon_0\left[B_{nm}E_n^m(a_2)\dot{F}_n^m(a_2) + C_{nm}F_n^m(a_2)\dot{F}_n^m(a_2)\right]$$
$$= \varepsilon_{nm}\left[B_{nm}\dot{E}_n^m(a_2)F_n^m(a_2) + C_{nm}F_n^m(a_2)\dot{F}_n^m(a_2)\right]. \qquad (3.205)$$

The above homogeneous equations have nonzero solutions only for such values of ε_{nm} that the determinant of these equations is equal to zero. This leads to the quadratic equation for ε_{nm}. We shall write this quadratic equation for the particular (but important) case of dipole modes ($n = 1$). In this case, by using the same line of reasoning as in the case of the ellipsoidal nanoparticle (see formulas (3.178)-(3.184)), the coefficients of this quadratic equation can be expressed in terms of depolarizing coefficients $N_m^{(1)}$ and $N_m^{(2)}$ corresponding to $\xi = a_1$ and $\xi = a_2$, respectively. The quadratic equation has the form

$$\varepsilon_{1m}^2\left[N_m^{(2)}\left(1 - f - N_m^{(1)} + fN_m^{(2)}\right)\right] + \varepsilon_{1m}\left[\varepsilon_0\left(1 - N_m^{(2)}\right)\left(1 - N_m^{(1)} + fN_m^{(2)}\right)\right.$$
$$\left. + \tilde{\varepsilon}N_m^{(2)}\left(f + N_m^{(1)} - fN_m^{(2)}\right)\right] + \tilde{\varepsilon}\varepsilon_0\left[\left(1 - N_m^{(2)}\right)\left(N_m^{(1)} - fN_m^{(2)}\right)\right] = 0,$$
$$(3.206)$$

where

$$f = \frac{a_1 b_1 c_1}{a_2 b_2 c_2}. \qquad (3.207)$$

For thin ellipsoidal shells, we have

$$f \approx 1, \qquad N_m^{(2)} \approx N_m^{(1)} = N_m, \qquad (3.208)$$

and from the quadratic equation (3.206) we derive

$$\boxed{\varepsilon_{1m}^- \approx -\frac{(1 - N_m)\varepsilon_0 + \tilde{\varepsilon}N_m}{(1 - f)(1 - N_m)N_m},} \qquad (3.209)$$

$$\boxed{\varepsilon_{1m}^+ \approx -\frac{\tilde{\varepsilon}\varepsilon_0(1 - f)(1 - N_m)N_m}{(1 - N_m)\varepsilon_0 + \tilde{\varepsilon}N_m}.} \qquad (3.210)$$

Figure 3.11

In the case when the ellipsoidal shell is reduced to the spherical shell of inner radius a and thickness δ we have

$$N_m = \frac{1}{3}, \qquad f \approx 1 - 3\frac{\delta}{a}, \qquad (3.211)$$

and the last two formulas coincide with formulas (3.155) and (3.156), respectively.

10. Plasmon modes in toroidal nano-rings

Plasmon modes in toroidal nano-rings with circular cross sections (see Figure 3.11) can be studied by using toroidal coordinates. These coordinates (see Figure 3.12) are obtained from bipolar coordinates (see Figure 3.4) by rotating them about the perpendicular bisector of the line connecting the two singular points P_1 and P_2 of the bipolar coordinates. The toroidal coordinates (μ, η, ϕ) are connected to rectangular coordinates (x, y, z) by the formulas

$$x = \frac{a \sinh \mu \cos \phi}{\cosh \mu - \cos \eta}, \qquad (3.212)$$

$$y = \frac{a \sinh \mu \sin \phi}{\cosh \mu - \cos \eta}, \qquad (3.213)$$

$$z = \frac{a \sin \eta}{\cosh \mu - \cos \eta}, \qquad (3.214)$$

and the corresponding metric coefficients are as follows:

$$h_\mu = h_\eta = \frac{a}{\cosh \mu - \cos \eta}, \qquad h_\phi = \frac{a \sinh \mu}{\cosh \mu - \cos \eta}. \qquad (3.215)$$

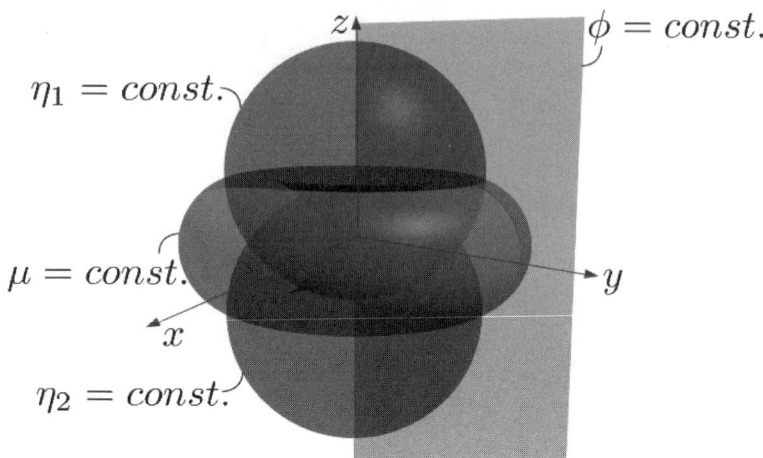

Figure 3.12

The utility of the toroidal coordinates for our problem stems from the fact that the boundary of the circular cross-section nano-ring coincides with coordinate surface $\mu = \mu_0$. The radius R of the cross section and x-coordinate of its center are related to μ_0 by the formulas

$$R = \frac{a}{\sinh \mu_0}, \qquad x_0 = a \coth \mu_0. \tag{3.216}$$

To analyze the plasmon modes in nano-rings, it is convenient to use the scalar potential ψ of the displacement field $\mathbf{D}$,

$$\mathbf{D} = -\operatorname{grad} \psi. \tag{3.217}$$

The plasmon modes are the nonzero solutions of the following homogeneous boundary value problem:

$$\nabla^2 \psi^\pm = 0 \quad \text{in } V^\pm, \tag{3.218}$$

$$\left. \frac{\psi^+}{\varepsilon} \right|_{\mu_0} = \left. \frac{\psi^-}{\varepsilon_0} \right|_{\mu_0}, \tag{3.219}$$

$$\left. \frac{\partial \psi^+}{\partial \mu} \right|_{\mu_0} = \left. \frac{\partial \psi^-}{\partial \mu} \right|_{\mu_0}, \tag{3.220}$$

$$\psi^-(\infty) = 0. \tag{3.221}$$

The method of separation of variables in toroidal coordinates leads to the following particular solutions of the Laplace equation which are regular inside and outside the nano-ring, respectively:

$$\sqrt{\cosh \mu - \cos \eta}\, e^{j(n\eta + m\phi)} Q^m_{n-\frac{1}{2}}\left(\cosh \mu\right), \tag{3.222}$$

$$\sqrt{\cosh \mu - \cos \eta}\, e^{j(n\eta + m\phi)} P^m_{n-\frac{1}{2}}\left(\cosh \mu\right), \tag{3.223}$$

where $j = \sqrt{-1}$, while $P^m_{n-\frac{1}{2}}$ and $Q^m_{n-\frac{1}{2}}$ are associated Legendre functions.

Due to the square-root factor, solutions (3.222) and (3.223) are not the product solutions of the Laplace equation. This complicates the analysis of plasmon modes and makes it impossible to get explicit analytical solutions. Nevertheless, some interesting results still can be obtained.

We fix m and look for the solutions of the boundary value problem (3.218)-(3.221) in the form

$$\psi^+_{(m)}(\mu, \eta, \phi) = \sqrt{\cosh \mu - \cos \eta} \sum_{n=-\infty}^{\infty} a^m_n e^{j(n\eta + m\phi)} Q^m_{n-\frac{1}{2}}\left(\cosh \mu\right), \tag{3.224}$$

$$\psi^-_{(m)}(\mu, \eta, \phi) = \sqrt{\cosh \mu - \cos \eta} \sum_{n=-\infty}^{\infty} b^m_n e^{j(n\eta + m\phi)} P^m_{n-\frac{1}{2}}\left(\cosh \mu\right). \tag{3.225}$$

By substituting formulas (3.224) and (3.225) into boundary condition (3.219), we find

$$a^m_n Q^m_{n-\frac{1}{2}}\left(\cosh \mu_0\right) = \frac{\varepsilon^{(m)}}{\varepsilon_0} b^m_n P^m_{n-\frac{1}{2}}\left(\cosh \mu_0\right). \tag{3.226}$$

By substituting formulas (3.224) and (3.225) into the boundary conditions (3.220), after somewhat lengthy but straightforward transformations we derive

$$\sum_{n=-\infty}^{\infty} \left[\alpha_{nm} a^m_n - \nu_{nm}\left(a^m_{n-1} + a^m_{n+1}\right)\right] e^{j(n\eta + m\phi)}$$

$$= \sum_{n=-\infty}^{\infty} \left[\beta_{nm} b^m_n - \gamma_{nm}\left(b^m_{n-1} + b^m_{n+1}\right)\right] e^{j(n\eta + m\phi)}, \tag{3.227}$$

where

$$\alpha_{nm} = \sinh \mu_0 Q^m_{n-\frac{1}{2}}\left(\cosh \mu_0\right) + 2\cosh \mu_0 \left.\frac{dQ^m_{n-\frac{1}{2}}\left(\cosh \mu\right)}{d\mu}\right|_{\mu_0}, \qquad (3.228)$$

$$\beta_{nm} = \sinh \mu_0 P^m_{n-\frac{1}{2}}\left(\cosh \mu_0\right) + 2\cosh \mu_0 \left.\frac{dP^m_{n-\frac{1}{2}}\left(\cosh \mu\right)}{d\mu}\right|_{\mu_0}, \qquad (3.229)$$

$$\nu_{nm} = \left.\frac{dQ^m_{n-\frac{1}{2}}\left(\cosh \mu\right)}{d\mu}\right|_{\mu_0}, \qquad (3.230)$$

$$\gamma_{nm} = \left.\frac{dP^m_{n-\frac{1}{2}}\left(\cosh \mu\right)}{d\mu}\right|_{\mu_0}. \qquad (3.231)$$

By equating the terms on both sides of formula (3.227) with identical exponents, we end up with the following linear equations:

$$\hat{A}_{(m)}\mathbf{a}^{(m)} = \hat{B}_{(m)}\mathbf{b}^{(m)}, \qquad (3.232)$$

where matrices $\hat{A}_{(m)}$ and $\hat{B}_{(m)}$ are tridiagonal, while $\mathbf{a}^{(m)}$ and $\mathbf{b}^{(m)}$ are column vectors whose components are a^m_n and b^m_n, respectively.

The infinite tridiagonal matrices $\hat{A}_{(m)}$ and $\hat{B}_{(m)}$ can be truncated and then the truncated tridiagonal matrix $\hat{A}_{(m)}$ can be easily inverted by using explicit formulas [7]. This leads to

$$\mathbf{a}^{(m)} = \hat{A}^{-1}_{(m)}\hat{B}_{(m)}\mathbf{b}^{(m)}. \qquad (3.233)$$

Now, by combining formulas (3.226) and (3.233), for any fixed m we arrive at the following eigenvalue problem:

$$\boxed{\frac{\varepsilon^{(m)}}{\varepsilon_0}\mathbf{b}^{(m)} = \hat{C}_{(m)}\mathbf{b}^{(m)},} \qquad (3.234)$$

where

$$\hat{C}_{(m)} = \hat{D}_{(m)}\hat{A}^{-1}_{(m)}\hat{B}_{(m)}, \qquad (3.235)$$

and $\hat{D}_{(m)}$ is the diagonal matrix whose diagonal elements are equal to the ratio of $P^m_{n-\frac{1}{2}}\left(\cosh \mu\right)$ to $Q^m_{n-\frac{1}{2}}\left(\cosh \mu\right)$. By solving the eigenvalue problem

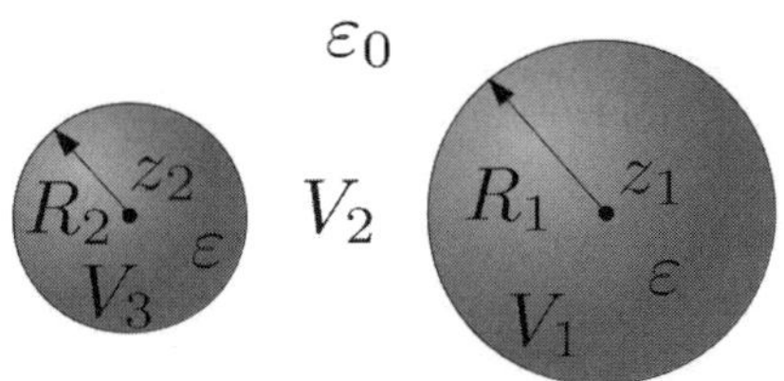

Figure 3.13

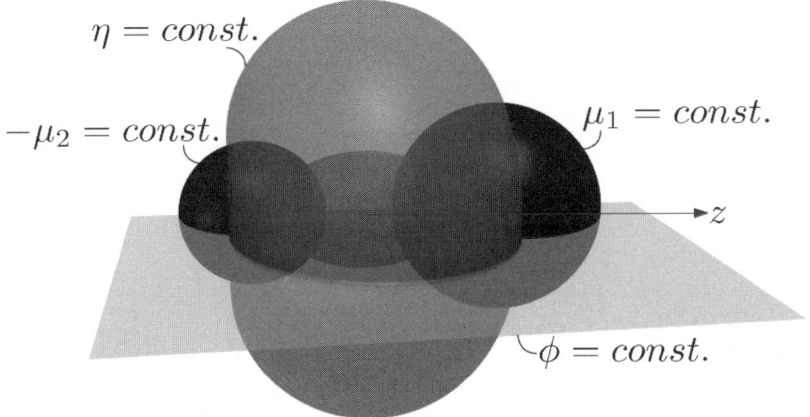

Figure 3.14

(3.234), eigenvalues $\varepsilon_k^{(m)}/\varepsilon_0$ and eigenvectors $\mathbf{b}_k^{(m)}$ can be found. In this way, plasmon modes can be separately computed (and classified) for any fixed m.

11. Plasmon modes in two adjacent spherical nanoparticles

Plasmon modes in two adjacent spherical nanoparticles (see Figure 3.13) can be studied by using bispherical coordinates (see Figure 3.14). These coordinates are obtained by rotating the plane of bipolar coordinates about the line which goes through the singular points P_1 and P_2 of the bipolar coordinates. The bispherical coordinates (μ, η, ϕ) are related to Cartesian coordinates by the formulas

$$x = \frac{a \sin \eta \cos \phi}{\cosh \mu - \cos \eta},$$

$$(3.236)$$

$$y = \frac{a \sin \eta \sin \phi}{\cosh \mu - \cos \eta}, \tag{3.237}$$

$$z = \frac{a \sinh \mu}{\cosh \mu - \cos \eta}, \tag{3.238}$$

and the corresponding metric coefficients are as follows:

$$h_\mu = h_\eta = \frac{a}{\cosh \mu - \cos \eta}, \qquad h_\phi = \frac{a \sin \eta}{\cosh \mu - \cos \eta}. \tag{3.239}$$

The bispherical coordinates can be chosen in such a way that the boundaries of the spherical nanoparticles coincide with coordinate surfaces $\mu = \mu_1$ and $\mu = -\mu_2$, while the radii of the nanoparticles and z-coordinates of their centers are given by the formulas

$$R_1 = \frac{a}{\sinh \mu_1}, \qquad R_2 = \frac{a}{\sinh \mu_2}, \tag{3.240}$$

$$z_1 = a \coth \mu_1, \qquad z_2 = -a \coth \mu_2. \tag{3.241}$$

To analyze plasmon modes in two adjacent spherical nanoparticles, we shall use the potential defined by formula (3.217). Plasmon modes are then the nonzero solutions of the following homogeneous boundary value problem:

$$\nabla^2 \psi_k = 0 \quad \text{in } V_k, \quad (k = 1, 2, 3), \tag{3.242}$$

$$\left. \frac{\psi_1}{\varepsilon} \right|_{\mu_1} = \left. \frac{\psi_2}{\varepsilon_0} \right|_{\mu_1}, \tag{3.243}$$

$$\left. \frac{\psi_2}{\varepsilon_0} \right|_{-\mu_2} = \left. \frac{\psi_3}{\varepsilon} \right|_{-\mu_2}, \tag{3.244}$$

$$\left. \frac{\partial \psi_1}{\partial \mu} \right|_{\mu_1} = \left. \frac{\partial \psi_2}{\partial \mu} \right|_{\mu_1}, \tag{3.245}$$

$$\left. \frac{\partial \psi_2}{\partial \mu} \right|_{-\mu_2} = \left. \frac{\partial \psi_3}{\partial \mu} \right|_{-\mu_2}. \tag{3.246}$$

The method of separation of variables in bispherical coordinates leads to the following particular solutions of the Laplace equations which are regular outside V_1 and V_3, respectively:

$$\sqrt{\cosh \mu - \cos \eta}\, e^{\left(n+\frac{1}{2}\right)\mu} P_n^m (\cos \eta)\, e^{jm\phi}, \tag{3.247}$$

$$\sqrt{\cosh \mu - \cos \eta}\, e^{-\left(n+\frac{1}{2}\right)\mu} P_n^m (\cos \eta)\, e^{jm\phi}. \tag{3.248}$$

It is apparent that these are not product solutions. As in the previous example, this complicates the analysis of plasmon modes and makes it impossible to obtain explicit analytical solutions. Nevertheless, some valuable results still can be obtained by using the same line of reasoning as in the previous example.

To simplify derivations, we shall consider the particular case $m = 0$ of rotationally symmetric plasmon modes. A general case $(m \neq 0)$ can be treated in a similar way. Thus, we shall look for the solutions of the boundary value problem (3.242)-(3.246) in the form

$$\psi_1(\mu, \eta) = \sqrt{\cosh \mu - \cos \eta} \sum_{n=0}^{\infty} a_n e^{-\left(n+\frac{1}{2}\right)\mu} P_n\left(\cos \eta\right), \tag{3.249}$$

$$\psi_2(\mu, \eta) = \sqrt{\cosh \mu - \cos \eta} \sum_{n=0}^{\infty} \left(b_n e^{-\left(n+\frac{1}{2}\right)\mu} + c_n e^{\left(n+\frac{1}{2}\right)\mu}\right) P_n\left(\cos \eta\right), \tag{3.250}$$

$$\psi_3(\mu, \eta) = \sqrt{\cosh \mu - \cos \eta} \sum_{n=0}^{\infty} d_n e^{\left(n+\frac{1}{2}\right)\mu} P_n\left(\cos \eta\right). \tag{3.251}$$

From the last three formulas and the boundary conditions (3.243) and (3.244) we derive

$$a_n = \frac{\varepsilon}{\varepsilon_0} \left(b_n + c_n e^{(2n+1)\mu_1}\right), \tag{3.252}$$

$$d_n = \frac{\varepsilon}{\varepsilon_0} \left(b_n e^{(2n+1)\mu_2} + c_n\right). \tag{3.253}$$

By substituting formulas (3.249) and (3.250) into the boundary condition (3.245) and by using the relation

$$\cos \eta \, P_n\left(\cos \eta\right) = \frac{n+1}{2n+1} P_{n+1}\left(\cos \eta\right) + \frac{n}{2n+1} P_{n-1}\left(\cos \eta\right), \tag{3.254}$$

after somewhat lengthy but straightforward transformations we derive

$$\sum_{n=0}^{\infty} \left[\alpha_n a_n + \nu_n a_{n-1} + \gamma_n a_{n+1}\right] P_n\left(\cos \eta\right)$$

$$= \sum_{n=0}^{\infty} \left[\alpha_n b_n + \nu_n b_{n-1} + \gamma_n b_{n+1} + \beta_n c_n + \chi_n c_{n-1} + \lambda_n c_{n+1}\right] P_n\left(\cos \eta\right), \tag{3.255}$$

where

$$\alpha_n = e^{-\left(n+\frac{1}{2}\right)\mu_1}\left[\sinh\mu_1 - (2n+1)\cosh\mu_1\right], \tag{3.256}$$

$$\nu_n = ne^{-\left(n-\frac{1}{2}\right)\mu_1}, \qquad \gamma_n = (n+1)e^{-\left(n+\frac{3}{2}\right)\mu_1}, \tag{3.257}$$

$$\beta_n = e^{\left(n+\frac{1}{2}\right)\mu_1}\left[\sinh\mu_1 + (2n+1)\cosh\mu_1\right], \tag{3.258}$$

$$\chi_n = -ne^{\left(n-\frac{1}{2}\right)\mu_1}, \qquad \lambda_n = -(n+1)e^{\left(n+\frac{3}{2}\right)\mu_1}. \tag{3.259}$$

By equating the terms on both sides of formula (3.255) with $P_n\left(\cos\eta\right)$, we end up with the following linear equations with tridiagonal matrices

$$\hat{A}_1\mathbf{a} = \hat{A}_1\mathbf{b} + \hat{B}_1\mathbf{c}, \tag{3.260}$$

where $\mathbf{a}$, $\mathbf{b}$ and $\mathbf{c}$ are column vectors whose components are a_n, b_n and c_n, respectively.

In a similar way, from the boundary condition (3.246) we derive

$$\hat{A}_2\mathbf{d} = \hat{B}_2\mathbf{b} + \hat{A}_2\mathbf{c}, \tag{3.261}$$

where tridiagonal matrices $\hat{A}_2$ and $\hat{B}_2$ have structure almost identical to matrices $\hat{A}_1$ and $\hat{B}_1$, respectively.

By inverting the tridiagonal matrices in equations (3.260) and (3.261), we find

$$\mathbf{a} = \mathbf{b} + \hat{A}_1^{-1}\hat{B}_1\mathbf{c}, \tag{3.262}$$

$$\mathbf{d} = \hat{A}_2^{-1}\hat{B}_2\mathbf{b} + \mathbf{c}. \tag{3.263}$$

Next, equations (3.252)-(3.253) can be written in the form

$$\frac{\varepsilon}{\varepsilon_0}\mathbf{b} = \hat{D}_1\mathbf{a} + \hat{D}_2\mathbf{d}, \tag{3.264}$$

$$\frac{\varepsilon}{\varepsilon_0}\mathbf{c} = \hat{D}_3\mathbf{a} + \hat{D}_4\mathbf{d}, \tag{3.265}$$

where $\hat{D}_k$, $(k = 1, 2, 3, 4)$, are diagonal matrices.

By substituting formulas (3.262) and (3.263) into formulas (3.264) and (3.265) we arrive at the following eigenvalue problem:

$$\boxed{\frac{\varepsilon}{\varepsilon_0}\begin{pmatrix}\mathbf{b}\\\mathbf{c}\end{pmatrix} = \begin{pmatrix}\hat{T}_1 & \hat{T}_2\\\hat{T}_3 & \hat{T}_4\end{pmatrix}\begin{pmatrix}\mathbf{b}\\\mathbf{c}\end{pmatrix}.} \tag{3.266}$$

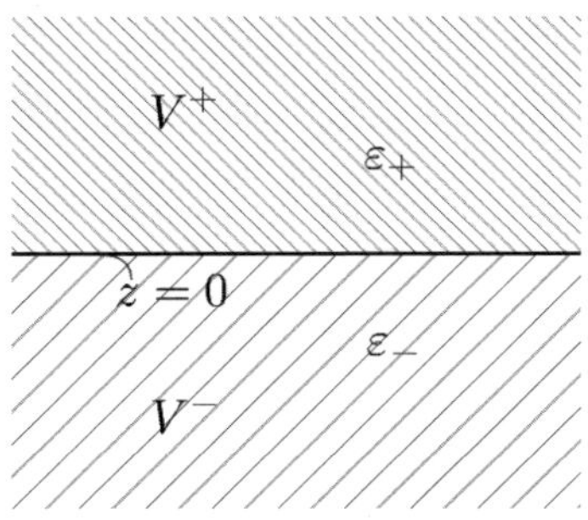

Figure 3.15

By solving this problem, eigenvalues $\varepsilon_k/\varepsilon_0$ and eigenvectors $\begin{pmatrix} \mathbf{b}_k \\ \mathbf{c}_k \end{pmatrix}$ can be found, which then can be used for the determination of resonance frequencies and corresponding plasmon fields. The analytical expressions for the fields will follow from formulas (3.249)-(3.251).

12. Plasmon modes in infinite flat structures

These structures were historically first studied (see [8, 9]) and the term "surface plasmons" was probably first coined in [9]. We begin with the discussion of plasmon modes at the flat interface between two media (see Figure 3.15). In the electrostatic approximation, plasmon modes must be nonzero solutions of the following homogeneous boundary value problem for electric potential:

$$\nabla^2 \varphi^\pm = 0 \quad \text{in } V^\pm, \tag{3.267}$$

$$\varphi^+\big|_{z=0} = \varphi^-\big|_{z=0}, \tag{3.268}$$

$$\varepsilon_+ \frac{\partial \varphi^+}{\partial z}\bigg|_{z=0} = \varepsilon_- \frac{\partial \varphi^-}{\partial z}\bigg|_{z=0}, \tag{3.269}$$

$$\varphi^+(\infty) = \varphi^-(-\infty) = 0. \tag{3.270}$$

The product solutions of the Laplace equation which satisfy the conditions (3.270) can be written in the form

$$\varphi^+(x, y, z) = A_+ e^{j\mathbf{k}\cdot\mathbf{r} - kz}, \tag{3.271}$$

$$\varphi^-(x, y, z) = A_- e^{j\mathbf{k}\cdot\mathbf{r} + kz}, \tag{3.272}$$

where the following notations are adopted:

$$\mathbf{r} = \mathbf{e}_x x + \mathbf{e}_y y, \tag{3.273}$$

$$\mathbf{k} = \mathbf{e}_x k_x + \mathbf{e}_y k_y, \tag{3.274}$$

$$k = |\mathbf{k}| = \sqrt{k_x^2 + k_y^2}. \tag{3.275}$$

By substituting formulas (3.271) and (3.272) into the boundary conditions (3.268) and (3.269), we end up with the following relations:

$$A_+ = A_-, \tag{3.276}$$

$$-\varepsilon_+ A_+ = \varepsilon_- A_-. \tag{3.277}$$

From the last two formulas we find that plasmon modes (3.271)-(3.272) may exist only under the condition

$$\boxed{\varepsilon_- = -\varepsilon_+.} \tag{3.278}$$

In the particular (and important) case when the permittivity ε_+ of the upper media is equal to ε_0, we find

$$\boxed{\varepsilon_- = -\varepsilon_0.} \tag{3.279}$$

By using the dispersion relation (1.8), we conclude that plasmon modes (3.271)-(3.272) can be excited at the frequency

$$\omega = \frac{\omega_p}{\sqrt{2}}. \tag{3.280}$$

There are at least two reasons why the plasmon modes (3.271)-(3.272) are called "surface plasmons." First, these plasmon modes are concentrated near the interface between plasmonic media and vacuum. This is evident from their exponential decay with respect to z, and this exponential decay is directly correlated to the intensity of oscillations of these plasmons along the interface. Second, the very existence of the plasmon modes (3.271)-(3.272) is due to the presence of the interface. Without it, only bulk plasma oscillations of the classical frequency ω_p may occur.

As was pointed out previously, the formula (3.279) reveals the physical reason why the resonance value of permittivity for higher-order plasmon modes in nanoparticles is close to $-\varepsilon_0$. These plasmon modes are highly oscillatory in nature and closely concentrated near the nanoparticle boundaries. As a result, these boundaries are "sensed" locally as almost flat interfaces.

It is interesting to point out that surface plasmon modes (3.271)-(3.272) may appear at interfaces between two conducting media. Historically, these plasmon modes have been discussed for flat interfaces between aluminum and magnesium [9]. These interface plasmon modes between two conducting media may exist under the condition (3.278). By using dispersion relations

$$\varepsilon_+ = \varepsilon_0 \left[1 - \left(\frac{\omega_p^+}{\omega} \right)^2 \right], \tag{3.281}$$

$$\varepsilon_- = \varepsilon_0 \left[1 - \left(\frac{\omega_p^-}{\omega} \right)^2 \right] \tag{3.282}$$

for these media, we find that interface plasmon modes can be excited at the frequency

$$\boxed{\omega = \sqrt{\frac{\left(\omega_p^+\right)^2 + \left(\omega_p^-\right)^2}{2}}.} \tag{3.283}$$

It is clear from the presented discussion that the spectrum of plasmon modes is continuous rather than discrete as is the case for nanoparticles. In other words, plasmon modes of any wave vector $\mathbf{k}$ can be excited at the frequencies (3.280) or (3.283). This means that actual electric potentials will be continuous superpositions of modes (3.271)-(3.272). These superpositions can be written in the following Fourier integral forms:

$$\varphi^+(x, y, z) = e^{-kz} \iint_{-\infty}^{\infty} A_+\left(\mathbf{k}\right) e^{j\mathbf{k}\cdot\mathbf{r}} dk_x dk_y, \tag{3.284}$$

$$\varphi^-(x, y, z) = e^{kz} \iint_{-\infty}^{\infty} A_+\left(\mathbf{k}\right) e^{j\mathbf{k}\cdot\mathbf{r}} dk_x dk_y. \tag{3.285}$$

It is understandable that function $A_+\left(\mathbf{k}\right)$ depends on excitation conditions. For specific excitation conditions, $A_+\left(\mathbf{k}\right)$ can be found by using Fourier transform analysis.

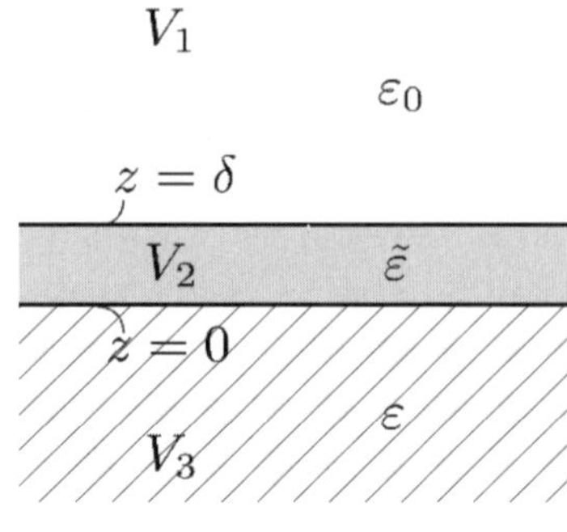

Figure 3.16

It is apparent from formula (3.279) that the resonance value of permittivity ε_- does not depend on the wave vector $\mathbf{k}$. This property is not valid if the conducting medium is separated from vacuum by a thin dielectric layer with permittivity $\tilde{\varepsilon}$ which can be formed as a result of oxidation. To demonstrate this, we consider plasmon modes in the structure shown in Figure 3.16. In this case, plasmon modes are nonzero solutions of the following homogeneous boundary value problem:

$$\nabla^2 \varphi_i = 0 \quad \text{in } V_i, \quad (i = 1, 2, 3), \tag{3.286}$$

$$\varphi_1\big|_{z=\delta} = \varphi_2\big|_{z=\delta}, \tag{3.287}$$

$$\varphi_2\big|_{z=0} = \varphi_3\big|_{z=0}, \tag{3.288}$$

$$\varepsilon_0 \left.\frac{\partial \varphi_1}{\partial z}\right|_{z=\delta} = \tilde{\varepsilon} \left.\frac{\partial \varphi_2}{\partial z}\right|_{z=\delta}, \tag{3.289}$$

$$\tilde{\varepsilon} \left.\frac{\partial \varphi_2}{\partial z}\right|_{z=0} = \varepsilon \left.\frac{\partial \varphi_3}{\partial z}\right|_{z=0}, \tag{3.290}$$

where zero conditions at $z = \pm\infty$ are assumed for φ_1 and φ_3, respectively.

The product solutions for the Laplace equation are given by the formulas

$$\varphi_1(x, y, z) = A e^{j\mathbf{k}\cdot\mathbf{r} - kz}, \tag{3.291}$$

$$\varphi_2(x, y, z) = \left(B e^{-kz} + C e^{kz}\right) e^{j\mathbf{k}\cdot\mathbf{r}}, \tag{3.292}$$

$$\varphi_3(x, y, z) = D e^{j\mathbf{k}\cdot\mathbf{r} + kz}. \tag{3.293}$$

It is clear that the boundary conditions (3.287)-(3.290) are satisfied if the following relations are valid:

$$Ae^{-k\delta} = Be^{-k\delta} + Ce^{k\delta}, \tag{3.294}$$

$$B + C = D, \tag{3.295}$$

$$\varepsilon_0 Ae^{-k\delta} = \tilde{\varepsilon}\left(Be^{-k\delta} - Ce^{k\delta}\right), \tag{3.296}$$

$$\tilde{\varepsilon}\left(B - C\right) = -\varepsilon D. \tag{3.297}$$

By substituting formulas (3.294) and (3.295) into equations (3.296) and (3.297), respectively, we end up with

$$\left(\tilde{\varepsilon} + \varepsilon\right)B + \left(\varepsilon - \tilde{\varepsilon}\right)C = 0, \tag{3.298}$$

$$\left(\tilde{\varepsilon} - \varepsilon_0\right)e^{-k\delta}B - \left(\tilde{\varepsilon} + \varepsilon_0\right)e^{k\delta}C = 0. \tag{3.299}$$

The above homogeneous equations have nonzero solutions only if their determinant is equal to zero. This leads to the following resonance value of dielectric permittivity:

$$\boxed{\varepsilon(k) = -\tilde{\varepsilon}\,\frac{\tilde{\varepsilon}\tanh k\delta + \varepsilon_0}{\tilde{\varepsilon} + \varepsilon_0\tanh k\delta}.} \tag{3.300}$$

Thus, contrary to formula (3.279), the resonance value of permittivity depends on the length of the wave vector. The direction of the wave vector does not matter because the problem is intrinsically isotropic, i.e., it is mathematically invariant with respect to rotations about the z-axis. It is also clear that in the limit of infinitely small δ, the last formula is reduced to formula (3.279).

As in the previous configuration (see Figure 3.15), the plasmon spectrum is continuous. It is worthwhile to note here that it is not always the case for flat infinite interfaces. For instance, if we consider the problem of a circular cross-section nanowire adjacent to a conducting (plasmonic) substrate (Figure 3.17), then localized plasmon modes of discrete spectrum appear. This problem, which is of interest in its own right, is easy to solve by using bipolar coordinates. Actually, the solution is a particular case of the problem of two adjacent circular cross-section nanowires which has been previously discussed. This particular case is realized when μ_2 is equal to zero. For this reason, by using formula (3.64), we find the following expression for discrete spectrum of permittivity:

$$\boxed{\varepsilon_n^{\pm} = -\varepsilon_0\left(\coth n\mu_1 \pm \sqrt{\coth^2 n\mu_1 - 1}\right).} \tag{3.301}$$

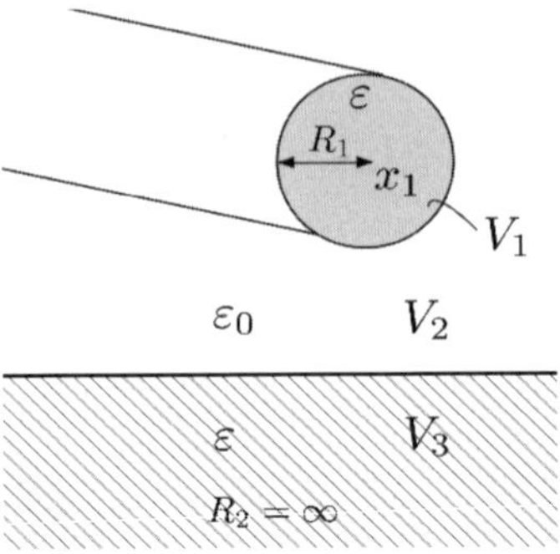

Figure 3.17

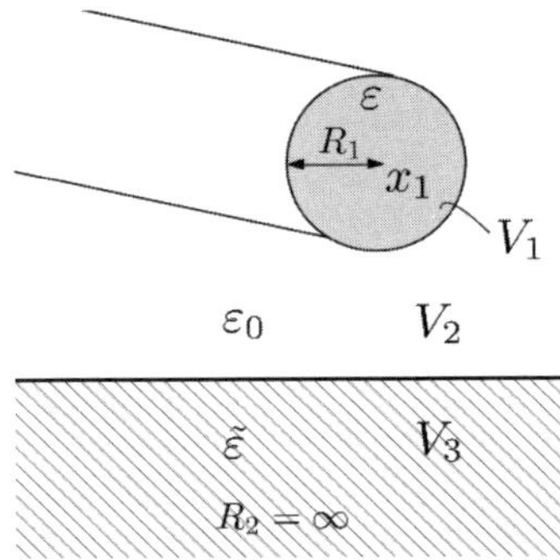

Figure 3.18

By using the bipolar coordinates, it is also easy to find the plasmon modes and their discrete spectrum in the case of circular cross-section nanowire adjacent to a flat dielectric substrate with permittivity $\tilde{\varepsilon}$ (see Figure 3.18). For this problem, the equations (3.61) and (3.62) should be modified as follows:

$$(\varepsilon_0 - \varepsilon_n) e^{-n\mu_1} B_n - (\varepsilon_0 + \varepsilon_n) e^{n\mu_1} C_n = 0, \tag{3.302}$$

$$(\varepsilon_0 + \tilde{\varepsilon}) B_n - (\varepsilon_0 - \tilde{\varepsilon}) C_n = 0. \tag{3.303}$$

The last two equations have nonzero solutions only when ε_n assumes the following values:

$$\varepsilon_n = -\varepsilon_0 \frac{(\tilde{\varepsilon} + \varepsilon_0) e^{2n\mu_1} + (\tilde{\varepsilon} - \varepsilon_0)}{(\tilde{\varepsilon} + \varepsilon_0) e^{2n\mu_1} - (\tilde{\varepsilon} - \varepsilon_0)}. \tag{3.304}$$

It is apparent that when $\tilde{\varepsilon} = \varepsilon_0$ the last formula is reduced to formula (3.9) derived for the case of a circular cross-section nanowire.

It is also clear that by using the same line of reasoning bispherical coordinates can be applied to the analysis of plasmon modes in a spherical nanoparticle adjacent to a dielectric substrate. Unfortunately, simple and explicit analytical results are out of reach in this case.

Finally, it must be noted that localized surface plasmon modes of discrete spectrum may appear around holes or apertures in conducting media. The simplest way to demonstrate this is to consider plasmon modes in infinite conducting media with a spherical hole. This problem can be solved by using the same mathematical machinery as for a spherical nanoparticle. The only change that must be made is to interchange ε_0 and ε_n. As apparent from formulas (3.125)-(3.127), this leads to the following expression for resonance values of permittivity:

$$\varepsilon_n = -\varepsilon_0 \frac{n}{n+1}. \tag{3.305}$$

As will be discussed later in the book, the plasmon modes localized around apertures in plasmonic media can be useful in applications. Apertures in thin plasmonic films can be produced by using ion beam milling.

3.2 Numerical Techniques for the Analysis of Plasmon Modes

It has been shown in the previous chapter that the analysis of plasmon modes in nanoparticles can be framed as the eigenvalue problem for the specific boundary integral equations. Namely, two dual formulations have been advanced in that chapter that require the solution of the following adjoint integral equations, respectively:

$$\sigma_k(Q) = \frac{\lambda_k}{2\pi} \oint_S \sigma_k(M) \frac{\mathbf{r}_{MQ} \cdot \mathbf{n}_Q}{r_{MQ}^3} ds_M, \tag{3.306}$$

$$\tau_k(Q) = \frac{\lambda_k}{2\pi} \oint_S \tau_k(M) \frac{\mathbf{r}_{QM} \cdot \mathbf{n}_M}{r_{QM}^3} ds_M. \tag{3.307}$$

As soon as the solutions of the above eigenvalue problems are found, the resonance values of dielectric permittivity and resonance frequencies can be

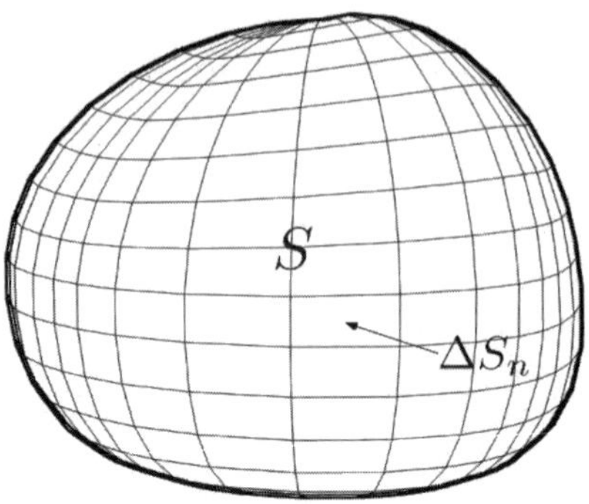

Figure 3.19

computed by using the relations

$$\lambda_k = \frac{\varepsilon_k - \varepsilon_0}{\varepsilon_k + \varepsilon_0}, \qquad \varepsilon'(\omega_k) = \varepsilon_k, \tag{3.308}$$

while the electric fields of plasmon modes can be determined as the fields created by surface charges $\sigma_k(M)$ or double layer charges $\tau_k(M)$. Thus, it is clear that the central issue in the analysis of plasmon modes is the numerical solution of eigenvalue problems (3.306) and (3.307). This issue is studied in detail in this section.

We begin with the discussion of numerical discretization of the integral equation (3.306). Here, we are confronted with two generic difficulties. First, the kernel of this integral equation is (weakly) singular, i.e., for sufficiently smooth surfaces S the kernel has an integrable in the usual sense singularity of $\frac{1}{r_{MQ}}$-type. Second, surface charges $\sigma_k(M)$ have singularities at corners and edges of nanoparticle boundaries. It turns out that these difficulties can be circumvented [10] by using the discretization technique presented below.

In this technique, some partition of the boundary S into N small pieces ΔS_n, ($n = 1, 2, ..., N$) is used (see Figure 3.19), and the integral equation (3.306) is first written in the form

$$\sigma_k(Q) = \frac{\lambda_k}{2\pi} \sum_{n=1}^{N} \int_{\Delta S_n} \sigma_k(M) \frac{\mathbf{r}_{MQ} \cdot \mathbf{n}_Q}{r_{MQ}^3} ds_M. \tag{3.309}$$

We shall next integrate both sides of the last formula over ΔS_m:

$$\int_{\Delta S_m} \sigma_k(Q)ds_Q = \frac{\lambda_k}{2\pi} \sum_{n=1}^{N} \int_{\Delta S_n} \sigma_k(M) \left(\int_{\Delta S_m} \frac{\mathbf{r}_{MQ} \cdot \mathbf{n}_Q}{r_{MQ}^3} ds_Q \right) ds_M,$$

$$(m = 1, 2, ..., N). \quad (3.310)$$

Now, we introduce the notation

$$\Omega_m(M) = \frac{1}{2\pi} \int_{\Delta S_m} \frac{\mathbf{r}_{MQ} \cdot \mathbf{n}_Q}{r_{MQ}^3} ds_Q. \quad (3.311)$$

It is apparent that $\Omega_m(M)$ is the solid angle normalized by 2π which ΔS_m subtends at point M. By using the notation (3.311), the formula (3.310) can be written as follows:

$$\int_{\Delta S_m} \sigma_k(Q)ds_Q = \lambda_k \sum_{n=1}^{N} \int_{\Delta S_n} \sigma_k(M)\Omega_m(M)ds_M,$$

$$(m = 1, 2, ..., N). \quad (3.312)$$

It is here that we introduce new variables (unknowns)

$$X_m^{(k)} = \int_{\Delta S_m} \sigma_k(Q)ds_Q, \quad (3.313)$$

and approximate the integrals in the right-hand side of equations (3.312) in the following manner:

$$\int_{\Delta S_n} \sigma_k(M)\Omega_m(M)ds_M \approx \Omega_m(M_n) \int_{\Delta S_n} \sigma_k(M)ds_M, \quad (3.314)$$

where M_n are some middle points of partitions ΔS_n. It is apparent (on intuitive grounds) that approximation (3.314) is more accurate than direct discretization of the integrals in (3.309). This is because normalized solid angles $\Omega_m(M)$ are smooth functions of M, while the kernel of integral equation (3.306) is weakly singular. It must be remarked that the approximation (3.314) is the only approximation made in the discretization of integral equation (3.306). By using the notation (3.313), the last formula can be written as follows:

$$\int_{\Delta S_n} \sigma_k(M)\Omega_m(M)ds_M \approx \Omega_{mn}X_n^{(k)}, \quad (3.315)$$

where

$$\Omega_{mn} = \Omega_m(M_n) = \frac{1}{2\pi} \int_{\Delta S_m} \frac{\mathbf{r}_{M_n Q} \cdot \mathbf{n}_Q}{r_{M_n Q}^3} ds_Q. \tag{3.316}$$

By substituting formulas (3.313) and (3.315) into formula (3.312) we arrive at the following finite-dimensional eigenvalue problem of linear algebra:

$$X_m^{(k)} = \lambda_k \sum_{n=1}^{N} \Omega_{mn} X_n^{(k)}, \quad (m = 1, 2, ..., N). \tag{3.317}$$

This eigenvalue problem can be written in the matrix form

$$\mathbf{X}^{(k)} = \lambda_k \hat{\Omega} \mathbf{X}^{(k)}. \tag{3.318}$$

Here, $\hat{\Omega}$ is an $N \times N$ matrix whose elements are Ω_{mn}, while $\mathbf{X}^{(k)}$ are column vectors whose elements are $X_n^{(k)}$.

Matrix elements can be computed by using formula (3.316). The evaluation of integrals in this formula in the case of off-diagonal elements Ω_{mn}, $(m \neq n)$ does not present any difficulties, because the integrands are smooth and finite. The evaluation of diagonal elements Ω_{nn} requires the evaluation of weakly singular integrals. However, this inconvenience can be completely circumvented. Indeed, according to the formula

$$\oint_S \frac{\mathbf{r}_{MQ} \cdot \mathbf{n}_Q}{r_{MQ}^3} ds_Q = 2\pi, \quad (M \in S) \tag{3.319}$$

and the definition (3.316) of the matrix elements Ω_{mn}, we have

$$\sum_{m=1}^{N} \Omega_{mn} = 1. \tag{3.320}$$

Consequently,

$$\Omega_{nn} = 1 - \sum_{\substack{m=1, \\ m \neq n}}^{N} \Omega_{mn}, \tag{3.321}$$

and diagonal elements can be found by computing off-diagonal elements.

Now, it is clear that the numerical solution of integral equation (3.306) requires the assembly of the matrix $\hat{\Omega}$ by using formulas (3.316) and (3.321) and the subsequent solution of the matrix eigenvalue problem (3.318). The latter can be accomplished by using standard (and easily available) linear algebra software packages. These packages are by and large based on iterative techniques which require repeated matrix-vector multiplications. The round-off errors in computations may corrupt the charge neutrality property of plasmon modes (see Chapter 2),

$$\oint_S \sigma_k(Q)ds_Q = 0, \tag{3.322}$$

which in terms of $X_m^{(k)}$ can be written as follows:

$$\sum_{m=1}^{N} X_m^{(k)} = 0. \tag{3.323}$$

It turns out that the matrix eigenvalue problem (3.318) can be modified to enforce charge neutrality condition (3.323). This "enforcement" also results in the matrix eigenvalue formulation that excludes the computation of the spurious eigenvalue $\lambda_0 = 1$ corresponding to the Robin problem (see Chapter 2). This "enforcement" is accomplished as follows. By summing up both sides of equations (3.317) with respect to m and taking into account formula (3.323), we obtain

$$\sum_{n=1}^{N} \left(\sum_{m=1}^{N} \Omega_{mn} \right) X_n = 0. \tag{3.324}$$

By using the last formula, equations (3.317) can be modified as follows:

$$X_m^{(k)} = \lambda_k \sum_{n=1}^{N} \left(\Omega_{mn} - \frac{1}{N} \sum_{m=1}^{N} \Omega_{mn} \right) X_n^{(k)}, \quad (m = 1, 2, ..., N). \tag{3.325}$$

It is apparent that equations (3.325) guarantee the fulfillment of charge neutrality condition (3.323) for any λ_k and the removal of spurious eigenvalue $\lambda_0 = 1$. This can be easily established by summing up both sides of equations (3.325).

Equations (3.325) can be written in the matrix form

$$\mathbf{X}^{(k)} = \lambda_k \hat{A} \mathbf{X}^{(k)}, \tag{3.326}$$

where $\hat{A}$ is the matrix with elements

$$a_{mn} = \Omega_{mn} - \frac{1}{N}\sum_{m=1}^{N}\Omega_{mn}. \tag{3.327}$$

Having solved the eigenvalue problem (3.318) or (3.326), the approximate, piece-wise constant values $\tilde{\sigma}_k(M)$ of $\sigma_k(M)$ can be found by using the formula

$$\tilde{\sigma}_k(M) = \frac{X_n^{(k)}}{A_n}, \quad \text{if } M \in \Delta S_n, \tag{3.328}$$

where A_n is the area of ΔS_n.

These piece-wise constant approximation solutions $\tilde{\sigma}_k(M)$ can be naturally smoothed by substituting them into integral equation (3.306). This leads to the expression

$$\tilde{\sigma}_k^{(s)}(Q) = \frac{\lambda_k}{2\pi}\sum_{n=1}^{N}\frac{X_n^{(k)}}{A_n}\int_{\Delta S_n}\frac{\mathbf{r}_{MQ}\cdot\mathbf{n}_Q}{r_{MQ}^3}ds_M, \tag{3.329}$$

where superscript (s) is used for the notation of smoothed values of $\tilde{\sigma}_k(M)$.

Having found $\mathbf{X}^{(k)}$ and $\tilde{\sigma}_k^{(s)}(M)$, the electric fields of the corresponding plasmon modes can be computed by discretizing the formula

$$\mathbf{E}^{(k)}(Q_i) = \frac{1}{4\pi\varepsilon_0}\oint_S \sigma_k(M)\frac{\mathbf{r}_{MQ_i}}{r_{MQ_i}^3}ds_M. \tag{3.330}$$

This discretization is performed as follows:

$$\mathbf{E}^{(k)}(Q_i) = \frac{1}{4\pi\varepsilon_0}\sum_{n=1}^{N}\int_{\Delta S_n}\sigma_k(M)\frac{\mathbf{r}_{MQ_i}}{r_{MQ_i}^3}ds_M$$

$$\approx \frac{1}{4\pi\varepsilon_0}\sum_{n=1}^{N}\frac{\mathbf{r}_{M_nQ_i}}{r_{M_nQ_i}^3}\int_{\Delta S_n}\sigma_k(M)ds_M. \tag{3.331}$$

By recalling formula (3.313), we end up with

$$\mathbf{E}^{(k)}(Q_i) \approx \frac{1}{4\pi\varepsilon_0}\sum_{n=1}^{N}X_n^{(k)}\frac{\mathbf{r}_{M_nQ_i}}{r_{M_nQ_i}^3}. \tag{3.332}$$

For instance, for the Cartesian x-component of the electric field we have

$$E_x^{(k)}(Q_i) = \sum_{n=1}^{N} b_{in} X_n^{(k)}, \tag{3.333}$$

where

$$b_{in} = \frac{x_{Q_i} - x_{M_n}}{4\pi\varepsilon_0 r_{M_n Q_i}^3}. \tag{3.334}$$

When normal components of plasmon electric fields are computed on the boundary S of the nanoparticle, then the following equation,

$$\mathbf{E}^{(k)\pm}(Q_i) \cdot \mathbf{n}_{Q_i} = \pm \frac{\sigma_k(Q_i)}{2\pi\varepsilon_0} + \frac{1}{4\pi\varepsilon_0} \oint_S \sigma_k(M) \frac{\mathbf{r}_{MQ_i} \cdot \mathbf{n}_{Q_i}}{r_{MQ_i}^3} ds_M, \tag{3.335}$$

must be discretized. This leads to the formula

$$\mathbf{E}^{(k)\pm}(Q_i) \cdot \mathbf{n}_{Q_i} \approx \pm \frac{\tilde{\sigma}_k^{(s)}(Q_i)}{2\pi\varepsilon_0} + \frac{1}{4\pi\varepsilon_0} \sum_{n=1}^{N} X_n^{(k)} \frac{\mathbf{r}_{M_n Q_i} \cdot \mathbf{n}_{Q_i}}{r_{M_n Q_i}^3}. \tag{3.336}$$

We shall next proceed to the discussion of discretization of integral equation (3.307). In contrast with $\sigma_k(M)$, the double layer charge density $\tau_k(M)$ is continuous on S and does not have singularities at corners and edges of S. For this reason, the discretization of integral equation (3.307) can be carried out as follows. In each partition ΔS_m we choose some middle points M_m and rewrite integral equation (3.307) in the form

$$\tau_k(M_m) = \frac{\lambda_k}{2\pi} \sum_{n=1}^{N} \int_{\Delta S_n} \tau_k(M) \frac{\mathbf{r}_{M_m M} \cdot \mathbf{n}_M}{r_{M_m M}^3} ds_M, \quad (m = 1, 2, ..., N). \tag{3.337}$$

We next use the approximation

$$\int_{\Delta S_n} \tau_k(M) \frac{\mathbf{r}_{M_m M} \cdot \mathbf{n}_M}{r_{M_m M}^3} ds_M \approx \tau_k(M_n) \int_{\Delta S_n} \frac{\mathbf{r}_{M_m M} \cdot \mathbf{n}_M}{r_{M_m M}^3} ds_M. \tag{3.338}$$

By substituting the last formula into equations (3.337) and by using the notation (3.316), we obtain

$$\boxed{\tau_k(M_m) \approx \lambda_k \sum_{n=1}^{N} \Omega_{nm} \tau_k(M_n), \quad (n = 1, 2, ..., N).} \tag{3.339}$$

The last equations can be written in the matrix form as follows:

$$\boldsymbol{\tau}^{(k)} = \lambda_k \hat{\Omega}' \boldsymbol{\tau}^{(k)}, \tag{3.340}$$

where $\hat{\Omega}'$ is the transpose of the matrix $\hat{\Omega}$, while $\boldsymbol{\tau}^{(k)}$ are column vectors whose elements are $\tau_k(M_n)$. It is clear that the assembly of matrix $\hat{\Omega}'$ is possible by using formulas (3.316) and (3.321). It is also clear that the following biorthogonality conditions are valid for the solutions of eigenvalue problems (3.318) and (3.340):

$$\left\langle \mathbf{X}^{(k)}, \boldsymbol{\tau}^{(i)} \right\rangle = \delta_{ki}. \tag{3.341}$$

Indeed, by recalling formula (3.313), the last conditions can be written as

$$\sum_{n=1}^{N} \tau_i(M_n) \int_{\Delta S_n} \sigma_k(M) ds_M = \delta_{ki}. \tag{3.342}$$

It is apparent that formula (3.342) is a discretized version of biorthogonality conditions (2.140).

The solution of the eigenvalue problem leads to piece-wise constant approximate solution

$$\tilde{\tau}_k(M) = \tau_k(M_n) \quad \text{if } M \in S_n \tag{3.343}$$

of integral equation (3.307). This solution can be naturally smoothed by substituting it in the integral equation (3.307). This leads to the expression

$$\tilde{\tau}_k^{(s)}(Q) = \frac{\lambda_k}{2\pi} \sum_{n=1}^{\infty} \tau_k(M_n) \int_{\Delta S_n} \frac{\mathbf{r}_{QM} \cdot \mathbf{n}_M}{r_{QM}^3} ds_M, \tag{3.344}$$

where as before superscript (s) is used for the notation of smoothed values of $\tilde{\tau}_k(M)$.

Next, we shall discuss the numerical calculation of plasmon electric displacement fields $\mathbf{D}_k$ created by double layer charges $\tau_k(M)$. These fields are given by the formula (see Chapter 2)

$$\mathbf{D}_k(Q) = -\text{grad}_Q \left(\frac{1}{4\pi} \oint_S \tau_k(M) \frac{\mathbf{r}_{QM} \cdot \mathbf{n}_M}{r_{QM}^3} ds_M \right). \tag{3.345}$$

The gradient operation cannot be brought inside the integral in the last formula because it will result in non-integrable singularities if $Q \in S$. This may create some difficulties in numerical computations. These difficulties can be circumvented by deriving another equivalent expression for $\mathbf{D}_k(Q)$. This expression is of interest in its own right because it establishes some interesting mathematical connections between electric fields created by double layer charges and magnetic fields created by surface currents. The derivation proceeds as follows. By using the fact that $\operatorname{grad}_M\left(\frac{1}{r_{MQ}}\right) = -\operatorname{grad}_Q\left(\frac{1}{r_{MQ}}\right)$, the last formula can be written in the form

$$\mathbf{D}_k(Q) = -\operatorname{grad}\operatorname{div}_Q\left(\frac{1}{4\pi}\oint_S \tau_k(M)\frac{\mathbf{n}_M}{r_{QM}}ds_M\right). \qquad (3.346)$$

Next, we shall use the identity

$$\operatorname{grad}\operatorname{div}\mathbf{a} = \operatorname{curl}\operatorname{curl}\mathbf{a} + \nabla^2\mathbf{a}. \qquad (3.347)$$

If $\mathbf{a}$ is identified with the integral in formula (3.346), then $\nabla^2\mathbf{a} = 0$ if $Q \notin S$. This means that formula (3.346) can be transformed as follows:

$$\mathbf{D}_k(Q) = -\operatorname{curl}\operatorname{curl}_Q\left(\frac{1}{4\pi}\oint_S \tau_k(M)\frac{\mathbf{n}_M}{r_{MQ}}ds_M\right). \qquad (3.348)$$

By bringing the curl operator inside of the integral, we obtain

$$\mathbf{D}_k(Q) = -\operatorname{curl}_Q\left[\frac{1}{4\pi}\oint_S \tau_k(M)\,\mathbf{n}_M \times \operatorname{grad}_M\left(\frac{1}{r_{MQ}}\right)ds_M\right]. \qquad (3.349)$$

The last integral can be written in the form

$$\oint_S \tau_k(M)\,\mathbf{n}_M \times \operatorname{grad}_M\left(\frac{1}{r_{MQ}}\right)ds_M$$

$$= \oint_S \mathbf{n}_M \times \operatorname{grad}_M\left(\frac{\tau_k(M)}{r_{MQ}}\right)ds_M - \oint_S \frac{[\mathbf{n}_M \times \operatorname{grad}_S\tau_k(M)]}{r_{MQ}}ds_M. \qquad (3.350)$$

But it is known (see [11]) that for any differentiable function ψ,

$$\oint_S (\mathbf{n}_M \times \operatorname{grad}_S\psi)\,ds = 0. \qquad (3.351)$$

This implies that formula (3.350) can be simplified as follows:

$$\oint_S \tau_k(M)\,\mathbf{n}_M \times \operatorname{grad}_M\left(\frac{1}{r_{MQ}}\right)ds_M = -\oint_S \frac{\mathbf{n}_M \times \operatorname{grad}_S\tau_k(M)}{r_{MQ}}ds_M. \qquad (3.352)$$

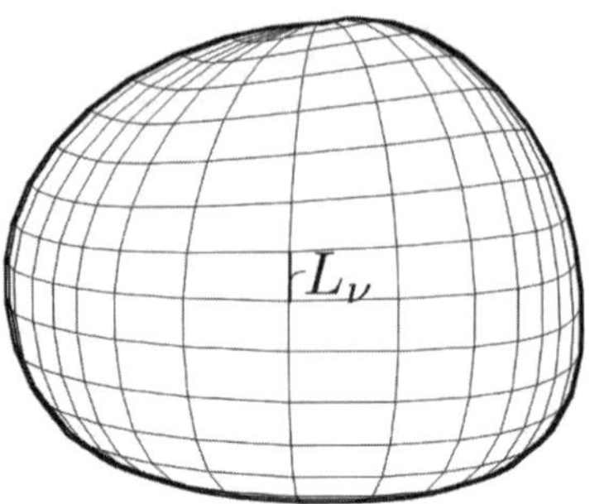

Figure 3.20

By substituting the last formula into equation (3.349), we derive

$$\mathbf{D}_k(Q) = \frac{1}{4\pi} \oint_S \frac{[\mathbf{n}_M \times \mathrm{grad}_S \tau_k(M)] \times \mathbf{r}_{MQ}}{r_{MQ}^3} ds_M. \tag{3.353}$$

It is clear that the last expression is mathematically similar to the Biot-Savart Law for surface distribution of currents. This similarity becomes transparent if we introduce virtual surface "magnetic" currents $\mathbf{i}_m$ by the formula

$$\mathbf{i}_m^{(k)} = \mathbf{n}_M \times \mathrm{grad}_S \tau_k(M) \tag{3.354}$$

and present the equation (3.353) in the form

$$\mathbf{D}_k(Q) = \frac{1}{4\pi} \oint_S \frac{\mathbf{i}_m^{(k)}(M) \times \mathbf{r}_{MQ}}{r_{MQ}^3} ds_M. \tag{3.355}$$

It is also clear from formula (3.354) that double layer density can be interpreted as a stream function for $\mathbf{i}_m^{(k)}(M)$ on S.

Formula (3.353) is convenient for numerical calculations of $\mathbf{D}_k(Q)$. Indeed, by using in formula (3.353) the piece-wise constant approximate expression (3.343) for $\tilde{\tau}_k(M)$ found through the solution of eigenvalue problem (3.340), we obtain the following expression for $\mathbf{D}_k(Q)$:

$$\mathbf{D}_k(Q) = \frac{1}{4\pi} \sum_\nu \int_{L_\nu} \left[\tilde{\tau}_k^+(P) - \tilde{\tau}_k^-(P)\right] \frac{d\mathbf{l}_P \times \mathbf{r}_{PQ}}{r_{PQ}^3}. \tag{3.356}$$

Here, L_ν are the edges of the mesh that partitions S (see Figure 3.20), while $\tilde{\tau}_k^+(P)$ and $\tilde{\tau}_k^-(P)$ are the values of $\tilde{\tau}_k(M)$ on opposite (with respect to L_ν) sides.

We further remark that formula (3.353) can be used for numerical computations of $\mathbf{D}_k(Q)$ in the case when $\tau_k(M)$ is replaced by smoothed solution $\tilde{\tau}_k^{(s)}(M)$ of the eigenvalue problem (3.340). This smoothed solution is given by formula (3.344), and $\operatorname{grad}_S \tilde{\tau}_k^{(s)}$ can be found through numerical differentiation. It is also understandable that (in contrast with expression (3.356)) formula (3.353) properly accounts for discontinuity of the tangential components of $\mathbf{D}_k(Q)$ on S. This is especially transparent from the physical interpretation of $\tau_k(M)$ in terms of surface "magnetic" currents. In other words, formula (3.356) is convenient to use for computations of normal components of $\mathbf{D}_k(Q)$ on S, while formula (3.353) must be used for the calculation of the tangential to S components of $\mathbf{D}_k(Q)$.

The presented discussion dealt with the discretizations of integral equations (3.306) and (3.307) for nanoparticles. The case of nanowires can be treated in a similar way. This case is actually much simpler because the kernels of the appropriate boundary integral equations (2.91) and (2.187) do not have singularities. Moreover, matrix coefficients Ω_{mn} are (plane) angles and can be computed by using exact and explicit formulas.

In our discussion, the case of single nanoparticles has been treated. The case of multiple nanoparticles can be treated in the same way as far as the discretization of integral equations is concerned. However, the matrices $\hat{\Omega}$ and $\hat{\Omega}'$ in the eigenvalue problems (3.318) and (3.340) become very large and the solution of these eigenvalue problems becomes very computationally expensive. Fortunately, since the fully populated matrices $\hat{\Omega}$ and $\hat{\Omega}'$ are generated through discretizations of integral operators whose kernels are derivatives of $\frac{1}{r}$-kernels, the computational cost of matrix-vector multiplications can be considerably reduced by using the fast multipole method (FMM) [12, 13] introduced and developed by V. Rokhlin and L. Greengard. In this method, the matrix-vector multiplications are performed at $O(N)$ computational cost for $N \times N$ fully populated matrices in contrast with $O(N^2)$ computational cost for standard matrix-vector multiplications. The fast multipole method is a very active research area in computer science and applied mathematics. As a result, many modifications, improvements and extensions of this method have been developed. This method employs very sophisticated hierarchical data structures and numerous essential mathematical details. For this reason, the complete description of this method is quite involved and it is beyond the scope of this book. Instead, we shall try to describe only the central idea of the fast multipole method.

One popular version of the fast multipole method which is most pertinent to the integral equation plasmon mode analysis is based on the "addition theorem." This theorem results in the following spherical harmonic expansion,

$$\frac{1}{r_{MQ}} = \frac{1}{|\mathbf{r}_Q - \mathbf{r}_M|} = \sum_{\ell=0}^{\infty} \sum_{m=-\ell}^{\ell} N_{\ell m}\left(\mathbf{r}_Q\right) F_{\ell m}\left(\mathbf{r}_M\right), \tag{3.357}$$

where the following notations are adopted:

$$N_{\ell m}\left(\mathbf{r}_Q\right) = \frac{r_Q^\ell P_{\ell m}\left(\cos\theta\right) e^{-jm\phi}}{(\ell+m)!}, \tag{3.358}$$

$$F_{\ell m}\left(\mathbf{r}_M\right) = \frac{(\ell-m)!\, P_{\ell m}\left(\cos\theta'\right) e^{-jm\phi'}}{r_M^{\ell+1}}. \tag{3.359}$$

Here, (r_Q, θ, ϕ) are spherical coordinates of point Q with respect to some chosen origin, while (r_M, θ', ϕ') are spherical coordinates of point M with respect to the same origin.

The expansion (3.357) is valid under the condition

$$r_Q < r_M. \tag{3.360}$$

The most computationally useful feature of the expansion (3.357) is the separation of variables, i.e., the representation of $\frac{1}{r_{MQ}}$ as the summation of terms which are the product of functions which separately depend on spherical coordinates of observation point Q and integration point M. The price which is paid for this is twofold. First, the sum in formula (3.357) is infinite. This difficulty is circumvented by truncating this infinite sum, and this truncation introduces some controllable errors. For this reason, the fast multipole method is approximate in nature in contrast with the fast Fourier transform technique. Second, the expansion (3.357) is not global and is only valid under the constraint (3.360). To circumvent this limitation, numerous translations of the origin of the spherical coordinates are used, and this eventually leads to the necessity of sophisticated hierarchical data structures.

By using formula (3.357) in its truncated form, the matrix-vector product $\hat{\Omega}\mathbf{X}^{(k)}$ can be represented as follows:

$$\left[\hat{\Omega}\mathbf{X}^{(k)}\right]_i = \sum_{\ell=0}^{L} \sum_{m,m'=-\ell}^{\ell} N_{\ell m'}\left(\mathbf{r}_{Q_i}\right) \sum_j a_j R_{\ell m}\left(\mathbf{r}_{M_j}\right) X_j^{(k)}. \tag{3.361}$$

Here, L is the truncation parameter, $R_{\ell m}\left(\mathbf{r}_{M_j}\right)$ are the appropriate derivatives of $F_{\ell m}\left(\mathbf{r}_M\right)$ needed for the calculation of the kernel of integral equation (3.306) and a_j are some coefficients used in the evaluation of integrals in formula (3.316) for matrix coefficients Ω_{mn} in terms of the kernel.

It is clear that the last sum in equation (3.361) can be computed only once and then used numerous times for various points Q_i. This is the main computational advantage rendered by the separation of variables in the expansion (3.357). Unfortunately, the use of precomputed data cannot be done for all points Q_i and M_j due to the limitation imposed by the inequality (3.360). Nevertheless, the appreciable portion of the last sum in (3.361) still can be precomputed and used for a certain set of points Q_i that are enclosed in some box. This portion consists of summation over all those points M_j that are outside the smallest sphere containing the above box.

This fact suggests to enclose nanoparticles in some large box and consider its hierarchical subdivisions into small boxes. When the calculations of $[\hat{\Omega}\mathbf{X}^{(k)}]_i$ are performed for mesh points in adjacent boxes, some appreciable part of the precomputed portion of the last sum in formula (3.361) that has been used for the previous box can be reused again at some small computational cost. This cost is due to the computations related to the translations of the origins of spherical coordinate systems used for different boxes. This extensive reusing of precomputed data is the key feature of the fast multipole method. This extensive reusing of precomputed data can be most efficiently realized by employing "hierarchical box" subdivisions [12, 13].

Finally, it can be noted that the fast multipole method can be effectively used not only for the solution of eigenvalue problems (3.318) and (3.340), but for calculation of plasmon electric fields as well. The latter is quite apparent from formulas (3.333) and (3.334).

3.3 Numerical Examples

The numerical analysis of plasmon modes in nanoparticles and nanowires discussed in the previous section has been software implemented. Namely, several codes have been developed. All these codes have three major components: a) pre-processing, which essentially performs the assembly of matrices $\hat{\Omega}$ and $\hat{\Omega}'$; b) processing, which consists in the numerical solution of eigenvalue problems (3.318) and (3.340); and c) post-processing, which includes the computation of plasmon mode electric fields as well as the graphical representation of computational results.

By using the developed software, plasmon modes in numerous nanoparticle structures have been computationally studied. This section presents some numerical examples of these computations. Some of these computations have been performed for nanowires and nanoparticles that have been analytically studied in the first section of this chapter. The comparison between analytical and computational results reveals the accuracy of the numerical techniques. Many computations have been carried out for nanoparticle arrangements that have been already studied experimentally due to their physical interest or possible technological applications. The comparison of computational results with available experimental data reveals coincidence which is mostly within five to seven percent. This agreement with experimental data is quite satisfactory and, by and large, within the accuracy of measuring techniques. This agreement can also be construed as some justification for using the macroscopically measured dispersion relation $\varepsilon(\omega)$ at the nanoscale, i.e., as the local constitutive relations for metallic nanoparticles.

We begin with simple examples of numerical analysis of plasmon modes in nanowires.

1. Plasmon modes in nanowires of elliptical cross section

This problem has been analytically studied in section 3.1 and simple analytical expressions (3.105) and (3.111) have been derived for resonance values of dielectric permittivity. These expressions lead to formulas (3.107) and (3.112), respectively, for eigenvalues of boundary integral equation (2.91). This integral equation has been solved numerically and the results of computations are presented in Table 3.1 for the case of semi-axes ratio $a/b = 5$. It is evident from this table that the numerical results are quite accurate and reproduce the mirror-symmetry property of twin spectrum of integral equation (2.91).

2. Plasmon resonances in nanowires of triangular cross section

Here, we present the results of numerical simulations for a nanowire whose cross section is an equilateral triangle with rounded corners (see Figure 3.21). In this case, the plasmon spectrum is not known *a priori*. However, certain qualitative features of this spectrum can be predicted on the basis of symmetry and then tested numerically. The above triangular cross section is invariant with respect to the transformation of the group C_{3v} [14]. This group has three inequivalent irreducible representations: two of dimension one and one of dimension two. This fact implies that the spectrum of integral equa-

Table 3.1

Numerical	Exact	Numerical	Exact
-1.500	-1.500	1.500	1.500
-2.250	-2.250	2.250	2.250
-3.375	-3.375	3.375	3.375
-5.064	-5.063	5.063	5.063
-7.598	-7.594	7.597	7.594
-11.404	-11.391	11.401	11.391
-17.130	-17.086	17.123	17.086
-25.774	-25.629	25.759	25.629
-38.921	-38.443	38.889	38.443
-59.240	-57.665	59.173	57.665

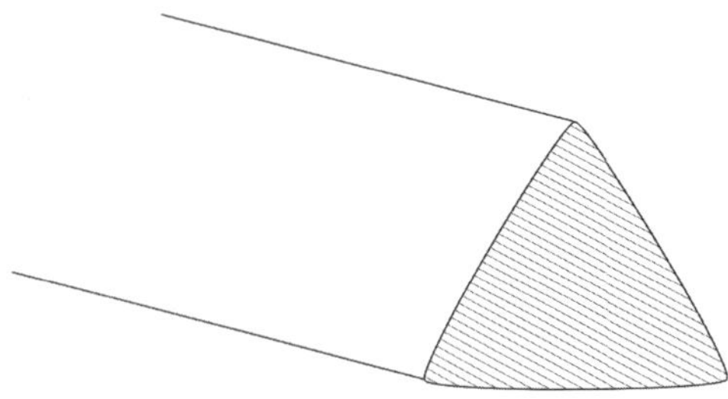

Figure 3.21

tion (2.91) may consist only of simple and twofold degenerate eigenvalues and that the dipole moments of resonance modes corresponding to simple eigenvalues are equal to zero. The computational results for this problem are summarized in Table 3.2. It is evident from this table that the numerical results are consistent with the qualitative features of the spectrum predicted on the basis of symmetry. Indeed, for the plasmon mode corresponding to the simple eigenvalue 2.05 the dipole moment is equal to zero, while for the eigenvalue 3.07 which has multiplicity two the dipole moments of corresponding plasmon modes are not equal to zero. It is also clear that similar qualitative spectral features are valid for three-dimensional prisms with triangular cross sections (see example below).

Table 3.2

Eigenvalue (λ)	p_x	p_y
-1.79	0.469	1.24
-1.79	1.22	-0.508
1.79	2.93	-1.95
1.79	2.16	2.77
-2.05	0.00	0.00
2.05	0.00	0.00
-3.07	0.492	0.568
-3.07	-0.550	0.512
3.07	0.869	-1.13
3.07	-1.09	-0.910
-4.01	0.00	0.00
4.02	0.00	0.00
-6.14	-0.275	-0.262
-6.14	0.286	-0.250
6.16	-0.378	0.376
6.16	0.385	0.369
-8.28	0.00	0.00
8.33	0.00	0.00
-12.32	0.167	0.103
-12.32	-0.097	0.171
12.44	-0.142	0.187
12.44	-0.188	-0.142

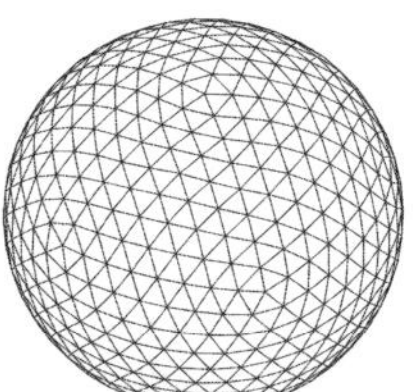

Figure 3.22

Table 3.3

Index	Eigenvalue	Theoretical	Index	Eigenvalue	Theoretical
1	2.999	3	21	9.039	9
2	2.999	3	22	9.049	9
3	2.999	3	23	9.049	9
			24	9.049	9
4	4.980	5			
5	4.980	5	25	10.863	11
6	4.980	5	26	10.863	11
7	5.023	5	27	10.863	11
8	5.023	5	28	10.941	11
			29	10.942	11
9	6.928	7	30	11.038	11
10	6.982	7	31	11.038	11
11	6.982	7	32	11.038	11
12	6.982	7	33	11.065	11
13	7.027	7	34	11.065	11
14	7.027	7	35	11.065	11
15	7.027	7			
			36	12.756	13
16	8.915	9	37	12.847	13
17	8.916	9	38	12.848	13
18	8.916	9	39	12.848	13
19	8.980	9	40	12.932	13
20	8.980	9	41	12.932	13

3. Plasmon modes in spherical nanoparticles

These modes have been studied analytically in sections 2.1 and 3.1, where simple expressions (2.53) and (2.54) have been derived for eigenvalues of integral equation (3.306) and resonance values of dielectric permittivity, respectively. To test the accuracy of three-dimensional code for analysis of plasmon modes in nanoparticles, the same problem has been solved numerically. The partition (mesh) of the spherical boundary used in computation is shown in Figure 3.22. The numerically computed eigenvalues and their comparison with analytical results are summarized in Table 3.3. It is apparent from this table that the numerical results are quite accurate even for plasmon modes of appreciably high orders. It is also evident from this table that the

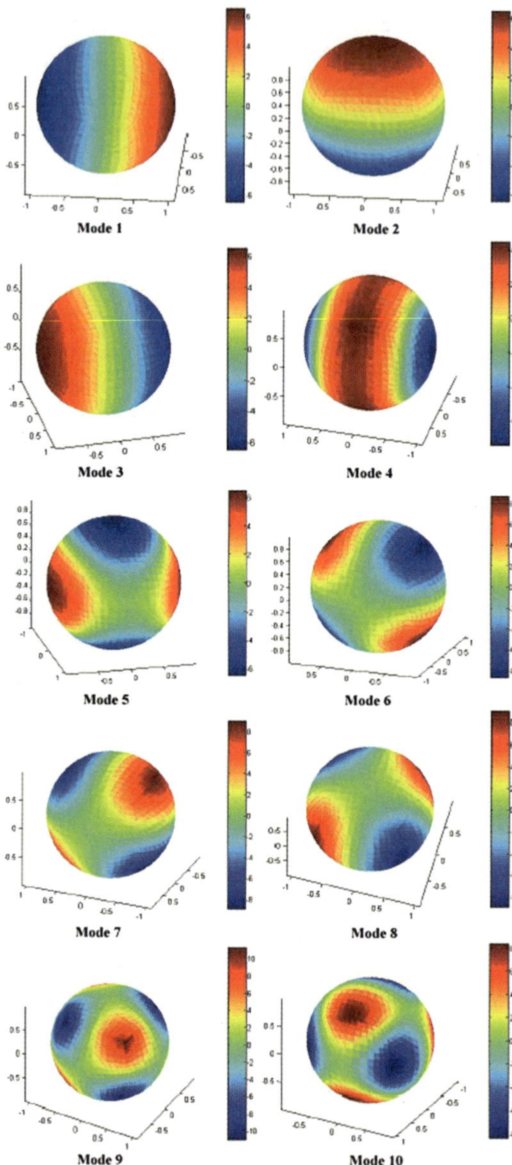

Figure 3.23

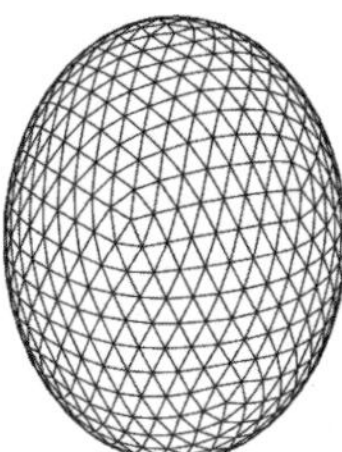

Figure 3.24

computations properly reveal the multiplicities of the eigenvalues of integral equation (3.306) that coincide with the dimensions of irreducible representations of the rotation group, which is the symmetry group of the sphere. The surface charge distributions $\sigma_k(M)$ for different plasmon modes are shown in Figure 3.23. These distributions mimic very well the distributions of spherical harmonics on the sphere and illustrate the generic feature of plasmon modes that their oscillatory nature is enhanced as the mode order is increased.

4. Plasmon modes in ellipsoidal nanoparticles

This problem has been studied analytically in sections 2.1 and 3.1 where simple formulas (2.65) and (3.184) have been derived for the resonance values of dielectric permittivity in the case of dipole (spatially) uniform plasmon modes. These formulas imply that the corresponding eigenvalues of integral equation (3.306) are equal to $\lambda_k = 1/(1 - 2N_k)$. This problem has also been solved numerically for ellipsoid of revolution with main axis ratio 1:1:1.55. The partition (mesh) of the ellipsoidal boundary used in computation is shown in Figure 3.24. The numerically computed eigenvalues of integral equation (3.306) are presented in Table 3.4 where they are compared with theoretical results in the case of dipole modes. Formulas (3.186)-(3.188) have been used in determining the theoretical values of λ_k for dipole modes. It is evident from this table that the numerical results are quite accurate. It is also apparent that for dipole modes $\sum_k (1/\lambda_k)$ is very close to one, as it must be. It is also clear that the eigenvalue 4.590 has multiplicity two, which is due to the rotational symmetry of ellipsoids of revolution. As will be discussed later in the book, this feature provides the opportunity for excitation of circularly polarized plasmon modes that may be instrumental in all-optical magnetic recording. The surface charge distributions $\sigma_k(M)$ for different plasmon modes are illustrated in Figure 3.25.

Table 3.4

Index	Eigenvalue	Theoretical	Index	Eigenvalue
1	1.977	1.972	21	8.549
2	3.335		22	9.316
3	4.081	4.057	23	9.481
4	4.081	4.057	24	9.612
5	4.590		25	9.633
6	4.590		26	9.634
7	4.830		27	10.639
8	5.646		28	10.772
9	5.646		29	10.835
10	6.315		30	11.046
11	6.929		31	11.054
12	6.929		32	11.244
13	7.670		33	11.258
14	7.690		34	11.453
15	7.811		35	11.461
16	7.826		36	11.824
17	7.879		37	11.873
18	8.268		38	11.873
19	8.268		39	12.074
20	8.422		40	12.191

5. Plasmon modes in nano-rings of circular cross sections

This problem has been studied analytically in section 3.1 by using the toroidal coordinate system and the method of separation of variables. The analysis of plasmon modes has been reduced to the eigenvalue problem (3.234) for specific matrix equations. The same problem has been studied numerically by solving the eigenvalue problem for integral equation (3.306). The comparison of the results of computations for different plasmon modes ($m = 1, 2, 3$) and for different ratio of $\frac{R}{x_0}$ is shown in Figure 3.26. It is apparent from this figure that the numerical and analytical results are practically the same. The surface charge distributions $\sigma_k(M)$ for these three plasmon modes are shown in Figure 3.27. We have also found numerically that resonance values of dielectric permittivity are not very sensitive to the actual shape of nano-ring cross section as long as the cross-sectional area remains the same.

Our calculations also reveal that the solution of eigenvalue problem (3.234) is computationally more efficient because only a small number of terms with respect to n is needed to achieve reasonably high accuracy. Furthermore, once the eigenvalue problem (3.234) is solved, the analytical form of the solution is easily obtained by using formulas (3.224) and (3.225).

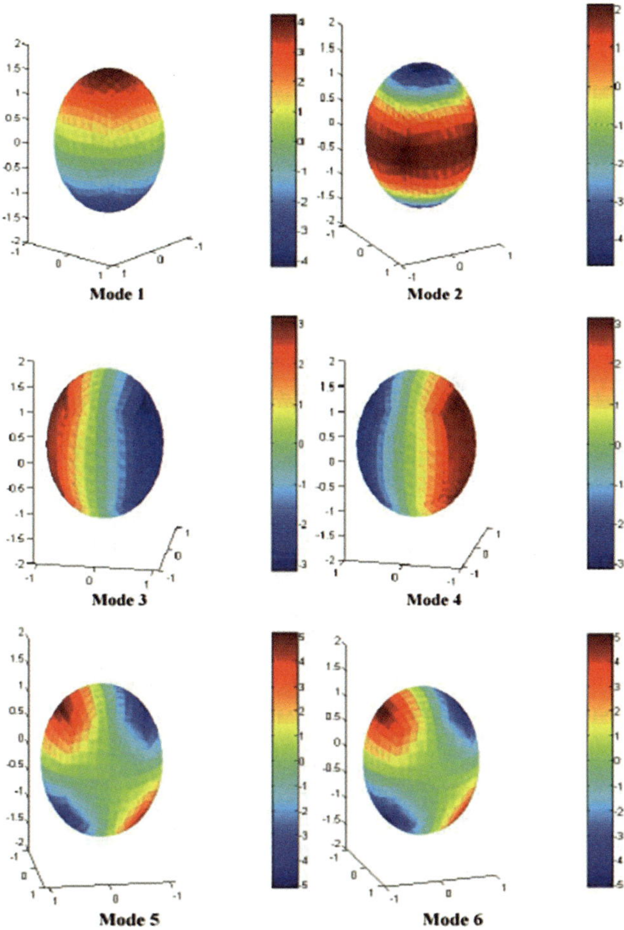

Figure 3.25

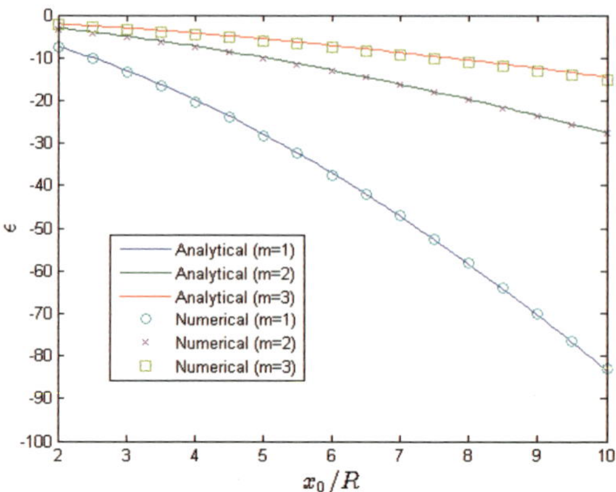

Figure 3.26

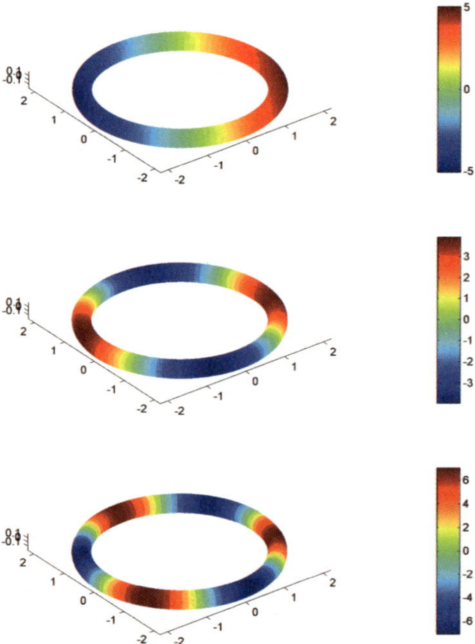

Figure 3.27

6. Plasmon modes in spherical nano-dimers

This problem has been studied analytically in section 3.1 by using bi-spherical coordinates and the method of separation of variables. In this way, the analysis of dipole plasmon modes has been reduced to the eigenvalue problem (3.266). The same problem has been studied numerically by solving the eigenvalue problem for boundary integral equations (2.80) and using for this purpose the discretization technique discussed in the previous section. The comparison of computational results for different ratios of nanoparticle radii and different gaps is presented in Figure 3.28 for dipole plasmon modes. It is evident from this figure that these computational results are practically the same. It is also clear from this figure that if the gap between the spherical nanoparticles is equal to or larger than the radius of the smallest nanoparticle then the resonance value of dielectric permittivity is about $-2\varepsilon_0$, i.e., the same as for dipole modes of a single spherical nanoparticle. This implies that for these separations of spherical nanoparticles their interaction can be neglected. The above figure also reveals that by reducing the separation (gap) between spherical nanoparticles the resonance value of permittivity and the resonance frequency (resonance wavelength) can be effectively controlled. However, accurate positioning of nanoparticles to guarantee their desired separation on the nanoscale is a challenging problem. Finally, the surface charge distribution $\sigma_k(M)$ for dipole plasmon modes of spherical dimers is illustrated by Figure 3.29.

7. Plasmon modes in spherical nano-dimers placed on dielectric substrates

The schematics of this nanostructure are shown in Figure 3.30. This problem has been studied numerically by solving the boundary integral equations (2.90) discretized in the manner described in the previous section. Figure 3.31 presents the computational results for resonance free-space wavelength of dipole plasmon modes as a function of relative separation (D/R_1) between gold spherical nanoparticles placed on dielectric substrate with $\tilde{\varepsilon} = 2.25\varepsilon_0$. The dispersion relation for gold from P. B. Johnson and R. W. Christy [15] has been used in calculations. It is clear (as in the previous example) that the separation between two spheres can be used for the tuning of plasmon resonances to desirable wavelengths. It can be also inferred from Figure 3.31 that the resonance wavelength of dipole modes of single gold nanospheres is about 500 nm. This inference is consistent with existing experimental data.

Plasmon Resonances in Nanoparticles

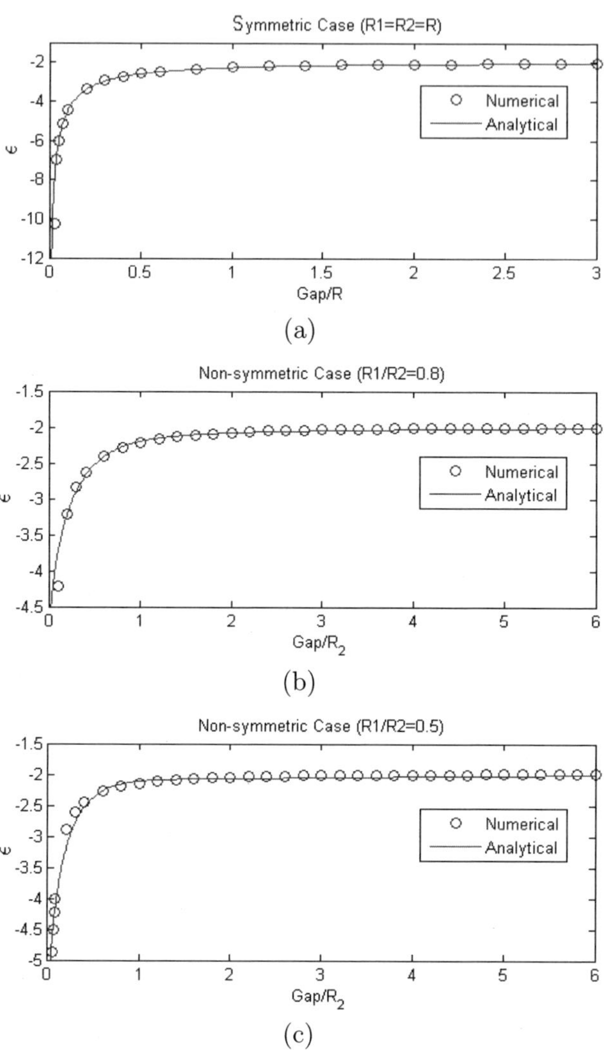

Figure 3.28

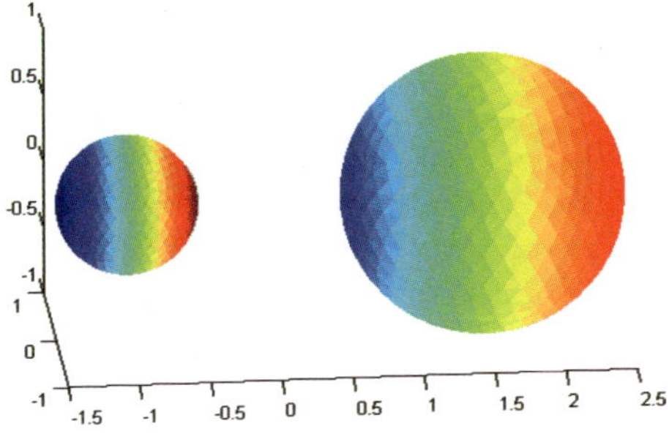

Figure 3.29

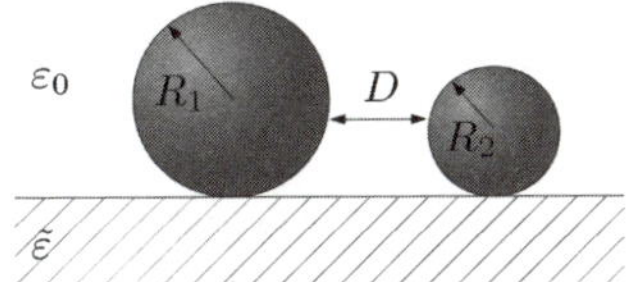

Figure 3.30

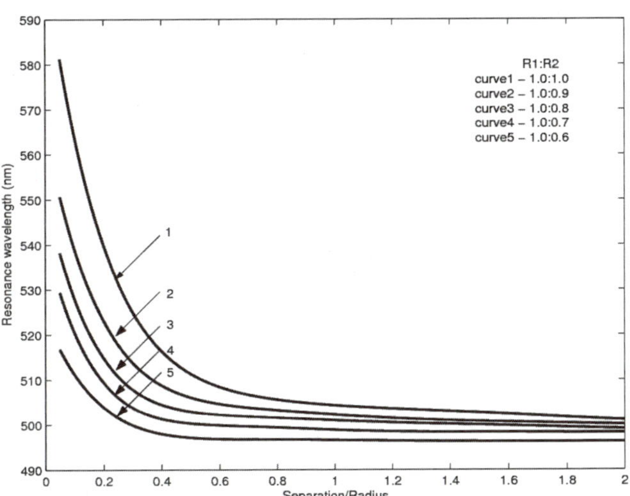

Figure 3.31

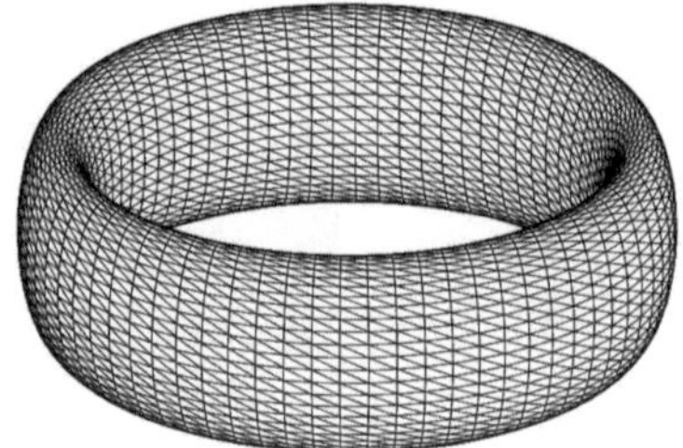

Figure 3.32

8. Plasmon modes in nano-rings placed on a dielectric substrate

Plasmon resonances in gold nano-rings prepared by colloidal lithography and placed on a dielectric substrate with permittivity $\tilde{\varepsilon} = 2.12\varepsilon_0$ have been studied in [16]. The plasmon resonance free-space wavelengths have been determined experimentally as corresponding to the peaks of measured extinction cross sections. The analysis of extinction cross sections of nanoparticles is discussed in the next chapter. Here, it suffices to state that extinction cross sections represent overall losses in nanoparticles. Since the absorption losses due to the imaginary parts of permittivity dominate, the peaks of extinction cross sections occur at wavelengths corresponding to peaks of absorption losses. The peaks of absorption losses coincide with peaks of electric fields inside nanoparticles. The latter peaks occur at resonances. This is the reason why the wavelengths corresponding to extinction cross-section peaks can be identified with plasmon resonance wavelengths.

We have studied plasmon resonances in gold nano-rings placed on a dielectric substrate with $\tilde{\varepsilon} = 2.12\varepsilon_0$ by solving numerically the eigenvalue problem for the boundary integral equation (2.87). Figure 3.32 presents the example of the partition (mesh) of the toroidal boundary used in our computations. Table 3.5 presents the comparison between computational and experimental results for dipole plasmon modes of gold nano-rings of various ring-wall thickness. The parenthetical number (2) in the table indicates the uncertainty in wall thickness measurements. It is evident from this table that the agreement between experimental and computational results is fairly good. It is shown in the next chapter that radiation corrections improve this agreement. It can also be observed that the resonance wavelengths are within the range where the ratio of the real part to the imaginary part of permittivity is most appreciable and, consequently, plasmon resonances can be most efficiently excited.

Table 3.5

	Ring 1	Ring 2	Ring 3
Outer radius (nm)	60	60	60
Height of the ring (nm)	40	40	40
Thickness of ring wall (nm)	14 (2)	10 (2)	9 (2)
Experimental resonance (nm)	1000	1180	1350
Computational resonance (nm)	940	1102	1159

Finally, Figure 3.33 presents the distribution of surface charges $\sigma_k(M)$ for various plasmon modes.

9. Plasmon modes in nanocubes

Plasmon resonances in silver cubic nanoparticles with edge length about 30 nm have been studied in [17]. The extinction cross sections of such nanocubes have been measured by using the dark-field microscopy technique, which allows for plasmon resonance characterization of single particles rather than their ensembles. The measurements have been performed for nanocubes immersed in water and for nanocubes placed on glass substrates. It has been found that there exist two distinct plasmon resonance peaks which are red-shifted and blue-shifted with respect to the peak of the nanocube ensemble. We have numerically studied the same problem by solving eigenvalue problems for boundary integral equations (2.36) for a nanocube in water and (2.87) for a nanocube on a glass substrate with $\tilde{\varepsilon} = 2.25\varepsilon_0$, respectively. The partition of nanocube boundary used in our calculations is shown in Figure 3.34. The comparison between the experimental and computational results is illustrated by Table 3.6. It is shown in the next chapter that this agreement can be further improved by taking into account radiation corrections.

10. Plasmon modes in gold nanocylinders (nanodisks) on a dielectric substrate

We have numerically studied plasmon resonances in short cylindrical dimers of elliptical cross sections placed on substrates with $\tilde{\varepsilon} = 2.25\varepsilon_0$ (see Figure 3.35). Figure 3.36 presents the dependence of resonance wavelength of dipole plasmon modes as a function of the ratio of the separation of cylinders

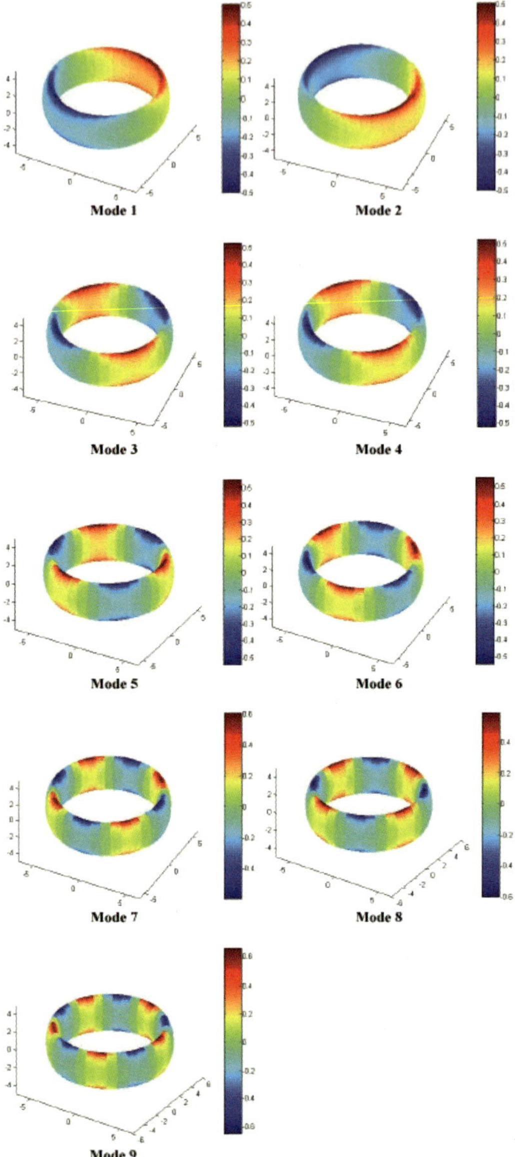

Figure 3.33

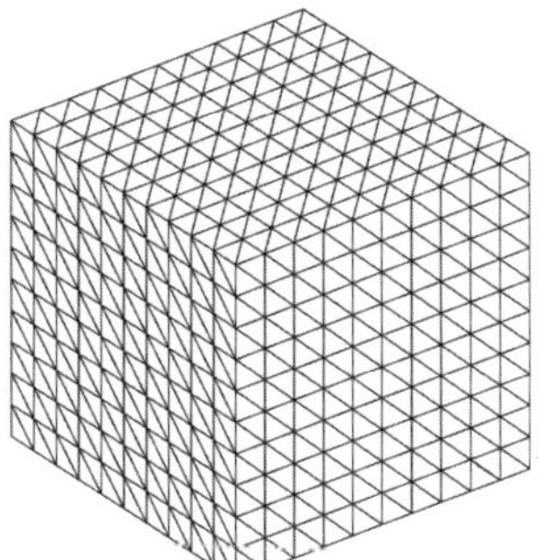

Figure 3.34

Table 3.6

Nanocube in water	Peak 1 (nm)	Peak 2 (nm)
Experimental data	432	500
Computational result	405	454
Nanocube on substrate	Peak 1 (nm)	Peak 2 (nm)
Experimental data	395	457
Computational result	383	421

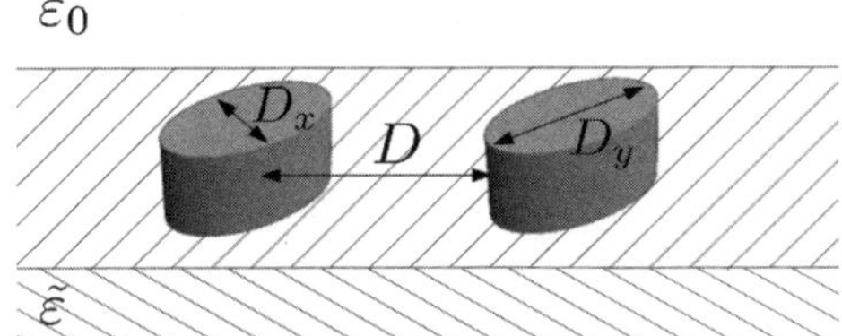

Figure 3.35

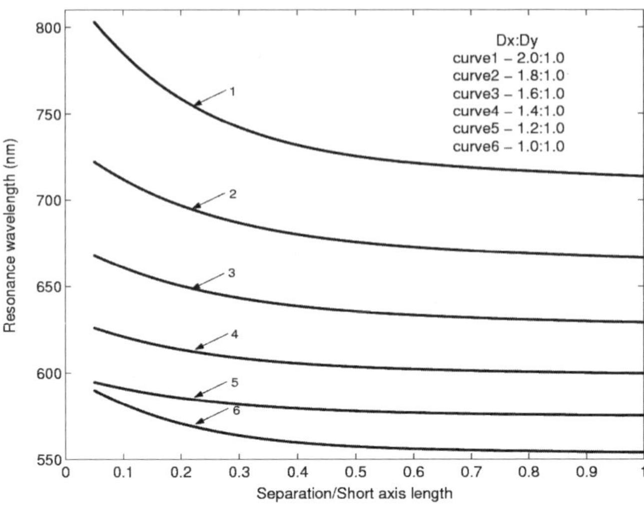

Figure 3.36

to their short axis. Computations have been performed by solving the eigen-value problem for boundary integral equation (2.87). It is instructive to compare these computational results with experimental results published in [18] for short (30 nm high) gold nanocylinders of almost elliptical cross section. These gold nanoparticles were prepared by using electron-beam lithography and lift-off process which permitted the accurate placement of nanoparticles in desired locations. The experimental results in [18] show that depending on nanocylinder dimensions and separation the resonance wavelength varies between 600 nm and 780 nm, which is consistent with computational results shown in Figure 3.36. We have also found numerically (and this can be inferred from Figure 3.36) that the resonance wavelength for a single cylinder with elliptical cross section (height 30 nm, long axis 130 nm and short axis 84 nm) placed on a substrate with $\tilde{\varepsilon} = 2.25\varepsilon_0$ is 622 nm, which is in good agreement with the experimental result of 645 nm.

11. Plasmon modes in a triangular prism

We have numerically studied plasmon modes in equilateral triangular prisms of various dimensions. Figure 3.37 presents a typical example of prism boundary partition used in our simulations. The computed eigenvalues of

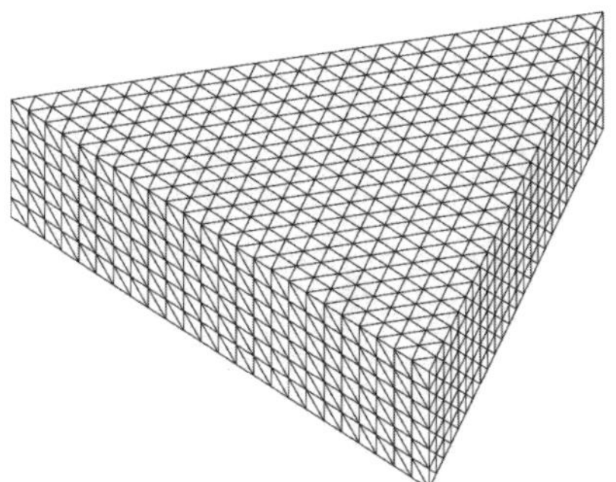

Figure 3.37

Table 3.7

Index	Eigenvalue	Index	Eigenvalue
1	1.260	11	2.020
2	1.260	12	-2.029
3	1.467	13	-2.030
4	1.522	14	-2.036
5	1.522	15	2.217
6	1.523	16	2.244
7	1.688	17	2.244
8	1.688	18	2.245
9	1.889	19	2.287
10	2.020	20	2.287

integral equation (2.36) are given in Table 3.7. It is apparent from this table that there are eigenvalues with multiplicity one and two. This is consistent with the fact that the prism is invariant with respect to the transformations of the group C_{3v}. The distributions of $\sigma_k(M)$ on the prism boundary are shown in Figure 3.38. We have tried to compare our computational results with the experiments discussed in paper [19], where tiny triangular prisms (edge length less than 10 nm) were used for the formation and growth of so-called "three-tipped" nanoparticles. The shape of these "three-tipped" nanoparticles somewhat deviates from the shape of the triangular prism. Nevertheless, we have found that the resonance wavelength 653 nm for the dipole plasmon

Mode 1
Mode 2
Mode 3
Mode 4
Mode 5
Mode 6
Mode 7
Mode 8
Mode 9
Mode 10

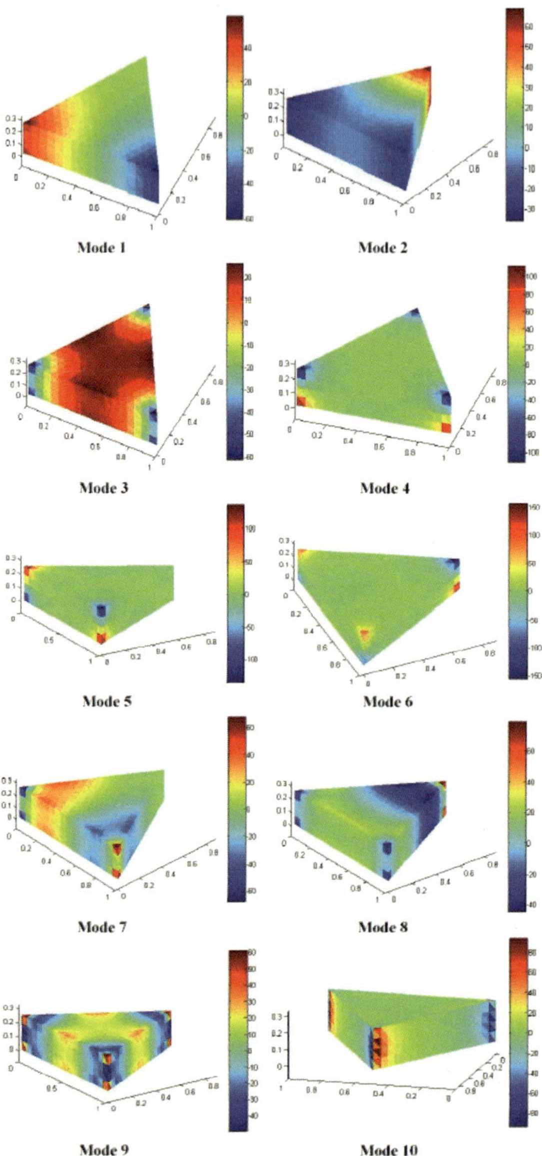

Figure 3.38

Table 3.8

	Shell 1	Shell 2
Inner/outer radius	60/80 nm	55/75 nm
Experimental resonance wavelength	752 nm	717 nm
Computational resonance wavelength	795 nm	764 nm

Table 3.9

	Shell 1	Shell 2
a:b:c	1.414:1:1	1.414:1:1
Resonance values of ε_1 (Theoretical)	-25.00	-21.04
	-0.61	-0.71
Resonance values of ε_1 (Computational)	-25.83	-22.08
	-0.59	-0.69

mode of triangular prisms (edge length of 48 nm and thickness of 14 nm) is reasonably close to the measured resonance wavelength 690 nm of the "three-tipped" nanoparticles of similar dimensions.

12. Plasmon modes in nanoshells

We have studied plasmon modes in spherical and ellipsoidal nanoshells both analytically (section 3.1) and numerically (see section 2.4) and compared the results of our theoretical studies with available experimental data [20]. This comparison has been performed for two spherical nanoshells with silicon cores ($\tilde{\varepsilon} = 15.2\varepsilon_0$) and the results of this comparison are summarized in Table 3.8. We have also compared the analytical results (see formulas (3.209)-(3.210)) with numerical results (see integral equations (2.349)-(2.350)) for ellipsoidal shells of revolution with silicon cores. The results of this comparison (see [31]) are presented in Table 3.9, while the mesh (surface partition) used in these computations is shown in Figure 3.39. Finally, Figure 3.40 presents the dependence of resonance wavelength for dipole plasmon modes as a function of the ratio of outer radius to inner radius of the spherical shells. It is apparent from this figure that the resonance wavelength can be effectively controlled by the thickness of the spherical nanoshells and this makes them very attractive in many applications.

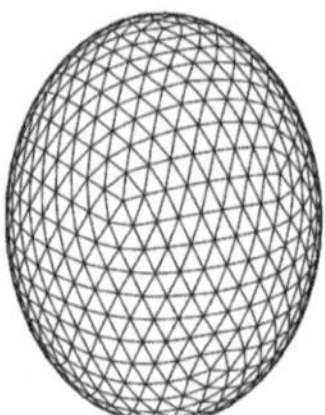

Figure 3.39

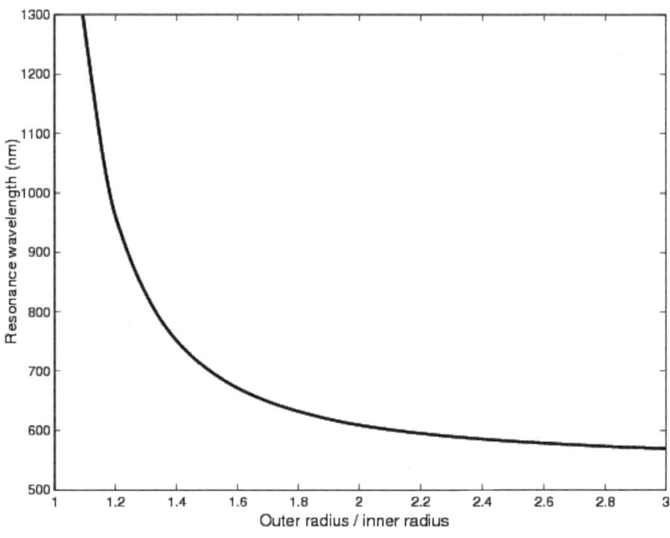

Figure 3.40

3.4 Universal Numerical Technique for the Solution of Boundary Integral Equations

It will be shown in the next chapter that the analysis of extinction cross sections of nanoparticles can be reduced to the solution of the following inhomogeneous integral equation:

$$\sigma(Q) - \frac{\lambda}{2\pi} \oint_S \sigma(M) \frac{\mathbf{r}_{MQ} \cdot \mathbf{n}_Q}{r_{MQ}^3} ds_M = f(Q), \qquad (3.362)$$

where λ is a complex number

$$\lambda = \frac{\varepsilon(\omega) - \varepsilon_0}{\varepsilon(\omega) + \varepsilon_0}, \quad (\varepsilon(\omega) = \varepsilon'(\omega) + j\varepsilon''(\omega)) \tag{3.363}$$

and the right-hand side $f(Q)$ depends on an incident electric field.

Since λ is complex and all eigenvalues of the integral operator in equation (3.362) are real, the boundary integral equation (3.362) is uniquely solvable for any right-hand side $f(Q)$. It turns out that many problems in electromagnetics can be reduced to uniquely solvable integral equations [21]. These integral equations may have weak or strong singularities. (The latter means that integrals in these equations exist in the sense of Cauchy's principal value.) It is very desirable to develop numerical techniques for the solution of these integral equations that are insensitive to the singular nature of integral equation kernels and are applicable to any (reasonable) geometry of S. The techniques with such features can be naturally termed "universal." Such techniques are desirable for the development of general purpose software. One such technique is presented in this section [22]. It is demonstrated below that unique solvability of discretized equations, convergence and the rate of convergence of this numerical technique can be established under only one natural condition of unique solvability of the integral equation to be solved.

Consider a general boundary integral equation

$$\sigma(Q) + \oint_S \sigma(M)K(Q, M)ds_M = f(Q), \tag{3.364}$$

which is assumed to be uniquely solvable in $\mathcal{L}_2(S)$ for any right-hand side $f(Q) \in \mathcal{L}_2(S)$.

For the sake of brevity, we shall write the above integral equation in the operator form

$$\hat{T}\sigma = f, \tag{3.365}$$

where $\hat{T}$ is a bounded operator in $\mathcal{L}_2(S)$. The latter means that

$$\|\hat{T}\| \leq C_1, \tag{3.366}$$

where C_1 is some constant. Since integral equation (3.364) is assumed to be uniquely solvable for any right-hand side, this implies (according to the Banach theorem from functional analysis [23]) the existence of bounded inverse operator $\hat{T}^{-1}$,

$$\left\| \hat{T}^{-1} \right\| \leq C_2, \tag{3.367}$$

where C_2 is some constant.

The numerical technique discussed below is fully justified under conditions (3.366) and (3.367) and no additional assumptions are needed. This technique can be described as follows.

Consider some partition of S into n small pieces $\Delta S_k^{(n)}$, $(k = 1, 2, ..., n)$, and introduce the following basis functions:

$$\psi_k^{(n)}(Q) = \begin{cases} \dfrac{1}{\sqrt{A_k^{(n)}}}, & \text{if } Q \in \Delta S_k^{(n)}, \\[2mm] 0, & \text{if } Q \notin \Delta S_k^{(n)}, \end{cases} \tag{3.368}$$

where $A_k^{(n)}$ is the area of $\Delta S_k^{(n)}$.

It is apparent that

$$\left\langle \psi_k^{(n)}, \psi_{k'}^{(n)} \right\rangle = \oint_S \psi_k^{(n)}(M)\psi_{k'}^{(n)}(M)ds_M = \delta_{kk'}, \tag{3.369}$$

where $\langle\ ,\ \rangle$ stands for the inner product in $\mathcal{L}_2(S)$ and $\delta_{kk'}$ is the Kronecker symbol.

We shall consider a sequence of "embedded" partitions. The latter means that for two partitions $\Delta S_k^{(n_1)}$, $(k = 1, 2, ..., n_1)$ and $\Delta S_k^{(n_2)}$, $(k = 1, 2, ..., n_2)$, the set of linear combinations of basis functions

$$M_{n_1} = \left\{ \sum_{k=1}^{n_1} a_k \psi_k^{(n_1)} \right\} \tag{3.370}$$

of the first partition is embedded in the similar set

$$M_{n_2} = \left\{ \sum_{k=1}^{n_2} b_k \psi_k^{(n_2)} \right\} \tag{3.371}$$

of the second partition if $n_2 > n_1$. This embedding can be written as

$$M_{n_1} \subset M_{n_2}, \quad \text{if } n_2 > n_1. \tag{3.372}$$

It is apparent that this embedding is realized if pieces $\Delta S_k^{(n_2)}$ are obtained by partition of pieces $\Delta S_k^{(n_1)}$.

We look for the approximate solution of integral equation (3.364) in the form

$$\sigma_n(Q) = \sum_{k=1}^{n} a_k^{(n)} \psi_k^{(n)}(Q). \tag{3.373}$$

The coefficients $a_k^{(n)}$ are found from the condition that

$$\left\| \hat{T}\sigma_n - f \right\|^2 = \left\| \hat{T}\left(\sum_{k=1}^{n} a_k^{(n)} \psi_k^{(n)} \right) - f \right\|^2 \tag{3.374}$$

is **minimal**.

By using simple transformations, we find

$$\left\| \hat{T}\sigma_n - f \right\|^2 = \sum_{k=1}^{n}\sum_{m=1}^{n} a_k^{(n)} a_m^{(n)} \left\langle \hat{T}\psi_k^{(n)}, \hat{T}\psi_m^{(n)} \right\rangle$$
$$- 2\sum_{k=1}^{n} a_k^{(n)} \left\langle \hat{T}\psi_k^{(n)}, f \right\rangle + \|f\|^2. \tag{3.375}$$

It is clear now that $\left\| \hat{T}\sigma_n - f \right\|^2$ is a quadratic function of coefficients $a_k^{(n)}$. The minimum condition for $\left\| \hat{T}\sigma_n - f \right\|^2$ can be written in the form

$$\frac{\partial}{\partial a_k^{(n)}} \left\| \hat{T}\sigma_n - f \right\|^2 = 0, \quad (k = 1, 2, ..., n), \tag{3.376}$$

which leads to the following linear simultaneous equations for $a_m^{(n)}$:

$$\sum_{m=1}^{n} a_m^{(n)} \left\langle \hat{T}\psi_k^{(n)}, \hat{T}\psi_m^{(n)} \right\rangle = \left\langle \hat{T}\psi_k^{(n)}, f \right\rangle, \quad (k = 1, 2, ..., n). \tag{3.377}$$

The last linear equations can be written in the matrix form

$$\hat{D}^{(n)}\mathbf{a}^{(n)} = \mathbf{f}^{(n)}, \tag{3.378}$$

where $\mathbf{a}^{(n)}$ is the column vector of unknown coefficients $a_m^{(n)}$, the symmetric matrix $\hat{D}^{(n)}$ is defined as

$$\hat{D}^{(n)} = \left\{ d_{km}^{(n)} \right\} = \left\langle \hat{T}\psi_k^{(n)}, \hat{T}\psi_m^{(n)} \right\rangle \tag{3.379}$$

and the column vector $\mathbf{f}^{(n)}$ has components equal to $\left\langle \hat{T}\psi_k^{(n)}, f \right\rangle$.

By solving linear equations (3.378), we find the approximate solution (3.373). It turns out that linear equations (3.378) are uniquely solvable for any partition of S. Namely, the following theorem is valid:

Theorem 1. If integral equation (3.364) is uniquely solvable, then discretized linear algebraic equations (3.378) are uniquely solvable for any partition of S.

Proof. For any n-dimensional vector $\mathbf{b}^{(n)}$ we find

$$\left\langle \hat{D}^{(n)} \mathbf{b}^{(n)}, \mathbf{b}^{(n)} \right\rangle_n = \sum_{m=1}^{n} \sum_{k=1}^{n} \left\langle \hat{T} \psi_k^{(n)}, \hat{T} \psi_m^{(n)} \right\rangle b_k^{(n)} b_m^{(n)}, \tag{3.380}$$

where $\langle\ ,\ \rangle_n$ stands for inner product in $\mathbb{R}^n$, while $b_k^{(n)}$ are elements of $\mathbf{b}^{(n)}$. The last formula can be written as

$$\left\langle \hat{D}^{(n)} \mathbf{b}^{(n)}, \mathbf{b}^{(n)} \right\rangle_n = \left\langle \hat{T} \left(\sum_{k=1}^{n} b_k^{(n)} \psi_k^{(n)} \right), \hat{T} \left(\sum_{m=1}^{n} b_m^{(n)} \psi_m^{(n)} \right) \right\rangle. \tag{3.381}$$

Next, we introduce the function

$$\xi_n(Q) = \sum_{k=1}^{n} b_k^{(n)} \psi_k^{(n)}(Q). \tag{3.382}$$

It follows from the orthogonality conditions (3.369) that

$$\left\| \xi_n \right\|_{\mathcal{L}_2(S)}^2 = \left\| \mathbf{b}^{(n)} \right\|_{\mathbb{R}^n}^2. \tag{3.383}$$

By using $\xi_n(Q)$, formula (3.381) can be represented in the form

$$\left\langle \hat{D}^{(n)} \mathbf{b}^{(n)}, \mathbf{b}^{(n)} \right\rangle_n = \left\| \hat{T} \xi^{(n)} \right\|_{\mathcal{L}_2(S)}^2. \tag{3.384}$$

Next, by using the identity

$$\xi_n = \hat{T}^{-1} \left(\hat{T} \xi_n \right), \tag{3.385}$$

we find

$$\left\| \xi_n \right\| \leq \left\| \hat{T}^{-1} \right\| \left\| \hat{T} \xi_n \right\|, \tag{3.386}$$

which, according to the inequality (3.367), leads to

$$\left\|\hat{T}\xi_n\right\|^2 \geq \frac{1}{C_2^2}\|\xi_n\|^2. \tag{3.387}$$

Now, by combining (3.383), (3.384) and (3.387) we find

$$\left\langle \hat{D}^{(n)}\mathbf{b}^{(n)}, \mathbf{b}^{(n)}\right\rangle_n \geq \frac{1}{C_2^2}\left\|\mathbf{b}^{(n)}\right\|_{\mathbb{R}^n}. \tag{3.388}$$

The latter means that for any partition of S the matrix $\hat{D}^{(n)}$ is strictly positive definite. From the last inequality immediately follows that the linear homogeneous equations

$$\hat{D}^{(n)}\mathbf{b}^{(n)} = 0 \tag{3.389}$$

must have only trivial solution $\mathbf{b}^{(n)} = 0$. This implies that discretized equations (3.378) are uniquely solvable for any partition of S.

We next proceed to the discussion of convergence of approximate solution σ_n to the exact solution σ. The main result is given by the following:

Theorem 2. If integral equation (3.364) is uniquely solvable for any right-hand side, then there are sequences of embedded partitions such that

$$\lim_{n\to\infty} \|\sigma - \sigma_n\| = 0. \tag{3.390}$$

Proof. The set of continuous functions on S is dense in $\mathcal{L}_2(S)$, and any continuous function on S can be approximated with any desired accuracy by a piece-wise constant function. This means that for any small $\varepsilon > 0$ we can always find such a number n_0 and such a partition of S into n_0 pieces $\Delta S_k^{(n_0)}$ as well as such numbers $\alpha_k^{(n_0)}$ that we have

$$\left\|\sigma - \sum_{k=1}^{n_0} \alpha_k^{(n_0)}\psi_k^{(n_0)}\right\| \leq \frac{\varepsilon}{\|\hat{T}\|\|\hat{T}^{-1}\|}, \tag{3.391}$$

where as before σ is the solution of integral equation (3.364).

From the last inequality we find

$$\left\|\hat{T}\sigma - \hat{T}\left(\sum_{k=1}^{n_0}\alpha_k^{(n_0)}\psi_k^{(n_0)}\right)\right\| \leq \|\hat{T}\|\left\|\sigma - \sum_{k=1}^{n_0}\alpha_k^{(n_0)}\psi_k^{(n_0)}\right\| \leq \frac{\varepsilon}{\|\hat{T}^{-1}\|}. \tag{3.392}$$

By taking into account that $\hat{T}\sigma = f$, we obtain

$$\left\|\hat{T}\left(\sum_{k=1}^{n_0} \alpha_k^{(n_0)}\psi_k^{(n_0)}\right) - f\right\| \leq \frac{\varepsilon}{\|\hat{T}^{-1}\|}. \tag{3.393}$$

According to (3.374), the last inequality will be strengthened if coefficients $\alpha_k^{(n_0)}$ are replaced by $a_k^{(n_0)}$ found by solving linear simultaneous equations (3.378),

$$\left\|\hat{T}\left(\sum_{k=1}^{n_0} a_k^{(n_0)}\psi_k^{(n_0)}\right) - f\right\| < \frac{\varepsilon}{\|\hat{T}^{-1}\|}. \tag{3.394}$$

This is true because the left-hand side of (3.393) achieves its minimum value when $\alpha_k^{(n_0)} = a_k^{(n_0)}$. Due to the embedding condition (3.372), this minimum does not increase if n_0 is replaced by any $n > n_0$. This implies that

$$\left\|\hat{T}\left(\sum_{k=1}^{n} a_k^{(n)}\psi_k^{(n)}\right) - f\right\| \leq \frac{\varepsilon}{\|\hat{T}^{-1}\|} \quad \text{for all } n > n_0. \tag{3.395}$$

By taking into account formulas (3.373) and (3.365), we find

$$\left\|\hat{T}\sigma_n - \hat{T}\sigma\right\| \leq \frac{\varepsilon}{\|\hat{T}^{-1}\|} \quad \text{for all } n > n_0, \tag{3.396}$$

which is equivalent to

$$\left\|\hat{T}(\sigma - \sigma_n)\right\| \leq \frac{\varepsilon}{\|\hat{T}^{-1}\|}. \tag{3.397}$$

Finally, from the above inequality we derive

$$\|\sigma - \sigma_n\| = \left\|\hat{T}^{-1}\left(\hat{T}(\sigma - \sigma_n)\right)\right\| \leq \|\hat{T}^{-1}\|\|\hat{T}(\sigma - \sigma_n)\| < \varepsilon \quad \text{for all } n > n_0. \tag{3.398}$$

This proves the convergence of the approximate solution σ_n to the exact one σ.

By using a similar line of reasoning, the following theorem which establishes the rate of convergence of approximate solution σ_n to the exact solution σ can be proved.

Theorem 3. The rate of convergence of σ_n to σ is characterized by the inequality

$$\|\sigma - \sigma_n\| \le \|\hat{T}\|\|\hat{T}^{-1}\|\|\sigma - \zeta_n^{(\sigma)}\|, \tag{3.399}$$

where $\zeta_n^{(\sigma)}$ is the best least-square approximation of σ by piece-wise constant functions from M_n, i.e.,

$$\min_{\alpha_k^{(n)}} \left\| \sigma - \sum_{k=1}^{n} \alpha_k^{(n)} \psi_k^{(n)} \right\| = \left\| \sigma - \zeta_n^{(\sigma)} \right\|. \tag{3.400}$$

Proof. Let $\beta_k^{(n)}$ be such values of $\alpha_k^{(n)}$ at which the minimum in the last formula is achieved. Then

$$\zeta_n^{(\sigma)} = \sum_{k=1}^{n} \beta_k^{(n)} \psi_k^{(n)}. \tag{3.401}$$

By using the last formula, we find

$$\left\| f - \hat{T}\left(\sum_{k=1}^{n} \beta_k^{(n)} \psi_k^{(n)} \right) \right\| = \left\| \hat{T}\sigma - \hat{T}\zeta_n^{(\sigma)} \right\| \le \|\hat{T}\|\|\sigma - \zeta_n^{(\sigma)}\|. \tag{3.402}$$

The last inequality is further strengthened if coefficients $\beta_k^{(n)}$ are replaced by $a_k^{(n)}$ at which the first term of the last inequality achieves its minimum value. Thus, we have

$$\left\| f - \hat{T}\left(\sum_{k=1}^{n} a_k^{(n)} \psi_k^{(n)} \right) \right\| \le \|\hat{T}\|\|\sigma - \zeta_n^{(\sigma)}\|. \tag{3.403}$$

By using the definition of σ_n, from the last inequality we derive

$$\left\| \hat{T}\sigma - \hat{T}\sigma_n \right\| = \left\| \hat{T}(\sigma - \sigma_n) \right\| \le \|\hat{T}\|\|\sigma - \zeta_n^{(\sigma)}\|. \tag{3.404}$$

It is evident that

$$\|\sigma - \sigma_n\| = \left\| \hat{T}^{-1}\left(\hat{T}(\sigma - \sigma_n) \right) \right\| \le \|\hat{T}^{-1}\|\|\hat{T}(\sigma - \sigma_n)\|. \tag{3.405}$$

Now, by using inequality (3.404) in the last formula, we obtain

$$\|\sigma - \sigma_n\| \le \|\hat{T}^{-1}\|\|\hat{T}\|\|\sigma - \zeta_n^{(\sigma)}\| \tag{3.406}$$

and the theorem is established.

The proven theorem suggests that the rate of convergence of approximate numerical solution σ_n to the exact one σ is the same as the rate of convergence of the best approximation $\zeta_n^{(\sigma)}$ to σ. In this sense, the presented numerical technique has the optimal rate of convergence. Inequality (3.406) clearly reveals that the accuracy of the approximate solution is mesh-dependent. The better the design of the mesh (partition of S), the smaller the norm $\|\sigma - \zeta_n^{(\sigma)}\|$ for a chosen n and the higher the accuracy of the numerical solution σ_n.

The approximate solution σ_n is piece-wise constant. Its "natural" smoothing can be achieved by using integral equation (3.364) and by replacing σ by σ_n in the integral of this equation. This leads to the following expression for the smoothed approximate solution $\tilde{\sigma}_n$:

$$\tilde{\sigma}_n(Q) = -\oint_S \sigma_n(M)K(Q,M)ds_M + f(Q). \tag{3.407}$$

The approximate solution $\tilde{\sigma}_n(Q)$ inherits its smoothness from the kernel and may be appreciably more accurate than $\sigma_n(Q)$. It can be proved that under some conditions $\tilde{\sigma}_n(Q)$ may converge uniformly to $\sigma(Q)$ if the kernel $K(Q,M)$ is not very singular. Indeed, from formulas (3.364) and (3.407) we find

$$\sigma(Q) - \tilde{\sigma}_n(Q) = \oint_S [\sigma_n(M) - \sigma(M)]\,K(Q,M)ds_M. \tag{3.408}$$

From the last formula we derive

$$\max_{Q \in S} |\sigma(Q) - \tilde{\sigma}_n(Q)| \leq \|\sigma - \sigma_n\|_{\mathcal{L}_2(S)} \max_{Q \in S} A(Q), \tag{3.409}$$

where

$$A(Q) = \left(\oint_S K^2(Q,M)ds_M\right)^{\frac{1}{2}}. \tag{3.410}$$

It is clear from formula (3.409) that the uniform convergence of $\tilde{\sigma}_n(Q)$ to $\sigma(Q)$ follows from the proven convergence of $\sigma_n(Q)$ to $\sigma(Q)$ in $\mathcal{L}_2(S)$.

Assembly of matrix $\hat{D}^{(n)}$ is computationally expensive because it requires the evaluation of integrals to compute $\hat{T}\psi_k^{(n)}$ and $\left\langle \hat{T}\psi_k^{(n)}, \hat{T}\psi_m^{(n)}\right\rangle$. The fact that $\psi_k^{(n)}$ are functions with local support (i.e., $\psi_k^{(n)}$ is nonzero only on $\Delta S_k^{(n)}$) facilitates the evaluation of integrals in $\hat{T}\psi_k^{(n)}$. The evaluation of

$\left\langle \hat{T}\psi_k^{(n)}, \hat{T}\psi_m^{(n)} \right\rangle$ and $\left\langle \hat{T}\psi_k^{(n)}, f \right\rangle$ is more computationally extensive because it requires integration over the entire surface S. These computations can be avoided by using the following reasoning. Suppose that the following numerical integration formulas are used for the evaluation of $d_{km}^{(n)}$ and $f_k^{(n)}$:

$$d_{km}^{(n)} \approx \sum_{i=1}^{n} \left(\hat{T}\psi_k^{(n)} \right)_i \left(\hat{T}\psi_m^{(n)} \right)_i A_i^{(n)}, \tag{3.411}$$

$$f_k^{(n)} \approx \sum_{i=1}^{n} f_i \left(\hat{T}\psi_k^{(n)} \right)_i A_i^{(n)}, \tag{3.412}$$

where $\left(\hat{T}\psi_k^{(n)} \right)_i$ and f_i are the values of functions $\hat{T}\psi_k^{(n)}$ and f, respectively, in some middle points of $\Delta S_i^{(n)}$ and, as before, $A_i^{(n)}$ is the area of $\Delta S_i^{(n)}$. By introducing matrices

$$\hat{R}^{(n)} = \left\{ r_{ki}^{(n)} = \left(\hat{T}\psi_k^{(n)} \right)_i A_i^{(n)} \right\}, \tag{3.413}$$

$$\hat{L}^{(n)} = \left\{ l_{im}^{(n)} = \left(\hat{T}\psi_m^{(n)} \right)_i \right\}, \tag{3.414}$$

and column vector $\mathbf{F}^{(n)}$ whose elements are f_i and taking into account formulas (3.411) and (3.412), we can represent the discretized linear equations (3.378) as follows:

$$\hat{R}^{(n)} \hat{L}^{(n)} \mathbf{a}^{(n)} \approx \hat{R}^{(n)} \mathbf{F}^{(n)}. \tag{3.415}$$

If the matrix $\hat{R}^{(n)}$ is invertible (which is most likely the case for sufficiently large n because $\hat{D}^{(n)}$ is invertible and $\hat{D}^{(n)} \approx \hat{R}^{(n)} \hat{L}^{(n)}$), then the discretized equations (3.415) can be reduced to

$$\hat{L}^{(n)} \mathbf{a}^{(n)} = \mathbf{F}^{(n)}. \tag{3.416}$$

The last linear equations are computationally simpler to assemble. These equations can be construed as "collocation" discretized equations for integral equation (3.364). Thus, under the condition of invertibility of matrix $\hat{R}^{(n)}$, we may obtain from Theorems 2 and 3 the convergence and the rate of convergence of the "collocation" technique presented by equations (3.415) and (3.416).

3.5 Absorbing Boundary Conditions for Finite-Difference Time-Domain Analysis of Scattering Problems

It may seem that this topic is somewhat out of place in this book, which advocates the alternatives to the finite-difference time-domain (FDTD) technique for the analysis of plasmon resonances in nanoparticles. There are, however, two reasons why the discussion of this topic is included: first, the unique features of the absorbing boundary conditions (ABCs) discussed in this section and second, the extensive use of ABCs in FDTD analysis of plasmon resonances.

One of the most notorious difficulties in finite-difference or finite-element analysis of scattering problems is the open-space nature of these problems, i.e., electromagnetic fields are distributed in the entire three-dimensional space. To circumvent these difficulties, regions of field distributions are truncated by introducing some artificial boundaries. This truncation introduces errors in electromagnetic field analysis due to scattering from these artificial boundaries. To eliminate or reduce these spurious electromagnetic wave reflections, special absorbing boundary conditions are introduced on the artificial boundaries. The proper design of these boundary conditions is one of the most important issues in numerical analysis of scattering problems that has been extensively studied during the past three decades. The main results achieved in this area can be classified in two conceptually distinct groups. The first group consists of various ABCs that are designed to eliminate the most dominant terms in the Wilcox far-field expansion for electromagnetic fields. These boundary conditions are mostly posed on spherical artificial boundaries. The origin of this approach to the design of ABCs can be traced back to the paper [24] of Bayliss and Turkel. The second group consists of ABCs which are posed on flat surfaces. The origin of this approach can be traced to the paper [25] of Engquist and Majda. This approach has received a new impetus in the paper [26] of Berenger on a perfectly matched layer (PML).

In this section, a conceptually different approach to the design of absorbing boundary conditions is discussed. It offers certain advantages over the existing approaches. The central idea of this approach was advanced in [27, 28]. The essence of this idea is to interpret the Kirchhoff-type formulas as explicit Dirichlet boundary conditions that can be posed on any

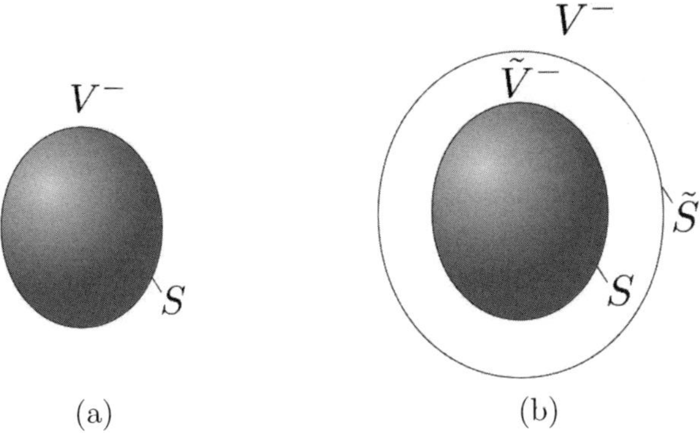

Figure 3.41

artificial boundary. These Dirichlet boundary conditions are not known beforehand and must be updated as computations proceed. This updating of the Dirichlet-type boundary conditions is possible due to the retardation phenomena. These boundary conditions are exact in principle. Their accuracy is only compromised and corrupted by numerical errors. In the case of multiple scatterers, separate artificial boundaries can be introduced around each scatterer and "updatable" Dirichlet boundary conditions can be posed on those boundaries.

To clearly reveal the essence of this idea, we first discuss scalar scattering problems. Consider scattering of incident scalar wave φ^0 by an ideal scatterer with boundary S (see Figure 3.41a). The scattered wave φ in the region V^- exterior to S satisfies the wave equation

$$\nabla^2 \varphi - \frac{1}{c^2}\frac{\partial^2 \varphi}{\partial t^2} = 0 \quad \text{in } V^- \tag{3.417}$$

and the following initial and boundary conditions:

$$\varphi(Q,t)\big|_{t=0} = 0 \quad \text{for } Q \in V^-, \tag{3.418}$$

$$\frac{\partial \varphi}{\partial t}(Q,t)\bigg|_{t=0} = 0 \quad \text{for } Q \in V^-, \tag{3.419}$$

$$\varphi(Q,t)\big|_{Q\in S} = -\varphi^0(Q,t)\big|_{Q\in S}. \tag{3.420}$$

We now introduce a closed surface $\tilde{S}$ which encloses the scattering object (see Figure 3.41b). It is our intention to demonstrate that the explicit (but "updatable") Dirichlet boundary condition

$$\varphi(Q,t)\big|_{Q\in\tilde{S}} = f(Q,t) \tag{3.421}$$

can be posed on $\tilde{S}$.

First, consider the time interval

$$0 \le t \le T, \tag{3.422}$$

where

$$T = \min_{M\in S, Q\in\tilde{S}} \frac{r_{MQ}}{c}. \tag{3.423}$$

It is clear that

$$\varphi(Q,t)\big|_{Q\in\tilde{S}} = 0 \quad \text{for } 0 \le t \le T. \tag{3.424}$$

The last boundary condition is valid due to the finite speed c of propagation of the scattered wave φ. Indeed, time longer than T is needed for the scattered wave to reach the artificial boundary $\tilde{S}$. If the solution of the initial boundary value problem defined by formulas (3.417)-(3.420) and (3.424) in the truncated region $\tilde{V}^-$ between S and $\tilde{S}$ is computed for the time interval (3.422), then the values of $\frac{\partial\varphi}{\partial n}(M,t)\big|_{M\in S}$ can be found for this time interval. By using these values, the Dirichlet boundary condition for φ on the artificial boundary $\tilde{S}$ can be found for the time interval

$$T \le t \le 2T \tag{3.425}$$

by using the well-known Kirchhoff formula

$$\varphi(Q,t) = \frac{1}{4\pi} \oint_S \left[\frac{1}{r_{MQ}} \frac{\partial\varphi}{\partial n_M} \left(M, t - \frac{r_{MQ}}{c}\right) - \varphi\left(M, t - \frac{r_{MQ}}{c}\right) \frac{\partial}{\partial n_M}\left(\frac{1}{r_{MQ}}\right) \right.$$
$$\left. + \frac{\partial\varphi}{\partial t}\left(M, t - \frac{r_{MQ}}{c}\right) \frac{1}{c\,r_{MQ}} \frac{\partial r_{MQ}}{\partial n_M} \right] ds_M, \tag{3.426}$$

where $Q \in \tilde{S}$.

Indeed, according to formula (3.423),

$$\frac{r_{MQ}}{c} \ge T \tag{3.427}$$

and it is clear that

$$t - \frac{r_{MQ}}{c} < T, \quad \text{if} \quad T < t < 2T. \tag{3.428}$$

Thus, $\frac{\partial \varphi}{\partial n}\left(M, t - \frac{r_{MQ}}{c}\right)$, $\varphi\left(M, t - \frac{r_{MQ}}{c}\right)$ and $\frac{\partial \varphi}{\partial t}\left(M, t - \frac{r_{MQ}}{c}\right)$ are known for the time interval $T < t < 2T$. Consequently, by using formula (3.426), the Dirichlet boundary condition for the scattered wave φ on $\tilde{S}$ can be computed for $T < t < 2T$. This updated boundary condition can be used to compute the scattered wave in the region $\tilde{V}^-$ for the time interval $T < t < 2T$. Thus, $\frac{\partial \varphi}{\partial n}(M, t)$ can be computed for the above time interval. This means that the values of $\frac{\partial \varphi}{\partial n}\left(M, t - \frac{r_{MQ}}{c}\right)$, $\varphi\left(M, t - \frac{r_{MQ}}{c}\right)$ and $\frac{\partial \varphi}{\partial t}\left(M, t - \frac{r_{MQ}}{c}\right)$ on S are known for the time interval $0 < t < 2T$. By using these values and the fact that

$$t - \frac{r_{MQ}}{c} < 2T, \quad \text{if} \quad 2T < t < 3T, \tag{3.429}$$

the Kirchhoff formula (3.426) can again be used to compute the Dirichlet boundary condition on $\tilde{S}$ for the time interval $2T < t < 3T$. By literally repeating the previous reasoning, it can be concluded that if the solution is computed in the region $\tilde{V}^-$ for some time interval $0 < t < t_0$, then the Kirchhoff formula (3.426) can be used to find the Dirichlet boundary condition for time interval $t_0 < t < t_0 + T$. Therefore, the solution of the initial-boundary value problem in the region $\tilde{V}^-$ can be extended to a wider time interval. In other words, the Kirchhoff formula (3.426) can be continuously used to update the Dirichlet boundary condition on the artificial boundary $\tilde{S}$ as the computations of the scattered wave φ in the truncated region $\tilde{V}^-$ proceed. It is apparent that this updating is possible due to the retardation phenomena.

It is important to point out that the only reason for the reduction of the initial-boundary value problem (3.417)-(3.420) in the unbounded region V^- to the initial-boundary value problem in the truncated region $\tilde{V}^-$ is the localization of computations. When the scattered wave φ is computed in the truncated region $\tilde{V}^-$, the computations of φ outside the artificial boundary $\tilde{S}$ can be carried out by using the same Kirchhoff formula (3.426).

Now, we turn to the discussion of electromagnetic scattering problems. We first present the "first-order" formulation where the first-order (in time and space) Maxwell differential equations are discretized and solved by using finite differences.

The scattering of incident electromagnetic field $\mathbf{E}^0$ by an ideal conductor (see Figure 3.41a) requires the solution of the Maxwell equations for scattered electric $\mathbf{E}$ and magnetic $\mathbf{H}$ fields

$$\operatorname{curl} \mathbf{E} = -\mu_0 \frac{\partial \mathbf{H}}{\partial t} \quad \text{in } V^-, \tag{3.430}$$

$$\operatorname{curl} \mathbf{H} = \varepsilon_0 \frac{\partial \mathbf{E}}{\partial t} \quad \text{in } V^-, \tag{3.431}$$

subject to the following initial and boundary conditions

$$\mathbf{E}(Q,t)|_{t=0} = 0 \quad \text{for } Q \in V^-, \tag{3.432}$$

$$\mathbf{H}(Q,t)|_{t=0} = 0 \quad \text{for } Q \in V^-, \tag{3.433}$$

$$\mathbf{n} \times \mathbf{E}(Q,t)|_{Q \in S} = -\,\mathbf{n} \times \mathbf{E}^0(Q,t)|_{Q \in S}, \tag{3.434}$$

where $\mathbf{n}$ is a unit vector of outward normal to S.

As before, we introduce a closed surface $\tilde{S}$ which encloses S (see Figure 3.41b). It is our intention to demonstrate that the explicit (but "updatable") Dirichlet-type boundary conditions

$$\mathbf{n} \times \mathbf{E}(Q,t)|_{Q \in \tilde{S}} = \mathbf{f}(Q,t) \tag{3.435}$$

can be posed on $\tilde{S}$.

It is clear that

$$\mathbf{f}(Q,t) = 0, \quad \text{if } 0 \le t \le T, \tag{3.436}$$

where T is defined by formula (3.423). Thus, we can compute the scattering field in $\tilde{V}^-$ for the time interval $0 < t < T$. Next, we need the counterpart of the Kirchhoff formula for the Maxwell equations (3.430) and (3.431) in order to update the function $\mathbf{f}(Q,t)$ in the boundary condition (3.435) for $t > T$. This counterpart is the time-domain version of the well-known Stratton-Chu formula. For time-harmonic electromagnetic fields, the Stratton-Chu formula can be written as follows [29]:

$$\hat{\mathbf{E}}(Q) = \frac{1}{4\pi} \oint_S \left[j\omega\mu_0 \left(\mathbf{n}_M \times \hat{\mathbf{H}}(M) \right) \frac{e^{jkr_{MQ}}}{r_{MQ}} + \mathbf{n}_M \cdot \hat{\mathbf{E}}(M)\,\operatorname{grad}_M \left(\frac{e^{jkr_{MQ}}}{r_{MQ}} \right) \right.$$
$$\left. + \left(\mathbf{n}_M \times \hat{\mathbf{E}} \right) \times \operatorname{grad}_M \left(\frac{e^{jkr_{MQ}}}{r_{MQ}} \right) \right] ds_M. \tag{3.437}$$

Here, $\hat{\mathbf{E}}$ and $\hat{\mathbf{H}}$ are the phasors of electric and magnetic fields, respectively, and $k = \frac{\omega}{c}$, while all other notations have their usual meaning.

To derive the time-domain version of the Stratton-Chu formula, we shall treat $\hat{\mathbf{E}}$ and $\hat{\mathbf{H}}$ in formula (3.437) as Fourier transforms of electric and magnetic fields, respectively. To recover the actual (time-dependent) electric and magnetic fields, we apply the inverse Fourier transform

$$\mathbf{E}(Q,t) = \frac{1}{2\pi} \int_{-\infty}^{\infty} \hat{\mathbf{E}}(Q) e^{-j\omega t} d\omega \tag{3.438}$$

to both sides of formula (3.437). Factors $e^{jkr_{MQ}} = e^{j\omega \frac{r_{MQ}}{c}}$ in each term of the integrand in (3.437) result in $\frac{r_{MQ}}{c}$-retardations in time for inverse Fourier transforms of each term of this integrand. As a result, we obtain the following time-domain version of the Stratton-Chu formula:

$$\begin{aligned}
\mathbf{E}(Q,t) = \frac{1}{4\pi} \oint_S \Bigg[&\left(\mathbf{n}_M \times \frac{\partial \mathbf{H}}{\partial t} \left(M, t - \frac{r_{MQ}}{c} \right) \right) \frac{\mu_0}{r_{MQ}} \\
&+ \mathbf{n}_M \cdot \mathbf{E} \left(M, t - \frac{r_{MQ}}{c} \right) \mathrm{grad}_M \left(\frac{1}{r_{MQ}} \right) \\
&+ \mathbf{n}_M \cdot \frac{\partial \mathbf{E}}{\partial t} \left(M, t - \frac{r_{MQ}}{c} \right) \frac{\mathrm{grad}_M (r_{MQ})}{c r_{MQ}} \\
&+ \left(\mathbf{n}_M \times \mathbf{E} \left(M, t - \frac{r_{MQ}}{c} \right) \right) \times \mathrm{grad}_M \left(\frac{1}{r_{MQ}} \right) \\
&+ \left(\mathbf{n}_M \times \frac{\partial \mathbf{E}}{\partial t} \left(M, t - \frac{r_{MQ}}{c} \right) \right) \times \frac{\mathrm{grad}_M (r_{MQ})}{c r_{MQ}} \Bigg] ds_M.
\end{aligned} \tag{3.439}$$

By using the same line of reasoning as before, it can be concluded that if the scattering fields are computed in the truncated region $\tilde{V}^-$ for some time interval $0 < t < t_0$, then the time-domain version (3.439) of the Stratton-Chu formula can be used to update function $\mathbf{f}(Q,t)$ in the boundary condition (3.435) for the time interval $t_0 < t < t_0 + T$. In this way, (3.439) can be continuously used to update the boundary condition (3.435) as computations proceed. Numerical computations are initiated by using "zero" boundary conditions (3.435)-(3.436).

It is worthwhile to remark that another (dual) Dirichlet-type boundary condition,

$$\mathbf{n} \times \mathbf{H}(Q,t)|_{\tilde{S}} = \mathbf{g}(Q,t), \tag{3.440}$$

can also be posed on the artificial boundary $\tilde{S}$. In this boundary condition, $\mathbf{g}(Q,T) = 0$ for $0 < t < T$, and it can be continuously updated for times $t > T$ by using the "magnetic" time-domain version of the Stratton-Chu formula.

The presented formulation of electromagnetic scattering problems is a "first-order" formulation because it requires numerical integration of coupled first-order Maxwell equations (3.430)-(3.431). This can be accomplished by using the Yee finite-difference scheme [30], which is currently very popular in the finite-difference time-domain analysis of electromagnetic scattering problems. In this scheme, electric and magnetic fields are defined on staggered grids and computer storage of six field components is required. An alternative "second-order" formulation is discussed below.

The electromagnetic scattering problem posed before can be formulated in terms of the scattered electric field $\mathbf{E}$ alone. Indeed, the scattered electric field $\mathbf{E}$ satisfies the equations

$$\nabla^2 \mathbf{E} - \frac{1}{c^2}\frac{\partial^2 \mathbf{E}}{\partial t^2} = 0 \quad \text{in } V^-, \tag{3.441}$$

$$\text{div } \mathbf{E} = 0 \quad \text{in } V^- \tag{3.442}$$

and the following initial and boundary conditions:

$$\mathbf{E}(Q,t)\big|_{t=0} = 0 \quad \text{in } V^-, \tag{3.443}$$

$$\frac{\partial \mathbf{E}}{\partial t}(Q,t)\bigg|_{t=0} = 0 \quad \text{in } V^-, \tag{3.444}$$

$$\mathbf{n} \times \mathbf{E}(Q,t)\big|_{Q\in S} = -\,\mathbf{n} \times \mathbf{E}^0(Q,t)\big|_{Q\in S}. \tag{3.445}$$

The initial-boundary value problem (3.441)-(3.445) has a peculiar mathematical structure where three unknown functions (Cartesian components of $\mathbf{E}$) satisfy four scalar independent equations (3.441) and (3.442) and two independent boundary conditions (3.445). From the point of view of the conventional partial differential equation (PDE) theory, the initial-boundary value problem has an "over-specified" number of partial differential equations and "under-specified" number of boundary conditions. Certain advantages may be gained by transforming this initial-boundary value problem to the form where the number of PDEs is equal to the number of boundary conditions.

This can be accomplished by replacing differential equation (3.442) by the boundary condition

$$\text{div } \mathbf{E}\big|_S = 0. \tag{3.446}$$

It is easy to demonstrate that imposing the last boundary condition is indeed equivalent to imposing differential equation (3.442).

By using the boundary condition (3.446) instead of differential equation (3.442), the analysis of the posed scattering problem can be reduced to the solution of the following three decoupled equations for Cartesian component E_{x_k} of $\mathbf{E}$,

$$\nabla^2 E_{x_k} - \frac{1}{c^2}\frac{\partial^2 E_{x_k}}{\partial t^2} = 0, \quad (k = 1, 2, 3), \tag{3.447}$$

subject to the following initial and boundary conditions:

$$E_{x_k}(Q,t)\big|_{t=0} = 0, \qquad \frac{\partial E_{x_k}}{\partial t}(Q,t)\bigg|_{t=0} = 0, \tag{3.448}$$

$$\sum_{k=1}^{3} E_{x_k}(Q,t)\cos(x_k,\tau_i) = -\sum_{k=1}^{3} E_{x_k}^0(Q,t)\cos(x_k,\tau_i), \tag{3.449}$$

$$\sum_{k=1}^{3} \frac{\partial E_{x_k}}{\partial n}(Q,t)\cos(x_k,n) = \mathrm{div}_S\,\mathbf{E}^0, \tag{3.450}$$

where $Q \in S$, $i = 1, 2$ and τ_i are orthogonal directions in a plane tangential to S, while $\mathrm{div}_S\,\mathbf{E}^0 = \partial E_{\tau_1}^0/\partial\tau_1 + \partial E_{\tau_2}^0/\partial\tau_2$.

It is apparent that in the initial-boundary value problem (3.447)-(3.450) the Cartesian components E_{x_k} are coupled only through boundary conditions (3.449)-(3.450). After discretization, this coupling can be easily removed by exploiting the fact that the matrix of directional cosines is an orthogonal one and its transpose coincides with its inverse.

Now, we consider an arbitrary artificial boundary $\tilde{S}$ that encloses the scatterer. We intend to specify the Dirichlet boundary condition on this boundary,

$$E_{x_k}(Q,t)\big|_{\tilde{S}} = f_k(Q,t). \tag{3.451}$$

It is clear that

$$f_k(Q,t) = 0, \quad \text{if} \quad 0 < t < T, \tag{3.452}$$

where T is specified by formula (3.423). For $t > T$, functions $f_k(Q,t)$ in the boundary conditions (3.451) can be updated by using the Kirchhoff formula (3.426) for each component E_{x_k}.

This completes the description of the "second-order" formulation. Numerical implementation of this formulation can be accomplished by using a standard explicit finite-difference scheme for the scalar wave equation. This scheme and its stability condition are well understood. It is clear that the presented second-order formulation may offer some advantages over the first-order formulation because computer storage of only three field components is required. In addition, Kirchhoff formula (3.426) has (computationally) simpler structure than the time-domain version (3.439) of the Stratton-Chu formula.

Up to this point, problems with a single scatterer have been discussed. However, the advocated approach may have appreciable advantages in the solution of problems with multiple scatterers. This may be especially the case when scatterers are quite remote from one another and the use of one artificial boundary that encloses all of them leads to a sizable truncated region. The approach discussed above enables one to introduce separate artificial boundaries for each scatterer and, in this way, substantially reduce the size of the overall truncated region.

To be specific, consider n (ideally conducting) scatterers with boundaries S_m, $(m = 1, 2, ..., n)$, subject to incident electromagnetic wave $\mathbf{E}^0$. An artificial boundary $\tilde{S}_m$ can be introduced around each scatterer. We intend to specify Dirichlet boundary conditions on these artificial boundaries. For instance, in the case of the second-order formulation, these boundary conditions are written as follows:

$$E_{x_k}(Q,t)|_{\tilde{S}_m} = f_k^{(m)}(Q,t). \tag{3.453}$$

It is clear that for each truncated region $\tilde{V}_m^-$ we have

$$f_k^{(m)}(Q,t) = 0, \quad \text{if} \quad 0 < t < T_m, \tag{3.454}$$

where

$$T_m = \min_{Q \in \tilde{S}_m, M \in S_m} \frac{r_{MQ}}{c}. \tag{3.455}$$

To update the boundary conditions (3.453) for times $t > T_m$, the Kirchhoff formula can be used with the only qualification that integration in this formula must be performed over all boundaries S_m. In other words, S in formula (3.426) is now the union of all S_m, $(m = 1, 2, ..., n)$. It is through this integration over all S_m that the electromagnetic coupling (interaction) between the scatterers is taken into account.

It is clear from the previous discussion that the Kirchhoff formula (3.426) enables one to specify Dirichlet-type boundary conditions on artificial boundaries that are exact in principle. This is in contrast with most ABCs that are approximate in nature. However, it must be kept in mind that the accuracy of the Kirchhoff formula can be compromised by numerical errors. These errors may be of various origins. One source of errors is the process of numerical integration itself. Another source of errors is that the electric field and its derivatives used in the Kirchhoff formula are found as a result of numerical solution of corresponding initial-boundary value problems. For this reason, their values used in integration are not exact but contain some discretization and round-off errors. Discretization errors may be especially appreciable when boundaries of scatterers are not smooth and contain corners and edges where the spatial derivatives of the scattered field have singularities. As a result of those singularities, discretization errors at the scatterer boundaries may be appreciably higher than outside those boundaries. This leads to the suggestion to use another surface S' between S and $\tilde{S}$ as the surface of integration in the Kirchhoff formula. It is apparent that there is considerable freedom in the choice of integration surfaces.

Numerical implementation of the Kirchhoff formula may be computationally expensive. However, this implementation may substantially benefit from the use of parallel computation. When an explicit finite-difference scheme is employed for the integration of the wave equation, the overall computational algorithm becomes intrinsically parallel and ideally suitable for implementation on parallel processors with single-instruction multiple-data architecture.

It is instructive to point out that the numerical implementation of the Kirchhoff formula for three-dimensional (3-D) problems will benefit from the absence of the "rear-front diffusion" ("tail") phenomenon that is present in two-dimensional (2-D) problems. Indeed, in 3-D problems any finite-in-time pulse of the incident field leads to a finite-in-time pulse of the scattered field at any observation point Q. In the case of 2-D problems, the scattered field will have at any observation point Q an infinite-in-time "tail" ("rear-front diffusion"). The rear-front diffusion phenomenon leads to the increase in computer storage requirements for the numerical implementation of the Kirchhoff formula.

The origin of the rear-front diffusion phenomenon can be clearly explained by the example of the field created by a 2-D "point" source. In three dimensions, this "point" source is an infinite line that can be subdivided into an infinite number of "dipole" sources. The electromagnetic field created by a 2-D point source can be represented as the superposition of the fields created by the above "dipoles." Now, it is clear that for any observation point Q and any instant of time t, there is always a sufficiently remote dipole whose field

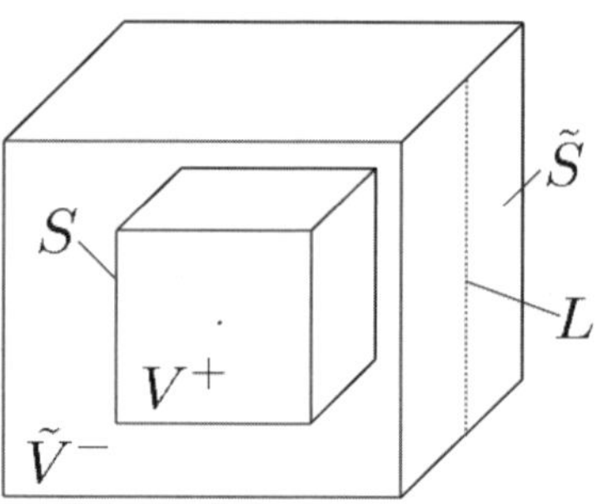

Figure 3.42

reaches the observation point Q at time t. These fields radiated by remote dipoles result in the "rear-front diffusion" phenomenon.

The presented technique has been numerically tested for scalar and electromagnetic wave propagation problems. The 3-D problems that admit simple analytical solutions have been used for numerical testing. In the case of scalar problems, the wave radiated by a point source has been used. This wave is given by

$$\varphi(Q,t) = \frac{F\left(t - \frac{r_{MQ}}{c}\right)}{r_{MQ}}, \qquad (3.456)$$

where $F(t)$ is the time "pulse-form" of the source, while M is the point of location of the source.

The calculations have been performed in the region exterior to a cube V^+ centered at the location of the source (see Figure 3.42). The boundary condition (3.420) on the surface of this cube has been specified by using formula (3.456). According to the uniqueness argument, the solution of initial-boundary value problem (3.417)-(3.420) coincides with (3.456) in the region V^- exterior to S. Artificial boundary $\tilde{S}$ has been chosen to be the surface of a larger cube (Figure 3.42). The initial-boundary value problem has been solved in the region $\tilde{V}^-$ between S and $\tilde{S}$ and the boundary condition (3.421) has been updated by using Kirchhoff formula (3.426). The comparison between this "updated" boundary condition and the exact boundary condition specified by formula (3.456) has been performed. The sampled results of this comparison are given in Figure 3.43 at the symmetry line L of a side face of $\tilde{S}$. These results were computed for the "pulse-form" $F(t) = A\sin(t/t_0)$ if $0 < t < t_0$ and $F(t) = 0$ if $t > t_0$.

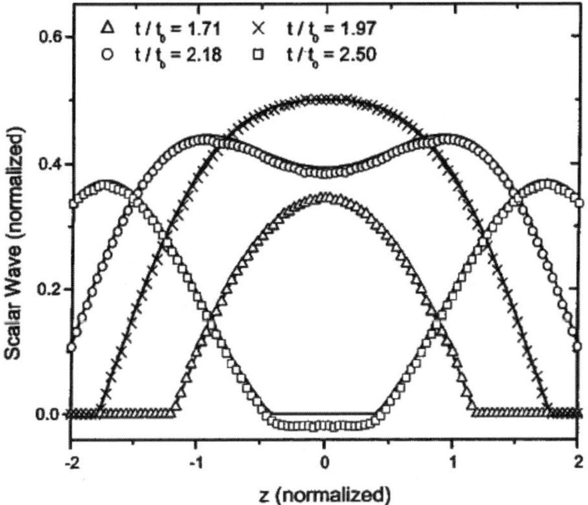

Figure 3.43

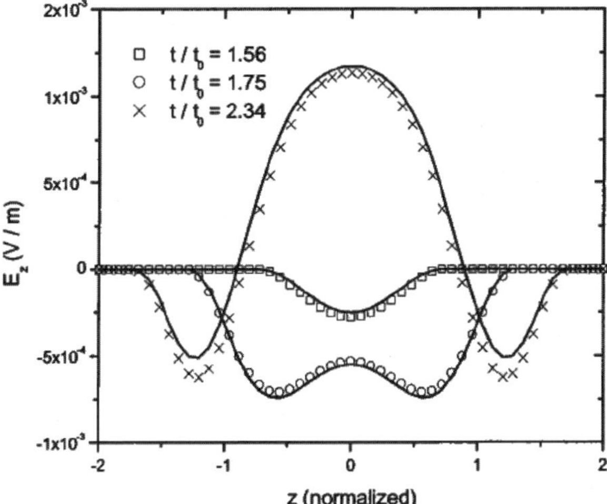

Figure 3.44

In the electromagnetic case, the point source has been replaced by a dipole and the problem was solved for the same geometry shown in Figure 3.42.

The electric field radiated by a dipole can be computed by using the following expression for the Hertz vector,

$$\mathbf{\Pi}(Q,t) = \mathbf{e}_z \frac{F\left(t - \frac{r_{MQ}}{c}\right)}{4\pi\epsilon_0 r_{MQ}},\tag{3.457}$$

and the formulas

$$E_r = \frac{2\cos\theta}{r}\frac{\partial \Pi_z}{\partial r}, \qquad E_\theta = \frac{\sin\theta}{r}\frac{\partial}{\partial r}\left(r\frac{\partial \Pi_z}{\partial r}\right).\tag{3.458}$$

The "second-order" formulation has been used in our computations and the boundary conditions on $\tilde{S}$ have been updated by using the Kirchhoff formula (3.426). The sampled results of the comparison between these updated boundary conditions and exact boundary conditions specified by formulas (3.457) and (3.458) are presented in Figure 3.44 for the E_z component. Figures 3.43 and 3.44 are indicative of the accuracy of Dirichlet-type absorbing boundary conditions generated by the Kirchhoff formula.

References

[1] P.M. Morse and H. Feshbach, Methods of Theoretical Physics, McGraw Hill, New York (1953).

[2] P.H. Moon and D.E. Spencer, Field Theory Handbook, Including Coordinate Systems, Differential Equations and Their Solutions, Springer Verlag, New York (1988).

[3] E.W. Hobson, The Theory of Spherical and Ellipsoidal Harmonics, The University Press, Cambridge (1931).

[4] W.E. Byerly, An Elementary Treatise on Fourier's Series and Spherical, Cylindrical and Ellipsoidal Harmonics, with Applications to Problems in Mathematical Physics, Ginn & Company, Boston (1893).

[5] I. Grigorenko and S. Haas, Physical Review Letters 97, 036806 (2006).

[6] R.D. Averitt, S.L. Westcott, and N.J. Halas, Journal of the Optical Society of America B 16, 1824 (1999).

[7] P. Schlegel, Mathematics of Computation 24, 665 (1970).

[8] R.H. Ritchie, Physical Review 106, 874 (1957).

[9] E.A. Stern and R.A. Ferrell, Physical Review 120, 130 (1960).

[10] I.D. Mayergoyz and Z. Zhang, Journal of Computational Electronics 4, 139 (2005).

[11] J.D. Jackson, Classical Electrodynamics, John Wiley, New York (1993).

[12] L. Greengard and V. Rokhlin, Journal of Computational Physics 73 (2), 325 (1987).

[13] N. Gumerov and R. Duraiswami, Fast Multipole Methods for the Helmholtz Equation in Three Dimensions, Elsevier, Amsterdam (2004).

[14] G.Y. Lyubarskii, Application of Group Theory in Physics, Pergamon Press, New York (1960).

[15] P.B. Johnson and R.W. Christy, Physical Review B 6, 4370 (1972).

[16] J. Aizpurua, P. Hanarp, D.S. Sutherland, M. Käll, G.W. Bryant, and F.J. García de Abajo, Physical Review Letters 90, 057401 (2003).

[17] L.J. Sherry, S.-H. Chang, B.J. Wiley, Y. Xia, G.C. Schatz, and R.P. Van Duyne, Nano Letters 5, 2034 (2005).

[18] K.-H. Su, Q.-H. Wei, X. Zhang, J.J. Mock, D.R. Smith, and S. Schultz, Nano Letters 3, 1087 (2003).

[19] E. Hao, G.C. Schatz, and J.T. Hupp, Journal of Fluorescence 14, 331 (2004).

[20] C.L. Nehl, N.K. Grady, G.P. Goodrich, F. Tam, N.J. Halas, and J.H. Hafner, Nano Letters 4, 2355 (2004).

[21] I.D. Mayergoyz, Iterative Techniques for the Analysis of Static Fields in Inhomogeneous, Anisotropic and Nonlinear Media, Naukova Dumka, Kyiv (1979).

[22] I. Mayergoyz, IEEE Transactions on Magnetics 38, 425 (2002).

[23] A. Friedman, Foundation of Modern Analysis, Dover, New York (1982).

[24] A. Bayliss and E. Turkel, Communications of Pure and Applied Mathematics **33**, 707 (1980).

[25] B. Engquist and A. Majda, Mathematics of Computation **31**, 629 (1977).

[26] J.P. Berenger, Journal of Computational Physics **114**, 185 (1994).

[27] I.D. Mayergoyz, Radiotechnics and Electronics USSR **1**, 8 (1977).

[28] I.D. Mayergoyz, P. Andrei, and B. Hakim, IEEE Transactions on Magnetics **38**, 327 (2002).

[29] J.A. Kong, Electromagnetic Wave Theory, Wiley, New York (1986).

[30] K.S. Yee, IEEE Transactions on Antennas and Propagation **14**, 302 (1966).

[31] I.D. Mayergoyz and Z. Zhang, IEEE Transactions on Magnetics **43**, 1689 (2007).

Chapter 4

Radiation Corrections, Excitation of Plasmon Modes and Selective Applications

4.1   Perturbation Technique

This final chapter deals with several topics. The first one is the radiation corrections to the electrostatic plasmon mode analysis presented in the previous chapters. These radiation corrections are mathematically treated as perturbations with respect to a small parameter which is defined as the ratio of particle dimensions (their diameters) to the free-space wavelengths. To discuss the radiation corrections we begin with homogeneous time-harmonic Maxwell equations and appropriate boundary conditions. For a nanoparticle (see Figure 4.1), these equations and boundary conditions are as follows:

$$\text{curl } \mathbf{E}^+ = -j\omega\mu_0\mathbf{H}^+ \quad \text{in } V^+, \tag{4.1}$$

$$\text{curl } \mathbf{H}^+ = j\omega\varepsilon\mathbf{E}^+ \quad \text{in } V^+, \tag{4.2}$$

$$\text{curl } \mathbf{E}^- = -j\omega\mu_0\mathbf{H}^- \quad \text{in } V^-, \tag{4.3}$$

$$\text{curl } \mathbf{H}^- = j\omega\varepsilon_0\mathbf{E}^- \quad \text{in } V^-, \tag{4.4}$$

$$\mathbf{n} \times \left(\mathbf{E}^+ - \mathbf{E}^-\right) = 0 \quad \text{on } S, \tag{4.5}$$

$$\mathbf{n} \times \left(\mathbf{H}^+ - \mathbf{H}^-\right) = 0 \quad \text{on } S. \tag{4.6}$$

Here, $\mathbf{E}^+$, $\mathbf{E}^-$ and $\mathbf{H}^+$, $\mathbf{H}^-$ are the phasors of electric and magnetic fields inside and outside the nanoparticle, respectively, $j = \sqrt{-1}$, while all other

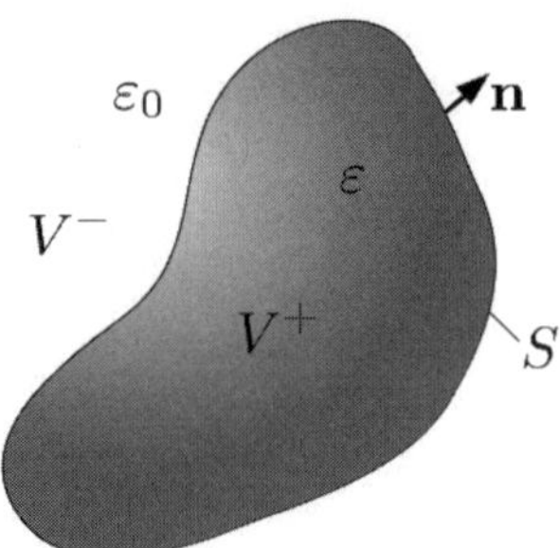

Figure 4.1

notations have their usual meaning. It is also tacitly assumed that appropriate outgoing radiation conditions are imposed on electromagnetic fields at infinity.

It is clear that equations

$$\operatorname{div} \mathbf{E}^{\pm} = 0 \quad \text{in } V^{\pm}, \tag{4.7}$$

$$\operatorname{div} \mathbf{H}^{\pm} = 0 \quad \text{in } V^{\pm} \tag{4.8}$$

and the boundary conditions

$$\mathbf{n} \cdot \left(\varepsilon \mathbf{E}^{+} - \varepsilon_0 \mathbf{E}^{-} \right) = 0 \quad \text{on } S, \tag{4.9}$$

$$\mathbf{n} \cdot \left(\mathbf{H}^{+} - \mathbf{H}^{-} \right) = 0 \quad \text{on } S, \tag{4.10}$$

are not independent but can be derived from equations (4.1)-(4.4) and boundary conditions (4.5) and (4.6), respectively.

We are interested in such negative values of ε for which source-free electromagnetic fields may exist provided that free-space wavelength is appreciably larger than particle dimensions. To find such permittivities, we shall write Maxwell equations (4.1)-(4.4) and the boundary conditions (4.5)-(4.6) in terms of the vectors

$$\mathbf{e}^{\pm} = \sqrt{\varepsilon_0}\, \mathbf{E}^{\pm}, \tag{4.11}$$

$$\mathbf{h}^{\pm} = \sqrt{\mu_0}\, \mathbf{H}^{\pm}, \tag{4.12}$$

and the spatial coordinates scaled by the diameter d of the nanoparticle

$$\tilde{x} = \frac{x}{d}, \quad \tilde{y} = \frac{y}{d}, \quad \tilde{z} = \frac{z}{d}. \tag{4.13}$$

This leads to the following boundary value problem with a small parameter β:

$$\operatorname{curl} \mathbf{e}^+ = -j\beta\mathbf{h}^+ \quad \text{in } \tilde{V}^+, \tag{4.14}$$

$$\operatorname{curl} \mathbf{h}^+ = j\frac{\varepsilon}{\varepsilon_0}\beta\mathbf{e}^+ \quad \text{in } \tilde{V}^+, \tag{4.15}$$

$$\operatorname{curl} \mathbf{e}^- = -j\beta\mathbf{h}^- \quad \text{in } \tilde{V}^-, \tag{4.16}$$

$$\operatorname{curl} \mathbf{h}^- = j\beta\mathbf{e}^- \quad \text{in } \tilde{V}^-, \tag{4.17}$$

$$\tilde{\mathbf{n}} \times \left(\mathbf{e}^+ - \mathbf{e}^-\right) = 0 \quad \text{on } \tilde{S}, \tag{4.18}$$

$$\tilde{\mathbf{n}} \times \left(\mathbf{h}^+ - \mathbf{h}^-\right) = 0 \quad \text{on } \tilde{S}, \tag{4.19}$$

where

$$\beta = \omega\sqrt{\mu_0\epsilon_0}d, \tag{4.20}$$

and $\tilde{V}^+$, $\tilde{V}^-$ and $\tilde{S}$ are scaled versions of V^+, V^- and S, respectively. It is also clear that the equations (4.7)-(4.8) and the boundary conditions (4.9)-(4.10) can be written in the form

$$\operatorname{div} \mathbf{e}^\pm = 0 \quad \text{in } \tilde{V}^\pm, \tag{4.21}$$

$$\operatorname{div} \mathbf{h}^\pm = 0 \quad \text{in } \tilde{V}^\pm, \tag{4.22}$$

$$\tilde{\mathbf{n}} \cdot \left(\frac{\varepsilon}{\varepsilon_0}\mathbf{e}^+ - \mathbf{e}^-\right) = 0 \quad \text{on } \tilde{S}, \tag{4.23}$$

$$\tilde{\mathbf{n}} \cdot \left(\mathbf{h}^+ - \mathbf{h}^-\right) = 0 \quad \text{on } \tilde{S}. \tag{4.24}$$

The last equations and boundary conditions formally follow from the boundary value problem (4.14)-(4.19).

In the case when the free-space wavelength is large in comparison with the nanoparticle dimensions, β can be treated as a small parameter, and source-free solutions $\mathbf{e}_k^\pm$, $\mathbf{h}_k^\pm$ of the boundary value problem (4.14)-(4.19) and permittivities ε_k at which they occur can be expanded in terms of β:

$$\mathbf{e}_k^\pm = \mathbf{e}_{0k}^\pm + \beta\mathbf{e}_{1k}^\pm + \beta^2\mathbf{e}_{2k}^\pm + \cdots, \tag{4.25}$$

$$\mathbf{h}_k^\pm = \mathbf{h}_{0k}^\pm + \beta\mathbf{h}_{1k}^\pm + \beta^2\mathbf{h}_{2k}^\pm + \cdots, \tag{4.26}$$

$$\varepsilon_k = \varepsilon_k^{(0)} + \beta\varepsilon_k^{(1)} + \beta^2\varepsilon_k^{(2)} + \cdots. \tag{4.27}$$

By substituting formulas (4.25)-(4.27) into equations (4.14)-(4.17) and boundary conditions (4.18)-(4.19) as well as into equations (4.21)-(4.22) and the boundary conditions (4.23)-(4.24) and by equating terms of the same powers of β, the boundary value problems for $\mathbf{e}_k^\pm$ and $\mathbf{h}_k^\pm$ can be derived. For zero-order terms, these boundary value problems can be written as follows:

$$\operatorname{curl} \mathbf{e}_{0k}^+ = 0 \quad \text{in } \tilde{V}^+, \tag{4.28}$$

$$\operatorname{div} \mathbf{e}_{0k}^+ = 0 \quad \text{in } \tilde{V}^+, \tag{4.29}$$

$$\operatorname{curl} \mathbf{e}_{0k}^- = 0 \quad \text{in } \tilde{V}^-, \tag{4.30}$$

$$\operatorname{div} \mathbf{e}_{0k}^- = 0 \quad \text{in } \tilde{V}^-, \tag{4.31}$$

$$\tilde{\mathbf{n}} \times \mathbf{e}_{0k}^+ = \tilde{\mathbf{n}} \times \mathbf{e}_{0k}^- \quad \text{on } \tilde{S}, \tag{4.32}$$

$$\varepsilon_k^{(0)} \tilde{\mathbf{n}} \cdot \mathbf{e}_{0k}^+ = \varepsilon_0 \tilde{\mathbf{n}} \cdot \mathbf{e}_{0k}^- \quad \text{on } \tilde{S} \tag{4.33}$$

and

$$\operatorname{curl} \mathbf{h}_{0k}^+ = 0 \quad \text{in } \tilde{V}^+, \tag{4.34}$$

$$\operatorname{div} \mathbf{h}_{0k}^+ = 0 \quad \text{in } \tilde{V}^+, \tag{4.35}$$

$$\operatorname{curl} \mathbf{h}_{0k}^- = 0 \quad \text{in } \tilde{V}^-, \tag{4.36}$$

$$\operatorname{div} \mathbf{h}_{0k}^- = 0 \quad \text{in } \tilde{V}^-, \tag{4.37}$$

$$\tilde{\mathbf{n}} \times \mathbf{h}_{0k}^+ = \tilde{\mathbf{n}} \times \mathbf{h}_{0k}^- \quad \text{on } \tilde{S}, \tag{4.38}$$

$$\tilde{\mathbf{n}} \cdot \mathbf{h}_{0k}^+ = \tilde{\mathbf{n}} \cdot \mathbf{h}_{0k}^- \quad \text{on } \tilde{S}. \tag{4.39}$$

It is apparent that the boundary value problem (4.28)-(4.33) is mathematically identical to the boundary value problem (2.1)-(2.6), which is the starting point for the electrostatic analysis of plasmon modes in nanoparticles in Chapter 2. Thus, nonzero solutions of the boundary value problem (4.28)-(4.33) indeed exist only for discrete values

$$\varepsilon_k^{(0)} = \varepsilon_k, \quad (k = 1, 2, ...), \tag{4.40}$$

which can be found by solving the eigenvalue problem for the boundary integral equation (2.36)-(2.37). The corresponding plasmon modes

$$\mathbf{E}_k^\pm = \frac{\mathbf{e}_{0k}^\pm}{\sqrt{\varepsilon_0}}, \quad (k = 1, 2, ...), \tag{4.41}$$

can be computed by using formula (2.39).

As far as the boundary value problem (4.34)-(4.39) is concerned, it is easy to prove that it has only the zero solution. Indeed, by introducing scalar magnetic potential φ_{mk},

$$\mathbf{h}_{0k}^{\pm} = -\text{grad } \varphi_{mk}^{\pm}, \tag{4.42}$$

the boundary value problem (4.34)-(4.39) can be reduced to the following boundary value problem for φ_{mk}:

$$\nabla^2 \varphi_{mk}^{+} = 0 \quad \text{in } \tilde{V}^{+}, \tag{4.43}$$

$$\nabla^2 \varphi_{mk}^{-} = 0 \quad \text{in } \tilde{V}^{-}, \tag{4.44}$$

$$\varphi_{mk}^{+} = \varphi_{mk}^{-} \quad \text{on } \tilde{S}, \tag{4.45}$$

$$\frac{\partial \varphi_{mk}^{+}}{\partial \tilde{n}} = \frac{\partial \varphi_{mk}^{-}}{\partial \tilde{n}} \quad \text{on } \tilde{S}. \tag{4.46}$$

It is also tacitly assumed here (as well as in the previous boundary value problems) that the "zero condition" at infinity must be satisfied. Now, by using the same line of reasoning as in the derivation of formula (2.19), from equations (4.43)-(4.46) we derive

$$\int_{\tilde{V}^{+}} \left|\text{grad } \varphi_{mk}^{+}\right|^2 d\tilde{v} + \int_{\tilde{V}^{-}} \left|\text{grad } \varphi_{mk}^{-}\right|^2 d\tilde{v} = 0. \tag{4.47}$$

The last relation implies that

$$\text{grad } \varphi_{mk}^{\pm} = 0, \tag{4.48}$$

and, consequently,

$$\mathbf{h}_{0k}^{\pm} = 0. \tag{4.49}$$

By summarizing the presented discussion, it can be concluded that the main (zero-order) terms of asymptotic expansions (4.25)-(4.27) can be obtained by using the electrostatic theory of plasmon resonances discussed in the previous two chapters. This can be construed as the mathematical justification of the electrostatic approximation for plasmon resonances in nanoparticles.

It is easy to establish that the boundary value problems for higher-order terms in power expansions (4.25)-(4.27) can be written in the form

$$\text{curl } \mathbf{e}^+_{mk} = -j\mathbf{h}^+_{m-1,k} \quad \text{in } \tilde{V}^+, \tag{4.50}$$

$$\text{div } \mathbf{e}^+_{mk} = 0 \quad \text{in } \tilde{V}^+, \tag{4.51}$$

$$\text{curl } \mathbf{e}^-_{mk} = -j\mathbf{h}^-_{m-1,k} \quad \text{in } \tilde{V}^-, \tag{4.52}$$

$$\text{div } \mathbf{e}^-_{mk} = 0 \quad \text{in } \tilde{V}^-, \tag{4.53}$$

$$\tilde{\mathbf{n}} \times \left(\mathbf{e}^+_{mk} - \mathbf{e}^-_{mk} \right) = 0 \quad \text{on } \tilde{S}, \tag{4.54}$$

$$\tilde{\mathbf{n}} \cdot \left(\frac{\varepsilon_k^{(0)}}{\varepsilon_0} \mathbf{e}^+_{mk} - \mathbf{e}^-_{mk} \right) = -\sum_{\nu=0}^{m-1} \frac{\varepsilon_k^{(m-\nu)}}{\varepsilon_0} \tilde{\mathbf{n}} \cdot \mathbf{e}^+_{\nu k} \quad \text{on } \tilde{S}, \tag{4.55}$$

and

$$\text{curl } \mathbf{h}^+_{mk} = j\sum_{\nu=0}^{m-1} \frac{\varepsilon_k^{(m-1-\nu)}}{\varepsilon_0} \mathbf{e}^+_{\nu k} \quad \text{in } \tilde{V}^+, \tag{4.56}$$

$$\text{div } \mathbf{h}^+_{mk} = 0 \quad \text{in } \tilde{V}^+, \tag{4.57}$$

$$\text{curl } \mathbf{h}^-_{mk} = j\mathbf{e}^-_{m-1,k} \quad \text{in } \tilde{V}^-, \tag{4.58}$$

$$\text{div } \mathbf{h}^-_{mk} = 0 \quad \text{in } \tilde{V}^-, \tag{4.59}$$

$$\tilde{\mathbf{n}} \times \left(\mathbf{h}^+_{mk} - \mathbf{h}^-_{mk} \right) = 0 \quad \text{on } \tilde{S}, \tag{4.60}$$

$$\tilde{\mathbf{n}} \cdot \left(\mathbf{h}^+_{mk} - \mathbf{h}^-_{mk} \right) = 0 \quad \text{on } \tilde{S}. \tag{4.61}$$

It is apparent from formulas (4.50)-(4.61) that, in principle, radiation corrections of all orders can be sequentially computed. This is illustrated in the next section for the first- and second-order corrections [1].

4.2   First- and Second-Order Radiation Corrections

From equations (4.56)-(4.61), we derive the following boundary value problem for the first-order radiation correction for plasmonic magnetic field:

$$\text{curl } \mathbf{h}^+_{1k} = j\frac{\varepsilon_k^{(0)}}{\varepsilon_0} \mathbf{e}^+_{0k}, \tag{4.62}$$

$$\text{div } \mathbf{h}^+_{1k} = 0, \tag{4.63}$$

$$\operatorname{curl} \mathbf{h}_{1k}^- = j\mathbf{e}_{0k}^-, \tag{4.64}$$

$$\operatorname{div} \mathbf{h}_{1k}^- = 0, \tag{4.65}$$

$$\tilde{\mathbf{n}} \times \left(\mathbf{h}_{1k}^+ - \mathbf{h}_{1k}^-\right) = 0, \tag{4.66}$$

$$\tilde{\mathbf{n}} \cdot \left(\mathbf{h}_{1k}^+ - \mathbf{h}_{1k}^-\right) = 0, \tag{4.67}$$

$$\mathbf{h}_{1k}^-(\infty) = 0. \tag{4.68}$$

It is clear from the above equations that $\mathbf{h}_{1k}$ can be interpreted as stationary magnetic field created by volume currents

$$\mathbf{J}_k^+ = j\frac{\varepsilon_k^{(0)}}{\varepsilon_0}\mathbf{e}_{0k}^+ \tag{4.69}$$

and

$$\mathbf{J}_k^- = j\mathbf{e}_{0k}^- \tag{4.70}$$

distributed in $\tilde{V}^+$ and $\tilde{V}^-$, respectively. Consequently, $\mathbf{h}_{1k}$ can be computed by using the Biot-Savart integral formula

$$\mathbf{h}_{1k}(Q) = \frac{j\varepsilon_k^{(0)}}{4\pi\varepsilon_0}\int_{\tilde{V}+}\frac{\mathbf{e}_{0k}^+(M) \times \tilde{\mathbf{r}}_{MQ}}{\tilde{r}_{MQ}^3}d\tilde{v}_M + \frac{j}{4\pi}\int_{\tilde{V}-}\frac{\mathbf{e}_{0k}^-(M) \times \tilde{\mathbf{r}}_{MQ}}{\tilde{r}_{MQ}^3}d\tilde{v}_M. \tag{4.71}$$

It turns out that the last formula can be appreciably simplified and reduced to an integral over the boundary S of the nanoparticle. This simplification is based on equation

$$\operatorname{curl} \mathbf{e}_{0k}^{\pm} = 0 \tag{4.72}$$

(see formulas (4.28) and (4.30)). Indeed, by using the last formula, equation (4.71) can be transformed as follows:

$$\mathbf{h}_{1k}(Q) = -\frac{j\varepsilon_k^{(0)}}{4\pi\varepsilon_0}\int_{\tilde{V}+}\operatorname{curl}_M\left(\frac{\mathbf{e}_{0k}^+(M)}{\tilde{r}_{MQ}}\right)d\tilde{v}_M - \frac{j}{4\pi}\int_{\tilde{V}-}\operatorname{curl}_M\left(\frac{\mathbf{e}_{0k}^-(M)}{\tilde{r}_{MQ}}\right)d\tilde{v}_M. \tag{4.73}$$

Now, by using the well-known integral identity from vector calculus

$$\int_V \operatorname{curl} \mathbf{a}\, dv = \oint_S \left(\mathbf{n} \times \mathbf{a}\right) ds \tag{4.74}$$

and by taking

$$\mathbf{a} = \frac{\mathbf{e}_{0k}(M)}{\tilde{r}_{MQ}}, \tag{4.75}$$

the volume integrals in formula (4.73) can be replaced by the appropriate surface integrals. This eventually leads to the following result:

$$\mathbf{h}_{1k}(Q) = -\frac{j\left(\varepsilon_k^{(0)} - \varepsilon_0\right)}{4\pi\varepsilon_0} \oint_{\tilde{S}} \frac{\tilde{\mathbf{n}}_M \times \mathbf{e}_{0k}(M)}{\tilde{r}_{MQ}} d\tilde{s}_M. \qquad (4.76)$$

It is apparent that formula (4.76) is appreciably simpler than formula (4.71) as far as calculations are concerned. It can also be observed that, from the mathematical point of view, formula (4.76) is identical to the formula for the vector magnetic potential of the magnetic field created by surface currents

$$\mathbf{i}_k(M) = -\frac{j\left(\varepsilon_k^{(0)} - \varepsilon_0\right)}{\varepsilon_0\mu_0}\tilde{\mathbf{n}}_M \times \mathbf{e}_{0k}(M) \qquad (4.77)$$

distributed over $\tilde{S}$. This observation suggests another derivation of formula (4.76) that can be accomplished without invoking the integral identity (4.74) and dealing with the singularities of $\mathbf{a}$ defined by equation (4.75). Indeed, by applying the curl-operation to both sides of equations (4.62) and (4.64), respectively, and by taking into account formula (4.72), the boundary value problem (4.62)-(4.68) can be written in the following equivalent form:

$$\operatorname{curl}\operatorname{curl}\mathbf{h}_{1k}^+ = 0, \qquad (4.78)$$
$$\operatorname{div}\mathbf{h}_{1k}^+ = 0, \qquad (4.79)$$
$$\operatorname{curl}\operatorname{curl}\mathbf{h}_{1k}^- = 0, \qquad (4.80)$$
$$\operatorname{div}\mathbf{h}_{1k}^- = 0, \qquad (4.81)$$
$$\tilde{\mathbf{n}} \times \left(\mathbf{h}_{1k}^- - \mathbf{h}_{1k}^+\right) = 0, \qquad (4.82)$$
$$\tilde{\mathbf{n}} \times \left(\operatorname{curl}\mathbf{h}_{1k}^- - \operatorname{curl}\mathbf{h}_{1k}^+\right) = \mu_0\mathbf{i}_k, \qquad (4.83)$$
$$\mathbf{h}_{1k}^-(\infty) = 0. \qquad (4.84)$$

It is apparent now that the boundary value problem (4.78)-(4.84) is mathematically identical to the boundary value problem for the magnetic vector potential of surface currents $\mathbf{i}_k$. This immediately leads to formula (4.76).

Next, we proceed to the analysis of the first-order radiation corrections $\mathbf{e}_{1k}$ and $\varepsilon_k^{(1)}$. From equations (4.50)-(4.55), we derive the following boundary value problem for $\mathbf{e}_{1k}$:

$$\operatorname{curl}\mathbf{e}_{1k}^{-} = 0, \tag{4.85}$$

$$\operatorname{div}\mathbf{e}_{1k}^{-} = 0, \tag{4.86}$$

$$\operatorname{curl}\mathbf{e}_{1k}^{+} = 0, \tag{4.87}$$

$$\operatorname{div}\mathbf{e}_{1k}^{+} = 0, \tag{4.88}$$

$$\tilde{\mathbf{n}}\times\left(\mathbf{e}_{1k}^{-} - \mathbf{e}_{1k}^{+}\right) = 0, \tag{4.89}$$

$$\tilde{\mathbf{n}}\cdot\left(\mathbf{e}_{1k}^{-} - \frac{\varepsilon_k^{(0)}}{\varepsilon_0}\mathbf{e}_{1k}^{+}\right) = \frac{\varepsilon_k^{(1)}}{\varepsilon_0}\tilde{\mathbf{n}}\cdot\mathbf{e}_{0k}^{+}, \tag{4.90}$$

$$\mathbf{e}_{1k}^{-}(\infty) = 0. \tag{4.91}$$

According to equations (4.85) and (4.87), the first-order electric field radiation correction $\mathbf{e}_{1k}$ can be described by electric potential φ_{1k}:

$$\mathbf{e}_{1k} = -\operatorname{grad}\varphi_{1k}. \tag{4.92}$$

It is apparent that this potential is the solution of the following boundary value problem:

$$\nabla^2\varphi_{1k}^{-} = 0, \tag{4.93}$$

$$\nabla^2\varphi_{1k}^{+} = 0, \tag{4.94}$$

$$\varphi_{1k}^{+} = \varphi_{1k}^{-}, \tag{4.95}$$

$$\frac{\partial\varphi_{1k}^{-}}{\partial\tilde{n}} - \frac{\varepsilon_k^{(0)}}{\varepsilon_0}\frac{\partial\varphi_{1k}^{+}}{\partial\tilde{n}} = -\frac{\varepsilon_k^{(1)}}{\varepsilon_0}\tilde{\mathbf{n}}\cdot\mathbf{e}_{0k}^{+}, \tag{4.96}$$

$$\varphi_{1k}^{-}(\infty) = 0. \tag{4.97}$$

We shall look for the solution of the boundary value problem (4.93)-(4.97) in the form of the "single layer" potential

$$\varphi_{1k}(Q) = \frac{1}{4\pi\varepsilon_0}\oint_{\tilde{S}}\frac{\sigma_{1k}(M)}{\tilde{r}_{MQ}}d\tilde{s}_M. \tag{4.98}$$

This potential satisfies equations (4.93) and (4.94) as well as the boundary condition (4.95) and the condition (4.97) at infinity. By using formula (2.32), we find that the boundary condition (4.96) is satisfied if $\sigma_{1k}(M)$ is the solution of the following inhomogeneous boundary integral equation:

$$\sigma_{1k}(Q) - \frac{\lambda_k}{2\pi}\oint_{\tilde{S}}\sigma_{1k}(M)\frac{\tilde{\mathbf{r}}_{MQ}\cdot\tilde{\mathbf{n}}_Q}{\tilde{r}_{MQ}^3}d\tilde{s}_M = \varepsilon_k^{(1)}\frac{2\varepsilon_0}{\varepsilon_k^{(0)} + \varepsilon_0}\tilde{\mathbf{n}}_Q\cdot\mathbf{e}_{0k}^{+}(Q), \tag{4.99}$$

where

$$\lambda_k = \frac{\varepsilon_k^{(0)} - \varepsilon_0}{\varepsilon_k^{(0)} + \varepsilon_0}. \tag{4.100}$$

It is clear that λ_k is one of the eigenvalues of the integral operator in equation (4.99). For this reason, a solution to the integral equation (4.99) exists only under the condition that the right-hand side of this equation is orthogonal on $\tilde{S}$ to a nonzero solution τ_{0k} of the homogeneous integral equation adjoint to equation (4.99). This is the so-called "normal solvability" condition of the Fredholm theory briefly discussed at the end of the first section of Chapter 2. It follows from equation (4.41) that this orthogonality condition is satisfied for nonzero $\varepsilon_k^{(1)}$ only if

$$\oint_S \tau_{0k}(Q)\mathbf{n}_Q \cdot \mathbf{E}_{0k}^+(Q)ds_Q = 0. \tag{4.101}$$

However, this is not the case. Indeed, from the relation

$$\mathbf{n}_Q \cdot \mathbf{E}_{0k}^+(Q) = -\frac{\partial \varphi_{0k}^+}{\partial n}(Q) \tag{4.102}$$

and formula (see (2.147))

$$\tau_{0k}(Q) = \left(\varepsilon_k^{(0)} - \varepsilon_0\right) \varphi_{0k}^+(Q), \tag{4.103}$$

we derive

$$\oint_S \tau_{0k}(Q)\mathbf{n}_Q \cdot \mathbf{E}_{0k}^+(Q)ds_Q = \left(\varepsilon_0 - \varepsilon_k^{(0)}\right) \oint_S \varphi_{0k}^+(Q)\frac{\partial \varphi_{0k}^+}{\partial n}(Q)ds_Q$$

$$= \left(\varepsilon_0 - \varepsilon_k^{(0)}\right) \int_{V^+} \left|\mathrm{grad}\, \varphi_{0k}^+\right|^2 dv \neq 0. \tag{4.104}$$

This means that the above normal solvability condition is satisfied only if

$$\boxed{\varepsilon_k^{(1)} = 0.} \tag{4.105}$$

As a consequence, the boundary integral equation (4.99) is reduced to the homogeneous boundary integral equation identical to equation (2.36). This implies that, up to a scale, $\sigma_{0k}(M)$ and $\sigma_{1k}(M)$ as well as $\mathbf{e}_{0k}$ and $\mathbf{e}_{1k}$ are identical. This justifies the equality

$$\boxed{\mathbf{e}_{1k}^\pm = 0.} \tag{4.106}$$

Thus, we have arrived at the conclusion that for *any shape* of nanoparticle the first-order radiation corrections for the resonance value of dielectric permittivity and plasmonic electric fields are equal to zero.

Now, we proceed to the discussion of second-order radiation corrections. From equations (4.56)-(4.61) as well as formulas (4.105) and (4.106), we derive the following boundary value problem for the second-order radiation corrections for plasmonic magnetic field:

$$\text{curl } \mathbf{h}_{2k}^{+} = 0, \tag{4.107}$$

$$\text{div } \mathbf{h}_{2k}^{+} = 0, \tag{4.108}$$

$$\text{curl } \mathbf{h}_{2k}^{-} = 0, \tag{4.109}$$

$$\text{div } \mathbf{h}_{2k}^{-} = 0, \tag{4.110}$$

$$\tilde{\mathbf{n}} \times \mathbf{h}_{2k}^{+} = \tilde{\mathbf{n}} \times \mathbf{h}_{2k}^{-}, \tag{4.111}$$

$$\tilde{\mathbf{n}} \cdot \mathbf{h}_{2k}^{+} = \tilde{\mathbf{n}} \cdot \mathbf{h}_{2k}^{-}, \tag{4.112}$$

$$\mathbf{h}_{2k}^{-}(\infty) = 0. \tag{4.113}$$

It is evident that the boundary value problem (4.107)-(4.113) is mathematically identical to the boundary value problem (4.34)-(4.39). For this reason, by literally repeating the same line of reasoning as in the first section of this chapter, we find that the second-order radiation corrections for the magnetic field are equal to zero,

$$\boxed{\mathbf{h}_{2k}^{\pm} = 0.} \tag{4.114}$$

Next, we turn to the discussion of second-order radiation corrections $\mathbf{e}_{2k}$ and $\varepsilon_{k}^{(2)}$. From equations (4.50)-(4.55) as well as equations (4.105) and (4.106), we derive the following boundary value problem for $\mathbf{e}_{2k}$:

$$\text{curl } \mathbf{e}_{2k}^{+} = -j\mathbf{h}_{1k}^{+}, \tag{4.115}$$

$$\text{div } \mathbf{e}_{2k}^{+} = 0, \tag{4.116}$$

$$\text{curl } \mathbf{e}_{2k}^{-} = -j\mathbf{h}_{1k}^{-}, \tag{4.117}$$

$$\text{div } \mathbf{e}_{2k}^{-} = 0, \tag{4.118}$$

$$\tilde{\mathbf{n}} \times \left(\mathbf{e}_{2k}^{-} - \mathbf{e}_{2k}^{+} \right) = 0, \tag{4.119}$$

$$\tilde{\mathbf{n}} \cdot \left(\mathbf{e}_{2k}^{-} - \frac{\varepsilon_{k}^{(0)}}{\varepsilon_{0}} \mathbf{e}_{2k}^{+} \right) = \frac{\varepsilon_{k}^{(2)}}{\varepsilon_{0}} \tilde{\mathbf{n}} \cdot \mathbf{e}_{0k}^{+}, \tag{4.120}$$

$$\mathbf{e}_{2k}^{-}(\infty) = 0. \tag{4.121}$$

We shall split $\mathbf{e}_{2k}$ into two components

$$\mathbf{e}_{2k} = \mathbf{e}'_{2k} + \mathbf{e}''_{2k} \tag{4.122}$$

that satisfy the following boundary value problems, respectively:

$$\operatorname{curl} \mathbf{e}'^{+}_{2k} = -j\mathbf{h}^{+}_{1k}, \tag{4.123}$$
$$\operatorname{div} \mathbf{e}'^{+}_{2k} = 0, \tag{4.124}$$
$$\operatorname{curl} \mathbf{e}'^{-}_{2k} = -j\mathbf{h}^{-}_{1k}, \tag{4.125}$$
$$\operatorname{div} \mathbf{e}'^{-}_{2k} = 0, \tag{4.126}$$
$$\tilde{\mathbf{n}} \times \left(\mathbf{e}'^{+}_{2k} - \mathbf{e}'^{-}_{2k}\right) = 0, \tag{4.127}$$
$$\tilde{\mathbf{n}} \cdot \left(\mathbf{e}'^{+}_{2k} - \mathbf{e}'^{-}_{2k}\right) = 0, \tag{4.128}$$
$$\mathbf{e}'^{-}_{2k}(\infty) = 0, \tag{4.129}$$

and

$$\operatorname{curl} \mathbf{e}''^{-}_{2k} = 0, \tag{4.130}$$
$$\operatorname{div} \mathbf{e}''^{-}_{2k} = 0, \tag{4.131}$$
$$\operatorname{curl} \mathbf{e}''^{+}_{2k} = 0, \tag{4.132}$$
$$\operatorname{div} \mathbf{e}''^{+}_{2k} = 0, \tag{4.133}$$
$$\tilde{\mathbf{n}} \times \left(\mathbf{e}''^{-}_{2k} - \mathbf{e}''^{+}_{2k}\right) = 0, \tag{4.134}$$
$$\tilde{\mathbf{n}} \cdot \left(\mathbf{e}''^{-}_{2k} - \frac{\varepsilon_k^{(0)}}{\varepsilon_0}\mathbf{e}''^{+}_{2k}\right) = \tilde{\mathbf{n}} \cdot \left[\frac{\varepsilon_k^{(2)}}{\varepsilon_0}\mathbf{e}^{+}_{0k} + \left(\frac{\varepsilon_k^{(0)}}{\varepsilon_0} - 1\right)\mathbf{e}'^{+}_{2k}\right], \tag{4.135}$$
$$\mathbf{e}''^{-}_{2k}(\infty) = 0. \tag{4.136}$$

It is clear that the mathematical structure of the boundary value problem (4.123)-(4.129) is identical to the mathematical structure of the boundary value problem (4.62)-(4.68). For this reason, the solution of the boundary value problem (4.123)-(4.129) can be written by using the Biot-Savart-type integral formula

$$\mathbf{e}'_{2k}(P) = -\frac{j}{4\pi} \int_{\mathbb{R}^3} \frac{\mathbf{h}_{1k}(Q) \times \tilde{\mathbf{r}}_{QP}}{\tilde{r}^3_{QP}} d\tilde{v}_Q, \tag{4.137}$$

where the integration over $\mathbb{R}^3$ is the integration over the entire three-dimensional space. By substituting formula (4.76) into the last formula and by changing the order of integration, we arrive at

$$\mathbf{e}'_{2k}(P) = \frac{\varepsilon_0 - \varepsilon_k^{(0)}}{16\pi^2\varepsilon_0} \oint_{\tilde{S}} [\tilde{\mathbf{n}}_M \times \mathbf{e}_{0k}(M)] \times \left(\int_{\mathbb{R}^3} \frac{\tilde{\mathbf{r}}_{QP}}{\tilde{r}_{QP}^3} \cdot \frac{1}{\tilde{r}_{QM}} d\tilde{v}_Q \right) d\tilde{s}_M. \quad (4.138)$$

The last formula can be substantially simplified by using the following reasoning. Consider the vector function

$$\mathbf{b}(P, M) = \frac{1}{4\pi\varepsilon_0} \int_{\mathbb{R}^3} \frac{\tilde{\mathbf{r}}_{QP}}{\tilde{r}_{QP}^3} \cdot \frac{1}{\tilde{r}_{QM}} d\tilde{v}_Q. \quad (4.139)$$

It is clear from the last formula that $\mathbf{b}(P, M)$ can be viewed as an electric field at point P created by electric charges of volume density $1/\tilde{r}_{QM}$. If point M is chosen as the coordinate origin, then this electric field is spherically symmetric and $\mathbf{b}(P, M)$ can be computed by using the Gauss law. The final result of these simple computations is the formula

$$\mathbf{b}(P, M) = \frac{\tilde{\mathbf{r}}_{MP}}{2\varepsilon_0\tilde{r}_{MP}}. \quad (4.140)$$

Now, it is apparent that by using relations (4.139) and (4.140), formula (4.138) can be simplified as follows:

$$\mathbf{e}'_{2k}(P) = \frac{\varepsilon_0 - \varepsilon_k^{(0)}}{8\pi\varepsilon_0} \oint_{\tilde{S}} \frac{[\tilde{\mathbf{n}}_M \times \mathbf{e}_{0k}(M)] \times \tilde{\mathbf{r}}_{MP}}{\tilde{r}_{MP}} d\tilde{s}_M. \quad (4.141)$$

Now, we proceed to the solution of boundary value problem (4.130)-(4.136) for $\mathbf{e}''_{2k}$. The mathematical structure of this boundary value problem is very similar to the structure of boundary value problem (4.85)-(4.91) for $\mathbf{e}_{1k}$. For this reason, we shall use the scalar potential φ_{2k},

$$\mathbf{e}''_{2k} = -\text{grad } \varphi_{2k}, \quad (4.142)$$

and look for this potential in the form

$$\varphi_{2k}(Q) = \frac{1}{4\pi\varepsilon_0} \oint_{\tilde{S}} \frac{\sigma_{2k}(M)}{\tilde{r}_{QM}} d\tilde{s}_M. \quad (4.143)$$

Then, by using formula (2.32), it can be shown as before that formulas (4.142)-(4.143) provide a solution to the boundary value problem (4.130)-(4.136) if $\sigma_{2k}(M)$ is a solution of the following boundary integral equation:

$$\sigma_{2k}(Q) - \frac{\lambda_k}{2\pi} \oint_{\tilde{S}} \sigma_{2k}(M)\frac{\tilde{\mathbf{r}}_{MQ} \cdot \tilde{\mathbf{n}}_Q}{\tilde{r}_{MQ}^3} d\tilde{s}_M = f(Q), \quad (4.144)$$

where

$$f(Q) = \frac{2\varepsilon_0}{\varepsilon_k^{(0)} + \varepsilon_0}\,\tilde{\mathbf{n}}_Q \cdot \left[\varepsilon_k^{(2)}\mathbf{e}_{0k}(Q) + \left(\varepsilon_k^{(0)} - \varepsilon_0\right)\mathbf{e}_{2k}'(Q)\right] \qquad (4.145)$$

and λ_k is given by formula (4.100).

Since λ_k is an eigenvalue of the integral operator in equation (4.144), a solution to this integral equation exists only under the condition that the right-hand side $f(Q)$ of this equation is orthogonal on $\tilde{S}$ to a nonzero solution $\tau_{0k}(Q)$ of the adjoint homogeneous integral equation, i.e.,

$$\oint_{\tilde{S}} f(Q)\tau_{0k}(Q)d\tilde{s}_Q = 0. \qquad (4.146)$$

By substituting formula (4.145) into orthogonality condition (4.146), we find that this orthogonality condition is satisfied if

$$\boxed{\varepsilon_k^{(2)} = \frac{\left(\varepsilon_0 - \varepsilon_k^{(0)}\right)\oint_{\tilde{S}} \tau_{0k}(Q)\tilde{\mathbf{n}}_Q \cdot \mathbf{e}_{2k}'(Q)d\tilde{s}_Q}{\oint_{\tilde{S}} \tau_{0k}(Q)\tilde{\mathbf{n}}_Q \cdot \mathbf{e}_{0k}(Q)d\tilde{s}_Q}.} \qquad (4.147)$$

Thus, in order to compute the second-order radiation correction $\varepsilon_k^{(2)}$ for resonance values of dielectric permittivity, we first solve the boundary integral equations for eigenfunctions σ_{0k} and τ_{0k} (this is the electrostatic approximation discussed in Chapter 2). Then, by using σ_{0k} we compute $\mathbf{e}_{0k}(Q)$ and $\mathbf{e}_{2k}'(Q)$ (see formula (4.141)). Finally, by using the last formula (4.147), $\varepsilon_k^{(2)}$ is calculated. As soon as $\varepsilon_k^{(2)}$ is found, the resonance values of dielectric permittivity are computed as follows:

$$\boxed{\varepsilon_k = \varepsilon_k^{(0)} + \beta^2\varepsilon_k^{(2)},} \qquad (4.148)$$

and the corresponding resonance frequencies of plasmon modes are determined by using the appropriate dispersion relation

$$\boxed{\varepsilon'(\omega_k) = \varepsilon_k = \varepsilon_k^{(0)} + \beta^2\varepsilon_k^{(2)}.} \qquad (4.149)$$

It is interesting to mention that the obtained results are consistent with the known analytical result for spherical nanoparticles that follows from the Mie theory [2]. Indeed, according to this theory, the following formula is valid for the resonance value ε_1 of dielectric permittivity for the first three (spatially uniform) plasmon modes:

$$\varepsilon_1 = -\left(2 + \frac{3}{5}\beta^2\right)\varepsilon_0 = -\left(2 + \frac{12}{5}\omega^2\mu_0\varepsilon_0 a^2\right)\varepsilon_0, \qquad (4.150)$$

where a is the radius of the spherical nanoparticle.

It is worthwhile to stress the general nature of our discussion which reveals that the first-order radiation corrections for resonance values of permittivity ε_k are equal to zero for any (not only spherical) shape of nanoparticles. Our discussion also results in the general formula (4.147) for the second-order radiation corrections for ε_k which are valid for any shape of nanoparticles and which are expressed only in terms of quantities obtained from the electrostatic analysis of plasmon modes.

It can be shown analytically that in the case of the spherical nanoparticle formula (4.150) can be derived from formulas (4.141) and (4.147)-(4.148). We shall omit these somewhat lengthy derivations and, instead, compare formula (4.150) with numerical computations performed by using formulas (4.141), (4.147) and (4.148). This comparison is presented in Figures 4.2 and 4.3, which reveals the coincidence of analytical (Mie theory) and computational results. These figures also reveal that the resonance wavelengths (resonance frequencies) are less sensitive to nanoparticle dimension variations than the resonance values of their dielectric permittivity, which is due to the nature of the permittivity dispersion relation.

As further illustrative examples, Tables 4.1, 4.2 and 4.3 present the results of computations of second-order corrections for resonance wavelengths of dipole plasmon modes in gold nanorings and gold nanocubes in water and on glass substrates, respectively. These second-order corrections have been computed by using formulas (4.141), (4.147) and (4.149). It is evident from these tables that the second-order corrections result in better agreement with experimental data than the results based on purely electrostatic analysis.

In the case of nanocubes in water, we have also compared the computational results based on the developed theory of second-order corrections with experimental results for three different dipole plasmon modes and different cube dimensions. The charge distributions corresponding to these different dipole plasmon modes are shown in Figure 4.4a, 4.4b and 4.4c, respectively. Figure 4.5 presents three continuous curves computed for resonance wavelengths of these three plasmon modes as well as their comparison with values

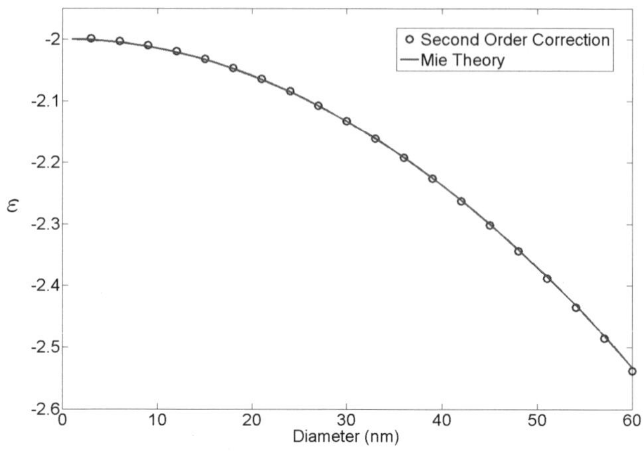

Figure 4.2

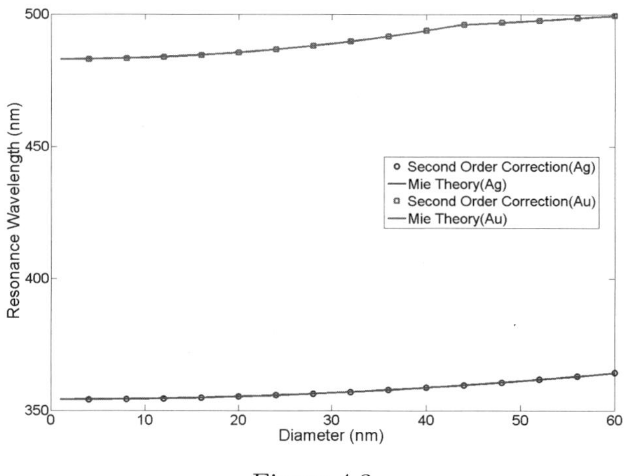

Figure 4.3

found from measurements of extinction spectra of nanocube suspensions. These measurements were performed by the group of Professor O. Rabin at the University of Maryland.

Finally, in the conclusion of this section, we outline how the analytical solution of integral equation (4.144) for $\sigma_{2k}(Q)$ (and, subsequently, for $\mathbf{e}''_{2k}(Q)$)

Table 4.1

Resonance Wavelength (nm)	Ring 1	Ring 2	Ring 3
Experimental Results [3]	1000	1180	1350
Computational Results (Zero-Order Solution)	940	1102	1159
Computational Results (Second-Order Solution)	987	1156	1214

Table 4.2

	Peak 1 (nm)	Peak 2 (nm)
Experimental Data [4]	432	500
Zero-Order Results	405	454
Second-Order Results	420	478

Table 4.3

	Peak 1 (nm)	Peak 2 (nm)
Experimental Data [4]	395	457
Zero-Order Results	383	421
Second-Order Results	410	443

can be obtained in terms of $\sigma_{0n}(Q)$ and $\tau_{0n}(Q)$. To this end, we shall look for $\sigma_{2k}(Q)$ in the form

$$\sigma_{2k}(Q) = \sum_{n \neq k}^{\infty} a_{kn} \sigma_{0n}(Q). \tag{4.151}$$

By substituting formula (4.151) into integral equation (4.144), we find

$$\sum_{n \neq k}^{\infty} a_{kn} \sigma_{0n}(Q) - \frac{\lambda_k}{2\pi} \sum_{n \neq k}^{\infty} a_{kn} \oint_{\tilde{S}} \sigma_{0n}(M) \frac{\tilde{\mathbf{r}}_{MQ} \cdot \tilde{\mathbf{n}}_Q}{\tilde{r}_{MQ}^3} d\tilde{s}_M = f(Q). \tag{4.152}$$

By taking into account that

$$\frac{1}{2\pi} \oint_{\tilde{S}} \sigma_{0n}(M) \frac{\tilde{\mathbf{r}}_{MQ} \cdot \tilde{\mathbf{n}}_Q}{\tilde{r}_{MQ}^3} d\tilde{s}_M = \frac{\sigma_{0n}(Q)}{\lambda_n}, \tag{4.153}$$

we derive

$$\sum_{n \neq k}^{\infty} a_{kn} \left(1 - \frac{\lambda_k}{\lambda_n} \right) \sigma_{0n}(Q) = f(Q). \tag{4.154}$$

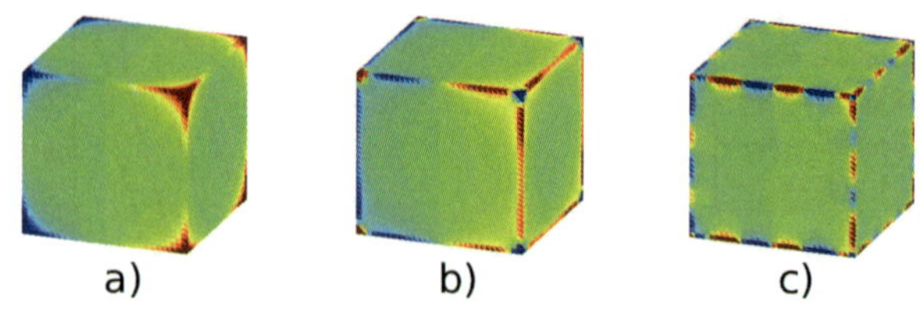

Figure 4.4

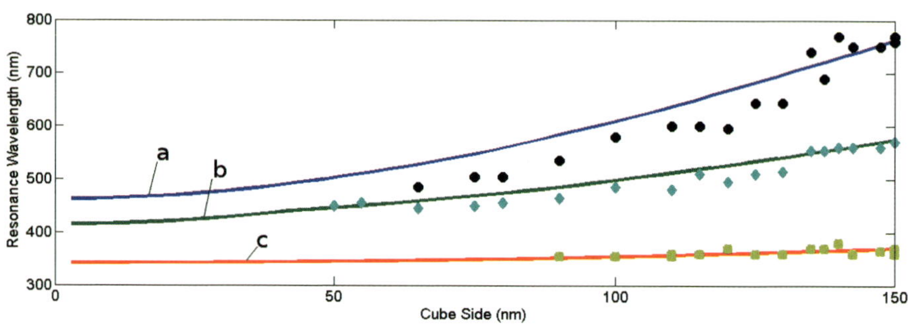

Figure 4.5

By multiplying both sides of the last formula by $\tau_{0m}(Q)$, then integrating over $\tilde{S}$ and taking into account the biorthogonality of $\sigma_{0n}(Q)$ and $\tau_{0m}(Q)$, we obtain

$$a_{kn}\left(1 - \frac{\lambda_k}{\lambda_n}\right) = \oint_{\tilde{S}} f(Q)\tau_{0n}(Q)d\tilde{s}_Q = \langle f, \tau_{0n} \rangle. \tag{4.155}$$

This implies that

$$a_{kn} = \frac{\lambda_n \langle f, \tau_{0n} \rangle}{\lambda_n - \lambda_k}. \tag{4.156}$$

By substituting the last formula into equation (4.151), we arrive at

$$\sigma_{2k}(Q) = \sum_{n \neq k}^{\infty} \frac{\lambda_n \langle f, \tau_{0n} \rangle}{\lambda_n - \lambda_k} \sigma_{0n}(Q). \tag{4.157}$$

Once again, $\sigma_{2k}(Q)$ is expressed only in terms of quantities computed in the electrostatic analysis of plasmon resonances.

4.3   Analysis of Extinction Cross Section

This section deals with the numerical analysis of extinction cross section of metallic nanoparticles as well as with the analysis of multiple nanoparticle structures used, for instance, in plasmon waveguides of light. The modeling of extinction cross section is of interest because the plasmon resonances in metallic nanoparticles are usually studied through the measurements of this cross section. The extinction (sometimes also called "total") cross section is usually defined as properly normalized overall losses which are due to absorption and scattering by metallic nanoparticles. The absorption losses are proportional to the square of electric field magnitude inside the nanoparticles. This magnitude is strongly peaked at resonance wavelengths of incident radiation. For this reason, plasmon resonances can be identified with the peaks in measured (or calculated) extinction cross sections. It turns out that the overall (absorption and scattering) losses are closely related to the behavior of far fields scattered in the forward direction. This general relationship is the essence of the "optical" (also called "forward scattering") theorem [5, 6]. This theorem is the foundation for the experimental measurements as well as calculation of extinction cross sections of metallic nanoparticles.

It is apparent from the above discussion that the analysis of extinction cross sections of nanoparticles requires the solution of the electromagnetic scattering problem and subsequent computations of forward scattered far fields. For this reason, we proceed with the discussion of the problem of electromagnetic scattering by nanoparticles and we shall use for this purpose the perturbation technique described in the first section of this chapter.

Consider an incident plane wave $\mathbf{E}_i$ scattered by a nanoparticle of arbitrary shape (see Figure 4.6) with complex permittivity

$$\varepsilon(\omega) = \varepsilon'(\omega) + j\varepsilon''(\omega). \tag{4.158}$$

It is clear that the total electromagnetic field is the sum of the incident and the scattered fields,

$$\mathbf{E} = \mathbf{E}_i + \mathbf{E}_s, \qquad \mathbf{H} = \mathbf{H}_i + \mathbf{H}_s, \tag{4.159}$$

where subscripts "i" and "s" are used for incident and scattered fields, respectively.

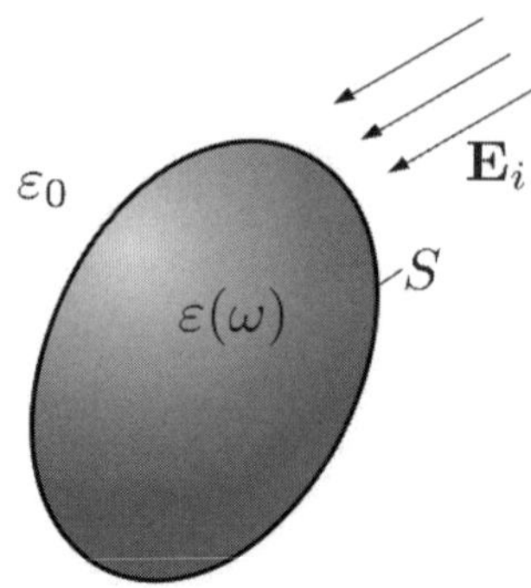

Figure 4.6

It is also apparent that the total electromagnetic field satisfies the following equations and boundary conditions:

$$\operatorname{curl}\mathbf{E}^+ = -j\omega\mu_0\mathbf{H}^+, \tag{4.160}$$

$$\operatorname{curl}\mathbf{H}^+ = j\omega\varepsilon\mathbf{E}^+, \tag{4.161}$$

$$\operatorname{curl}\mathbf{E}^- = -j\omega\mu_0\mathbf{H}^-, \tag{4.162}$$

$$\operatorname{curl}\mathbf{H}^- = j\omega\varepsilon_0\mathbf{E}^-, \tag{4.163}$$

$$\operatorname{div}\mathbf{H}^+ = 0, \tag{4.164}$$

$$\operatorname{div}\mathbf{H}^- = 0, \tag{4.165}$$

$$\operatorname{div}\mathbf{E}^+ = 0, \tag{4.166}$$

$$\operatorname{div}\mathbf{E}^- = 0, \tag{4.167}$$

$$\mathbf{n}\times\left(\mathbf{E}^+ - \mathbf{E}^-\right) = 0, \tag{4.168}$$

$$\mathbf{n}\times\left(\mathbf{H}^+ - \mathbf{H}^-\right) = 0, \tag{4.169}$$

$$\mathbf{n}\cdot\left(\varepsilon\mathbf{E}^+ - \varepsilon_0\mathbf{E}^-\right) = 0, \tag{4.170}$$

$$\mathbf{n}\cdot\left(\mathbf{H}^+ - \mathbf{H}^-\right) = 0. \tag{4.171}$$

The incident electromagnetic field satisfies the same equations and boundary conditions with the exception of equation (4.161) and boundary condition (4.170), which are modified, respectively, as follows:

$$\operatorname{curl}\mathbf{H}_i^+ = j\omega\varepsilon_0\mathbf{E}_i^+, \tag{4.172}$$

$$\mathbf{n}\cdot\left(\mathbf{E}_i^+ - \mathbf{E}_i^-\right) = 0. \tag{4.173}$$

By using this fact and formulas (4.159), we arrive at the following equations and boundary conditions for the scattered field:

$$\operatorname{curl} \mathbf{E}_s^+ = -j\omega\mu_0 \mathbf{H}_s^+, \tag{4.174}$$

$$\operatorname{curl} \mathbf{H}_s^+ = j\omega\varepsilon \mathbf{E}_s^+ + j\omega\left(\varepsilon - \varepsilon_0\right) \mathbf{E}_i^+, \tag{4.175}$$

$$\operatorname{curl} \mathbf{E}_s^- = -j\omega\mu_0 \mathbf{H}_s^-, \tag{4.176}$$

$$\operatorname{curl} \mathbf{H}_s^- = j\omega\varepsilon_0 \mathbf{E}_s^-, \tag{4.177}$$

$$\operatorname{div} \mathbf{H}_s^+ = 0, \tag{4.178}$$

$$\operatorname{div} \mathbf{H}_s^- = 0, \tag{4.179}$$

$$\operatorname{div} \mathbf{E}_s^+ = 0, \tag{4.180}$$

$$\operatorname{div} \mathbf{E}_s^- = 0, \tag{4.181}$$

$$\mathbf{n} \times \left(\mathbf{E}_s^+ - \mathbf{E}_s^-\right) = 0, \tag{4.182}$$

$$\mathbf{n} \times \left(\mathbf{H}_s^+ - \mathbf{H}_s^-\right) = 0, \tag{4.183}$$

$$\mathbf{n} \cdot \left(\varepsilon \mathbf{E}_s^+ - \varepsilon_0 \mathbf{E}_s^-\right) = \left(\varepsilon_0 - \varepsilon\right) \mathbf{n} \cdot \mathbf{E}_i, \tag{4.184}$$

$$\mathbf{n} \cdot \left(\mathbf{H}_s^+ - \mathbf{H}_s^-\right) = 0. \tag{4.185}$$

It is also tacitly assumed that the scattered electromagnetic field satisfies the outgoing radiation conditions at infinity.

In the case when nanoparticle dimensions are small in comparison with the free-space wavelength of incident radiation, the small parameter β defined by formula (4.20) can be introduced in the formulation of the scattering problem. This is done by using the spatial coordinate scaling (4.13) and by writing the boundary value problem (4.174)-(4.185) in terms of vectors

$$\mathbf{e}_s^\pm = \sqrt{\varepsilon_0}\mathbf{E}_s^\pm, \quad \mathbf{h}_s^\pm = \sqrt{\mu_0}\mathbf{H}_s^\pm, \quad \mathbf{e}_i^\pm = \sqrt{\varepsilon_0}\mathbf{E}_i^\pm. \tag{4.186}$$

This results in the following perturbative form of the boundary value problem (4.174)-(4.185):

$$\operatorname{curl} \mathbf{e}_s^+ = -j\beta \mathbf{h}_s^+, \tag{4.187}$$

$$\operatorname{curl} \mathbf{h}_s^+ = j\beta \frac{\varepsilon}{\varepsilon_0} \mathbf{e}_s^+ + j\beta \left(\frac{\varepsilon}{\varepsilon_0} - 1\right) \mathbf{e}_i^+, \tag{4.188}$$

$$\operatorname{curl} \mathbf{e}_s^- = -j\beta \mathbf{h}_s^-, \tag{4.189}$$

$$\operatorname{curl} \mathbf{h}_s^- = j\beta \mathbf{e}_s^-, \tag{4.190}$$

$$\operatorname{div} \mathbf{h}_s^+ = 0, \tag{4.191}$$

$$\operatorname{div}\mathbf{h}_s^- = 0, \tag{4.192}$$

$$\operatorname{div}\mathbf{e}_s^+ = 0, \tag{4.193}$$

$$\operatorname{div}\mathbf{e}_s^- = 0, \tag{4.194}$$

$$\tilde{\mathbf{n}} \times \left(\mathbf{e}_s^+ - \mathbf{e}_s^-\right) = 0, \tag{4.195}$$

$$\tilde{\mathbf{n}} \times \left(\mathbf{h}_s^+ - \mathbf{h}_s^-\right) = 0, \tag{4.196}$$

$$\tilde{\mathbf{n}} \cdot \left(\varepsilon\mathbf{e}_s^+ - \varepsilon_0\mathbf{e}_s^-\right) = \left(\varepsilon_0 - \varepsilon\right)\tilde{\mathbf{n}} \cdot \mathbf{e}_i, \tag{4.197}$$

$$\tilde{\mathbf{n}} \cdot \left(\mathbf{h}_s^+ - \mathbf{h}_s^-\right) = 0. \tag{4.198}$$

Next, we shall use the following expansions:

$$\mathbf{e}_s^\pm = \mathbf{e}_{s0}^\pm + \beta\mathbf{e}_{s1}^\pm + \beta^2\mathbf{e}_{s2}^\pm + \cdots, \tag{4.199}$$

$$\mathbf{h}_s^\pm = \mathbf{h}_{s0}^\pm + \beta\mathbf{h}_{s1}^\pm + \beta^2\mathbf{h}_{s2}^\pm + \cdots. \tag{4.200}$$

By substituting expansions (4.199) and (4.200) into formulas (4.187)-(4.198) and by equating the terms of the same powers of β, the boundary value problems for $\mathbf{e}_{sk}$ and $\mathbf{h}_{sk}$ can be derived. For zero-order terms, these boundary value problems are as follows:

$$\operatorname{curl}\mathbf{h}_{s0}^+ = 0, \tag{4.201}$$

$$\operatorname{curl}\mathbf{h}_{s0}^- = 0, \tag{4.202}$$

$$\operatorname{div}\mathbf{h}_{s0}^+ = 0, \tag{4.203}$$

$$\operatorname{div}\mathbf{h}_{s0}^- = 0, \tag{4.204}$$

$$\tilde{\mathbf{n}} \times \left(\mathbf{h}_{s0}^+ - \mathbf{h}_{s0}^-\right) = 0, \tag{4.205}$$

$$\tilde{\mathbf{n}} \cdot \left(\mathbf{h}_{s0}^+ - \mathbf{h}_{s0}^-\right) = 0, \tag{4.206}$$

and

$$\operatorname{curl}\mathbf{e}_{s0}^+ = 0, \tag{4.207}$$

$$\operatorname{curl}\mathbf{e}_{s0}^- = 0, \tag{4.208}$$

$$\operatorname{div}\mathbf{e}_{s0}^+ = 0, \tag{4.209}$$

$$\operatorname{div}\mathbf{e}_{s0}^- = 0, \tag{4.210}$$

$$\tilde{\mathbf{n}} \times \left(\mathbf{e}_{s0}^+ - \mathbf{e}_{s0}^-\right) = 0, \tag{4.211}$$

$$\tilde{\mathbf{n}} \cdot \left(\varepsilon\mathbf{e}_{s0}^+ - \varepsilon_0\mathbf{e}_{s0}^-\right) = \left(\varepsilon_0 - \varepsilon\right)\tilde{\mathbf{n}} \cdot \mathbf{e}_i. \tag{4.212}$$

The boundary value problem (4.201)-(4.206) is mathematically identical to the boundary value problem (4.34)-(4.39). Thus, by using the same line of reasoning as in the first section of this chapter, it can be established that

$$\mathbf{h}_{s0}^\pm = 0. \tag{4.213}$$

As far as the boundary value problem (4.207)-(4.212) is concerned, it is clear that the electric field $\mathbf{e}_{s0}^{\pm}$ can be described by a scalar potential

$$\mathbf{e}_{s0} = -\operatorname{grad} \varphi_{s0} \qquad (4.214)$$

that, in turn, can be viewed as the potential created by surface charges $\sigma_0(M)$ distributed over $\tilde{S}$

$$\varphi_{s0}(Q) = \frac{1}{4\pi\varepsilon_0} \oint_{\tilde{S}} \frac{\sigma_0(M)}{\tilde{r}_{MQ}} d\tilde{s}_M. \qquad (4.215)$$

The electric field created by these surface charges satisfies equations (4.207)-(4.210) and the boundary condition (4.211). By using formulas (2.30) and (2.31), it is easy to show that the boundary condition (4.212) is satisfied as well if the surface charge density $\sigma_0(M)$ is the solution to the following integral equation:

$$\sigma_0(M) - \frac{\lambda}{2\pi} \oint_{\tilde{S}} \sigma_0(M) \frac{\tilde{\mathbf{r}}_{MQ} \cdot \tilde{\mathbf{n}}_Q}{\tilde{r}_{MQ}^3} d\tilde{s}_M = -2\lambda\varepsilon_0 \tilde{\mathbf{n}}_Q \cdot \mathbf{e}_i(Q), \qquad (4.216)$$

where

$$\lambda = \frac{\varepsilon - \varepsilon_0}{\varepsilon + \varepsilon_0}. \qquad (4.217)$$

According to formulas (4.158) and (4.217), λ is a complex number. It has been proved in Chapter 2 that all the eigenvalues of integral equation (4.216) are real. This implies that the homogeneous integral equation corresponding to integral equation (4.216) has only the trivial zero solution. By using the Fredholm theory (see section 2.1), we conclude that there exists the unique solution of integral equation (4.216) for any right-hand side. For this reason, the numerical technique discussed in section 4 of Chapter 3 can be used for the numerical solution of equation (4.216).

Next, through the substitution of expansions (4.199)-(4.200) into formulas (4.187)-(4.198), we derive the following boundary value problems for $\mathbf{e}_{s1}$ and $\mathbf{h}_{s1}$:

$$\operatorname{curl} \mathbf{e}_{s1}^+ = 0, \qquad (4.218)$$
$$\operatorname{div} \mathbf{e}_{s1}^+ = 0, \qquad (4.219)$$
$$\operatorname{curl} \mathbf{e}_{s1}^- = 0, \qquad (4.220)$$
$$\operatorname{div} \mathbf{e}_{s1}^- = 0, \qquad (4.221)$$
$$\tilde{\mathbf{n}} \times \left(\mathbf{e}_{s1}^+ - \mathbf{e}_{s1}^- \right) = 0, \qquad (4.222)$$
$$\tilde{\mathbf{n}} \cdot \left(\varepsilon \mathbf{e}_{s1}^+ - \varepsilon_0 \mathbf{e}_{s1}^- \right) = 0, \qquad (4.223)$$

and

$$\operatorname{curl} \mathbf{h}_{s1}^{+} = j \frac{\varepsilon}{\varepsilon_0} \mathbf{e}_{s0}^{+} + j \left(\frac{\varepsilon}{\varepsilon_0} - 1 \right) \mathbf{e}_{i}^{+}, \tag{4.224}$$

$$\operatorname{div} \mathbf{h}_{s1}^{+} = 0, \tag{4.225}$$

$$\operatorname{curl} \mathbf{h}_{s1}^{-} = j \mathbf{e}_{s0}^{-}, \tag{4.226}$$

$$\operatorname{div} \mathbf{h}_{s1}^{-} = 0, \tag{4.227}$$

$$\tilde{\mathbf{n}} \times \left(\mathbf{h}_{s1}^{+} - \mathbf{h}_{s1}^{-} \right) = 0, \tag{4.228}$$

$$\tilde{\mathbf{n}} \cdot \left(\mathbf{h}_{s1}^{+} - \mathbf{h}_{s1}^{-} \right) = 0. \tag{4.229}$$

By introducing the scalar potential

$$\mathbf{e}_{s1} = -\operatorname{grad} \varphi_{s1} \tag{4.230}$$

and by using the same line of reasoning as in the derivation of formula (2.19), we obtain

$$\varepsilon \int_{\tilde{V}+} \left| \operatorname{grad} \varphi_{s1}^{+} \right|^2 d\tilde{v} + \varepsilon_0 \int_{\tilde{V}-} \left| \operatorname{grad} \varphi_{s1}^{-} \right|^2 d\tilde{v} = 0. \tag{4.231}$$

Since ε is assumed to be complex (see formula (4.158)), we conclude that

$$\mathbf{e}_{s1}^{\pm} = 0. \tag{4.232}$$

It is clear from formulas (4.224)-(4.229) that $\mathbf{h}_{s1}$ can be viewed as the stationary magnetic field created by currents

$$\mathbf{J}^{+} = j \frac{\varepsilon}{\varepsilon_0} \mathbf{e}_{s0}^{+} + j \left(\frac{\varepsilon}{\varepsilon_0} - 1 \right) \mathbf{e}_{i}^{+}, \tag{4.233}$$

$$\mathbf{J}^{-} = j \mathbf{e}_{s0}^{-} \tag{4.234}$$

distributed over $\tilde{V}^{+}$ and $\tilde{V}^{-}$, respectively. For this reason, $\mathbf{h}_{s1}$ can be computed by using the Biot-Savart integral formula

$$\mathbf{h}_{s1}(Q) = \frac{j}{4\pi} \int_{\tilde{V}+} \frac{\left[\frac{\varepsilon}{\varepsilon_0} \mathbf{e}_{s0}^{+}(M) + \left(\frac{\varepsilon}{\varepsilon_0} - 1 \right) \mathbf{e}_{i}^{+}(M) \right] \times \tilde{\mathbf{r}}_{MQ}}{\tilde{r}_{MQ}^{3}} d\tilde{v}_M$$

$$+ \frac{j}{4\pi} \int_{\tilde{V}-} \frac{\mathbf{e}_{s0}^{-}(M) \times \tilde{\mathbf{r}}_{MQ}}{\tilde{r}_{MQ}^{3}} d\tilde{v}_M. \tag{4.235}$$

Now, by taking into account that for any smooth vector $\mathbf{a}(M)$

$$\frac{\mathbf{a}(M) \times \tilde{\mathbf{r}}_{MQ}}{\tilde{r}_{MQ}^3} = -\mathrm{curl}\left(\frac{\mathbf{a}(M)}{\tilde{r}_{MQ}}\right) + \frac{\mathrm{curl}\,\mathbf{a}(M)}{\tilde{r}_{MQ}} \tag{4.236}$$

as well as the equations (4.207), (4.208) and

$$\mathrm{curl}\,\mathbf{e}_i^+ = -j\beta\mathbf{h}_i^+, \tag{4.237}$$

through the help of identity (4.74) the integral formula (4.235) can be transformed as follows:

$$\begin{aligned}
\mathbf{h}_{s1}(Q) = {}& \frac{(\varepsilon - \varepsilon_0)\,\beta}{4\pi\varepsilon_0} \int_{\tilde{V}+} \frac{\mathbf{h}_i^+(M)}{\tilde{r}_{MQ}} d\tilde{v}_M \\
& + \frac{j\,(\varepsilon - \varepsilon_0)}{4\pi\varepsilon_0} \oint_{\tilde{S}} \frac{\tilde{\mathbf{n}}_M \times [\mathbf{e}_{s0}(M) + \mathbf{e}_i(M)]}{\tilde{r}_{MQ}} d\tilde{s}_M.
\end{aligned} \tag{4.238}$$

It is apparent that formula (4.238) is simpler than formula (4.235) as far as calculations are concerned.

We next proceed to the analysis of the second-order terms $\mathbf{h}_{s2}$ and $\mathbf{e}_{s2}$ in power expansions (4.199) and (4.200). By substituting these expansions into equations (4.187)-(4.194) as well as boundary conditions (4.195)-(4.198) and by equating the terms of second power of β we arrive at the following boundary value problems:

$$\mathrm{curl}\,\mathbf{h}_{s2}^+ = 0, \tag{4.239}$$

$$\mathrm{div}\,\mathbf{h}_{s2}^+ = 0, \tag{4.240}$$

$$\mathrm{curl}\,\mathbf{h}_{s2}^- = 0, \tag{4.241}$$

$$\mathrm{div}\,\mathbf{h}_{s2}^- = 0, \tag{4.242}$$

$$\tilde{\mathbf{n}} \times \left(\mathbf{h}_{s2}^+ - \mathbf{h}_{s2}^-\right) = 0, \tag{4.243}$$

$$\tilde{\mathbf{n}} \cdot \left(\mathbf{h}_{s2}^+ - \mathbf{h}_{s2}^-\right) = 0, \tag{4.244}$$

and

$$\mathrm{curl}\,\mathbf{e}_{s2}^+ = -j\mathbf{h}_{s1}^+, \tag{4.245}$$

$$\mathrm{div}\,\mathbf{e}_{s2}^+ = 0, \tag{4.246}$$

$$\mathrm{curl}\,\mathbf{e}_{s2}^- = -j\mathbf{h}_{s1}^-, \tag{4.247}$$

$$\mathrm{div}\,\mathbf{e}_{s2}^- = 0, \tag{4.248}$$

$$\tilde{\mathbf{n}} \times \left(\mathbf{e}_{s2}^+ - \mathbf{e}_{s2}^-\right) = 0, \tag{4.249}$$

$$\tilde{\mathbf{n}} \cdot \left(\varepsilon\mathbf{e}_{s2}^+ - \varepsilon_0\mathbf{e}_{s2}^-\right) = 0. \tag{4.250}$$

The mathematical structure of the boundary value problem (4.239)-(4.244) is identical to the structure of the boundary value problem (4.34)-(4.39). Thus, by using the same line of reasoning as in the first section of this chapter, it can be established that

$$\mathbf{h}_{s2}^{\pm} = 0. \tag{4.251}$$

Next, we shall split $\mathbf{e}_{s2}$ into two components,

$$\mathbf{e}_{s2} = \mathbf{e}'_{s2} + \mathbf{e}''_{s2}, \tag{4.252}$$

which are the solutions of the following boundary value problems, respectively:

$$\operatorname{curl}\mathbf{e}'^{\pm}_{s2} = -j\mathbf{h}^{\pm}_{s1}, \tag{4.253}$$

$$\operatorname{div}\mathbf{e}'^{\pm}_{s2} = 0, \tag{4.254}$$

$$\tilde{\mathbf{n}} \times \left(\mathbf{e}'^{+}_{s2} - \mathbf{e}'^{-}_{s2}\right) = 0, \tag{4.255}$$

$$\tilde{\mathbf{n}} \cdot \left(\mathbf{e}'^{+}_{s2} - \mathbf{e}'^{-}_{s2}\right) = 0, \tag{4.256}$$

and

$$\operatorname{curl}\mathbf{e}''^{\pm}_{s2} = 0, \tag{4.257}$$

$$\operatorname{div}\mathbf{e}''^{\pm}_{s2} = 0, \tag{4.258}$$

$$\tilde{\mathbf{n}} \times \left(\mathbf{e}''^{+}_{s2} - \mathbf{e}''^{-}_{s2}\right) = 0, \tag{4.259}$$

$$\tilde{\mathbf{n}} \cdot \left(\varepsilon\mathbf{e}''^{+}_{s2} - \varepsilon_0\mathbf{e}''^{-}_{s2}\right) = (\varepsilon_0 - \varepsilon)\,\tilde{\mathbf{n}} \cdot \mathbf{e}'_{s2}. \tag{4.260}$$

It is clear that $\mathbf{e}'_{s2}$ can be viewed as created by magnetic currents $-j\mathbf{h}_{s1}$. For this reason, it can be represented by the following integral formula:

$$\mathbf{e}'_{s2}(P) = -\frac{j}{4\pi} \int_{\mathbb{R}^3} \frac{\mathbf{h}_{s1}(Q) \times \tilde{\mathbf{r}}_{QP}}{\tilde{r}^3_{QP}} d\tilde{v}_Q. \tag{4.261}$$

By substituting formula (4.238) into the last equation and by changing the order of integration, we find

$$\mathbf{e}'_{s2}(P) = -\frac{j\,(\varepsilon - \varepsilon_0)\,\beta}{16\pi^2\varepsilon_0} \int_{\tilde{V}+} \mathbf{h}^{+}_{i}(M) \times \left(\int_{\mathbb{R}^3} \frac{\tilde{\mathbf{r}}_{QP}}{\tilde{r}^3_{QP}} \cdot \frac{1}{\tilde{r}_{MQ}} d\tilde{v}_Q\right) d\tilde{v}_M$$

$$+ \frac{\varepsilon_0 - \varepsilon}{16\pi^2\varepsilon_0} \oint_{\tilde{S}} (\tilde{\mathbf{n}}_M \times [\mathbf{e}_{s0}(M) - \mathbf{e}_i(M)]) \times \left(\int_{\mathbb{R}^3} \frac{\tilde{\mathbf{r}}_{QP}}{\tilde{r}^3_{QP}} \cdot \frac{1}{\tilde{r}_{MQ}} d\tilde{v}_Q\right) d\tilde{s}_M.$$

$$\tag{4.262}$$

Finally, by using relations (4.139) and (4.140), the last formula can be simplified as follows:

$$\mathbf{e}'_{s2}(P) = -\frac{j\,(\varepsilon - \varepsilon_0)\,\beta}{8\pi\varepsilon_0} \int_{\tilde{V}+} \frac{\mathbf{h}_i^+(M) \times \tilde{\mathbf{r}}_{MP}}{\tilde{r}_{MP}} d\tilde{v}_M$$
$$+ \frac{\varepsilon_0 - \varepsilon}{8\pi\varepsilon_0} \oint_{\tilde{S}} \frac{(\tilde{\mathbf{n}}_M \times [\mathbf{e}_{s0}(M) - \mathbf{e}_i(M)]) \times \tilde{\mathbf{r}}_{MP}}{\tilde{r}_{MP}} d\tilde{s}_M. \qquad (4.263)$$

As far as the solution of the boundary value problem (4.257)-(4.260) is concerned, it can be accomplished by introducing the potential

$$\mathbf{e}''_{s2} = -\operatorname{grad} \varphi_{s2}, \qquad (4.264)$$

which can be viewed as the potential created by surface charges $\sigma_2(M)$ distributed over $\tilde{S}$

$$\varphi_{s2}(Q) = \frac{1}{4\pi\varepsilon_0} \oint_{\tilde{S}} \frac{\sigma_2(M)}{\tilde{r}_{MQ}} d\tilde{s}_M. \qquad (4.265)$$

The electric field created by these charges naturally satisfies equations (4.257)-(4.258) and the boundary condition (4.259). As before, by using formulas (2.30)-(2.31), it can be easily shown that the boundary condition (4.260) is satisfied as well if the surface charge density $\sigma_2(M)$ is the solution of the following integral equation:

$$\sigma_2(Q) - \frac{\lambda}{2\pi} \oint_{\tilde{S}} \sigma_2(M) \frac{\tilde{\mathbf{n}}_Q \cdot \tilde{\mathbf{r}}_{MQ}}{\tilde{r}_{MQ}^3} d\tilde{s}_M = -2\lambda\varepsilon_0\tilde{\mathbf{n}}_Q \cdot \mathbf{e}'_{s2}(Q), \qquad (4.266)$$

where λ is complex and defined by formula (4.217). Since λ is complex, this integral equation is uniquely solvable for any right-hand side and its solution can be found by using the numerical technique discussed in section 4 of Chapter 3. Having computed $\mathbf{e}_{s0}$ and $\mathbf{e}_{s2}$, the scattered electric field can be found as follows:

$$\mathbf{E}_s^- \simeq \varepsilon_0^{-\frac{1}{2}} \left(\mathbf{e}_{s0}^- + \beta^2 \mathbf{e}_{s2}^- \right). \qquad (4.267)$$

Thus, the algorithm for the solution of the scattering problem consists of computing $\mathbf{e}_{s0}$ and $\mathbf{e}_{s2}$ by using formula (4.263) and solving identical (up to the form of the right-hand side) integral equations (4.216) and (4.266). Having performed these computations, the dipole moment of the field $\mathbf{E}_s^-$ can be found and used for the calculation of far fields scattered in the forward direction. By using these fields, the extinction cross section can be determined. Indeed, considering linearly polarized incident electric field

$$\mathbf{E}_i = \mathbf{a}_i E_i e^{j\mathbf{k}\cdot\mathbf{r}}, \qquad (4.268)$$

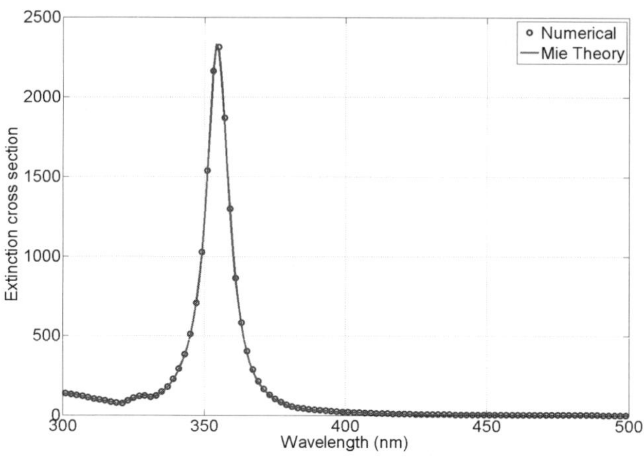

Figure 4.7

then the normalized forward scattering amplitude $\mathbf{f}$ can be defined as

$$\mathbf{f} = \lim_{r \to \infty} \left[\left(\frac{e^{jkr}}{r} \right)^{-1} \frac{\mathbf{E}_s^-}{E_i} \right], \qquad (4.269)$$

where $\mathbf{E}_s^-$ is the forward scattered (measured in $\mathbf{k}$-direction) far field. As soon as $\mathbf{f}$ is determined, by using the "optical theorem" the extinction cross section C_{ext} can be computed by using the formula

$$C_{ext} = \frac{4\pi}{k} \mathrm{Im} \left(\mathbf{f} \cdot \mathbf{a}_i \right). \qquad (4.270)$$

The described technique for the solution of the scattering problem and subsequent computation of extinction cross section have been software implemented. By using this software, numerous computations have been performed. Below, some examples of these computations are presented.

To test the accuracy of our technique and the developed software, the extinction cross sections of silver and gold nanospheres with diameters of 20 nm have been computed and compared with the Mie theory [2]. The results of these comparisons are presented in Figures 4.7 and 4.8, respectively, and they reveal sufficiently high accuracy of the numerical technique. The dispersion relations based on experimental data of P. B. Johnson and R. W. Christy [7] were used in these calculations as well as in other calculations presented

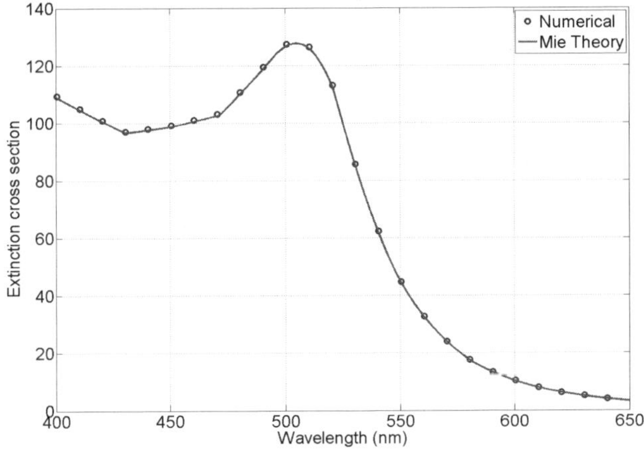

Figure 4.8

in this section. Computational results for extinction cross sections of silver ellipsoidal nanoparticles of five different aspect (major axes) ratios (1:1:0.8, 1:1:1, 1:1:1.2, 1:1:1.4 and 1:1:1.6) are presented in Figure 4.9, respectively. It is apparent from this figure that peaks (resonances) are red-shifted with the increase in aspect ratio. Figure 4.10 presents the extinction cross sections of three gold nanorings placed on a substrate with permittivity $2.21\varepsilon_0$ (see [52]). The geometrical dimensions of these rings are the same as in example 8 of section 3 of Chapter 3. It is apparent from Figure 4.10 that peaks in extinction cross sections occur at practically the same wavelengths as the resonance wavelengths computed in the above-mentioned example by using the eigenvalue technique. It is worthwhile to point out that some nanoparticles exhibit extinction cross sections with two (or more) adjacent peaks. This is usually the case when there are two (or more) dipole plasmon modes whose resonance wavelengths are somewhat close to one another. A typical and interesting example of such extinction cross sections is shown in Figures 4.11 and 4.12. These figures present the results of computations of extinction cross sections of nanocubes immersed in water and on glass substrates, respectively. These extinction cross sections are similar to those experimentally observed and published in [4] and the "peak" wavelengths are in good quantitative agreement (see also example 9 in section 3 of Chapter 3). Furthermore, Figures 4.13 and 4.14 present the extinction cross sections for

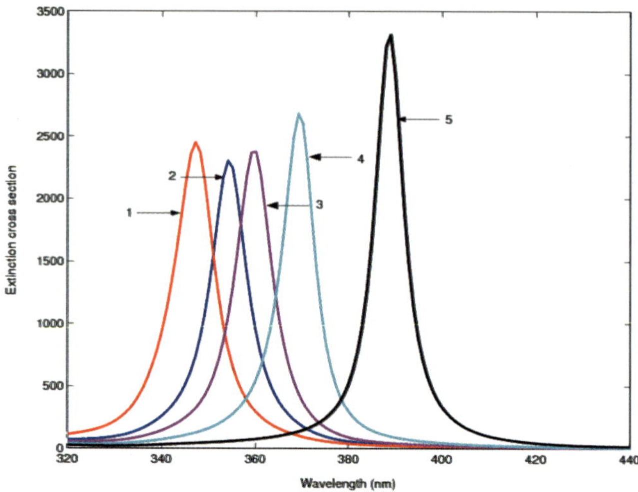

Figure 4.9

silver nanospheres placed on dielectric (Si) substrates. These extinction cross sections were computed for electric field polarizations parallel and perpendicular to the substrates, respectively. Remarkably, the presence of substrates manifests itself in the appearance of new peaks in extinction cross sections. This effect can be attributed to the symmetry breaking introduced by the substrate presence. This symmetry breaking results in splitting of resonance wavelength of the dipole modes of the spherical nanoparticle as well as in creation of new dipole modes from previously non-dipole modes. As a consequence, the existence of dipole modes of various resonance wavelengths may result in the appearance of additional peaks in extinction cross sections. Finally, Figure 4.15 presents the extinction cross section of oblate spheroidal silver nanoparticles with axes ratio 1:0.3 placed on a quartz ($\varepsilon = 4\varepsilon_0$) substrate. The wavelengths of peak values of this extinction cross section are practically the same as experimentally measured (see [8]).

Up to this point, the extinction cross section of a single nanoparticle has been discussed. However, in many applications, multiple closely spaced nanoparticles are utilized. Important examples are SERS multi-nanoparticle arrangements and nanoparticle-structured plasmon waveguides of light. To be specific, our subsequent discussion is concerned with the latter example, while SERS-related issues will be discussed in the last section of this chapter.

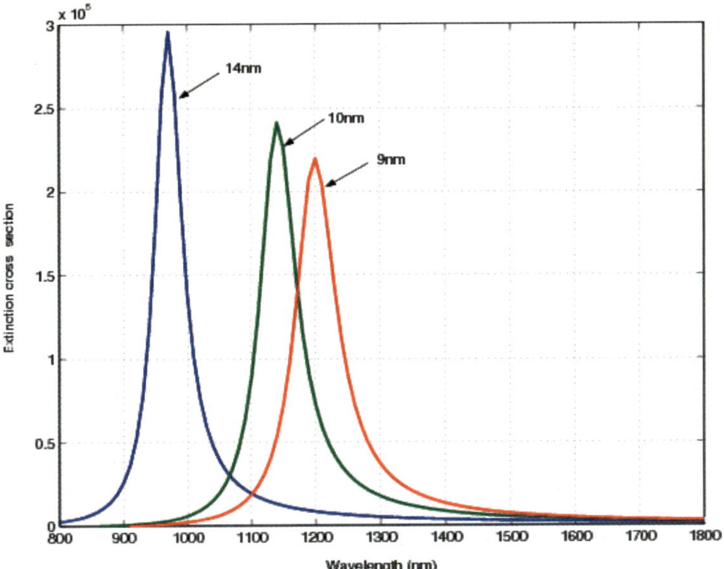

Figure 4.10

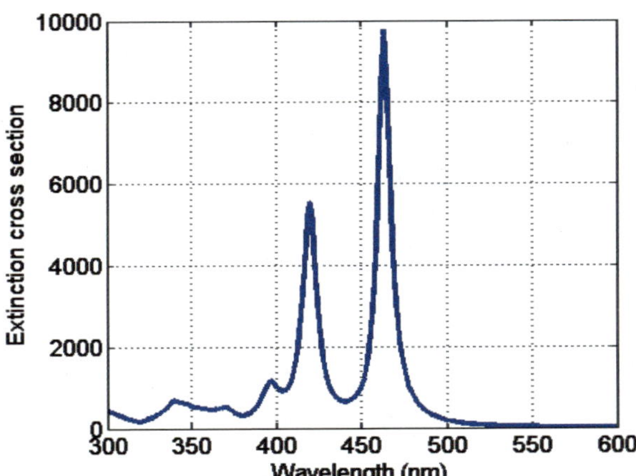

Figure 4.11

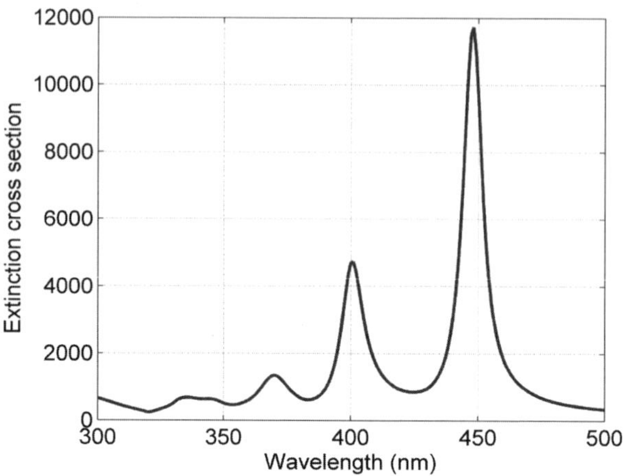

Figure 4.12

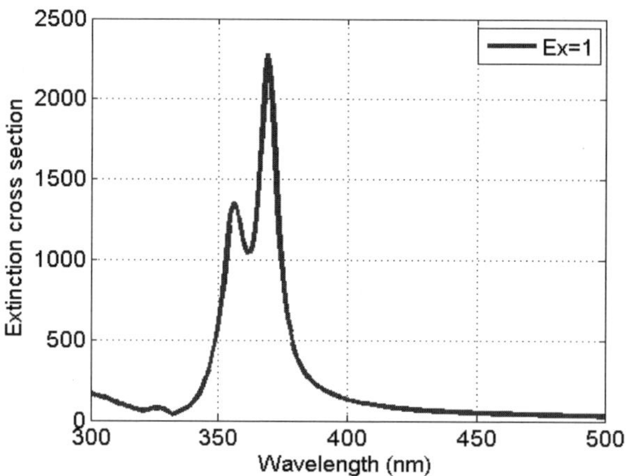

Figure 4.13

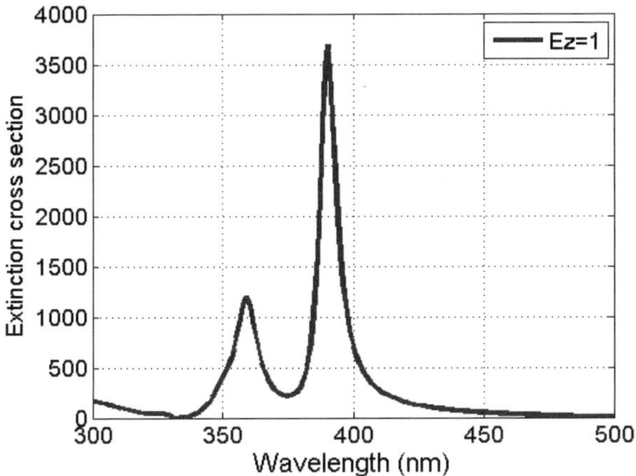

Figure 4.14

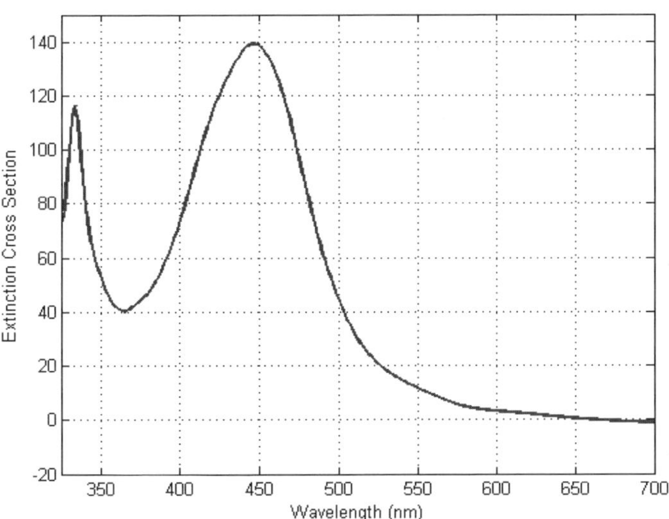

Figure 4.15

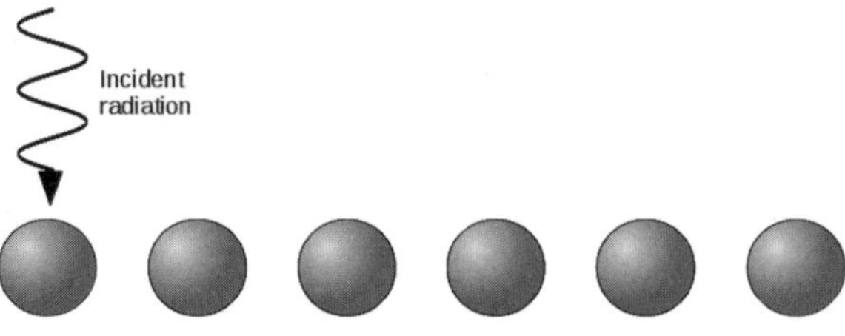

Figure 4.16

Plasmon waveguides of light have been of interest lately [9]-[11] because these waveguides hold unique promise for light guiding and bending at the nanoscale. These waveguides consist of a linear array of metallic nanoparticles (see Figure 4.16). The geometric dimensions of each nanoparticle are usually much smaller than a free-space wavelength of incident light. However, the overall dimensions of the nanoparticle array may be comparable to or exceed this wavelength. Under such circumstances, the electrostatic approximation can be applied locally to any chosen nanoparticle of the array, while for the electric field created by other nanoparticles in the vicinity of the chosen nanoparticle retardation effects must be accounted for [12]. The latter can be done by using the following formula:

$$
\mathbf{E}^{(p)}(Q) = \frac{1}{4\pi\varepsilon_0}\left[\oint_{S_p}\sigma_p(M)\frac{\mathbf{r}_{MQ}}{r_{MQ}^3}ds_M \right.
$$
$$
\left. + \sum_{m\neq p}^{N}\oint_{S_m}\sigma_m(M)\frac{\mathbf{r}_{MQ}e^{jkr_{MQ}}}{r_{MQ}^3}\left(1 - jkr_{MQ}\right)ds_M\right], \qquad (4.271)
$$

where $\mathbf{E}^{(p)}(Q)$ is the total electric field in the vicinity of the p-th nanoparticle of the array, while N is the total number of nanoparticles in the array.

Now, by using the same line of reasoning as before, we arrive at the following integral equations for electric charge densities:

$$
\sigma_p(Q) - \frac{\lambda}{2\pi}\left[\oint_{S_p}\sigma_p(M)\frac{\mathbf{r}_{MQ}\cdot\mathbf{n}_Q}{r_{MQ}^3}ds_M \right.
$$
$$
\left. + \sum_{m\neq p}^{N}\oint_{S_m}\sigma_m(M)\frac{\mathbf{r}_{MQ}\cdot\mathbf{n}_Q}{r_{MQ}^3}\left(1 - jkr_{MQ}\right)e^{jkr_{MQ}}ds_M\right]
$$
$$
= -2\lambda\mathbf{n}_Q\cdot\mathbf{E}_i(Q), \qquad (p = 1, 2, ..., N). \qquad (4.272)
$$

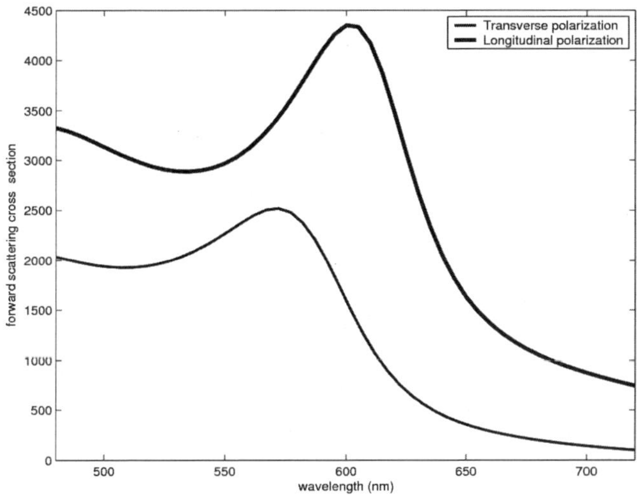

Figure 4.17

Table 4.4

	Experimental Results	Computational Results
Transverse polarization	585 nm	534 nm
Longitudinal polarization	604 nm	572 nm

This approach has been tested by computing extinction cross sections for an array of seven spherical gold nanoparticles on a glass substrate and by comparing the computational results with available experimental data [10, 11]. The spherical nanoparticles are chosen to be 50 nm in diameter with 75 nm center-to-center spacing (or 25 nm particle-to-particle spacing). The extinction cross sections have been computed for incident light propagating normal to the substrate with two distinct polarizations of light electric field: longitudinal polarization with electric field directed along the waveguide axis and transverse polarization with electric field perpendicular to this axis. The results of computations are shown in Figure 4.17. The comparison between the wavelengths corresponding to the peaks of measured [10, 11] and computed extinction cross section is given in Table 4.4, which reveals fairly good agreement. In the case of silver seven-nanosphere waveguides, the computed

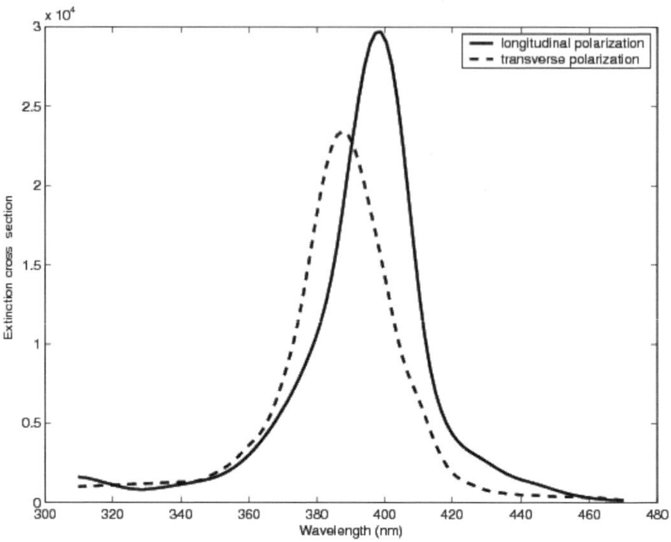

Figure 4.18

extinction cross sections are presented in Figure 4.18. The calculations have also been performed for the waveguides consisting of eleven gold spherical nanoparticles of chain, T-shape and L-shape configurations (see Figure 4.19a, 4.19b and 4.19c, respectively). The corresponding extinction cross sections are shown in Figures 4.20, 4.21 and 4.22. It is worthwhile to point out that for the T-shape waveguide the transverse polarization excitation is applied at the short arm. As plasmon fields get coupled to the long arm, the polarization becomes longitudinal. Similar polarization transformation occurs in the L-shape waveguide.

It can be remarked that by setting the right-hand sides of integral equations (4.272) to zero, we arrive at the following eigenvalue problem:

$$\sigma_p(Q) - \frac{\lambda}{2\pi} \left[\oint_{S_p} \sigma_p(M) \frac{\mathbf{r}_{MQ} \cdot \mathbf{n}_Q}{r_{MQ}^3} ds_M \right.$$

$$\left. + \sum_{m \neq p}^{N} \oint_{S_m} \sigma_m(M) \frac{\mathbf{r}_{MQ} \cdot \mathbf{n}_Q}{r_{MQ}^3} \left(1 - jkr_{MQ}\right) e^{jkr_{MQ}} ds_M \right] = 0, \quad (p = 1, 2, ..., N).$$

$$(4.273)$$

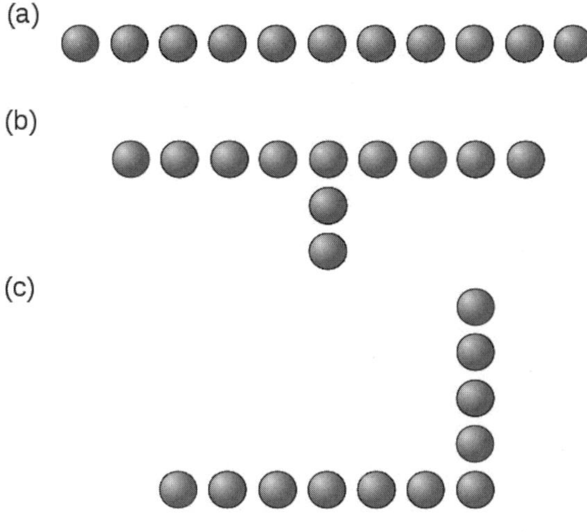

Figure 4.19

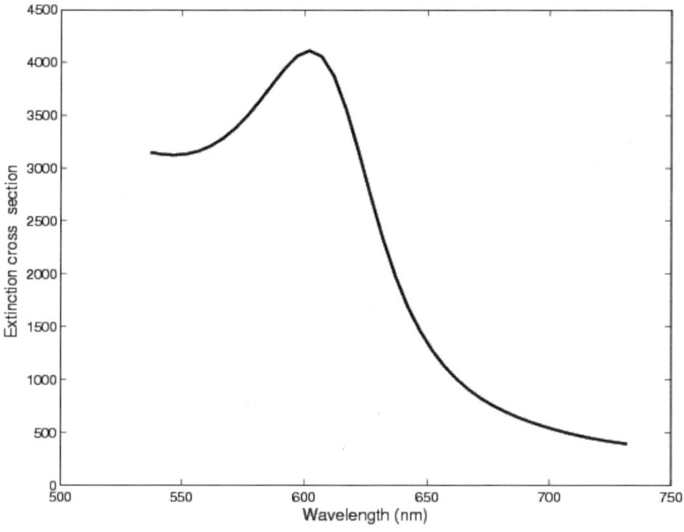

Figure 4.20

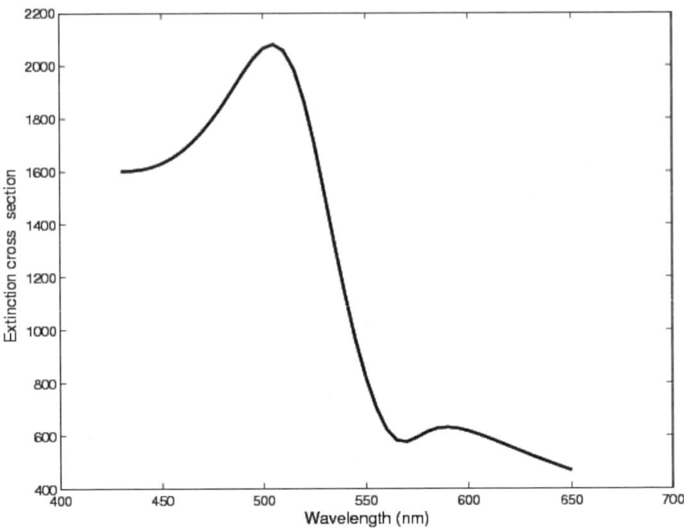

Figure 4.21

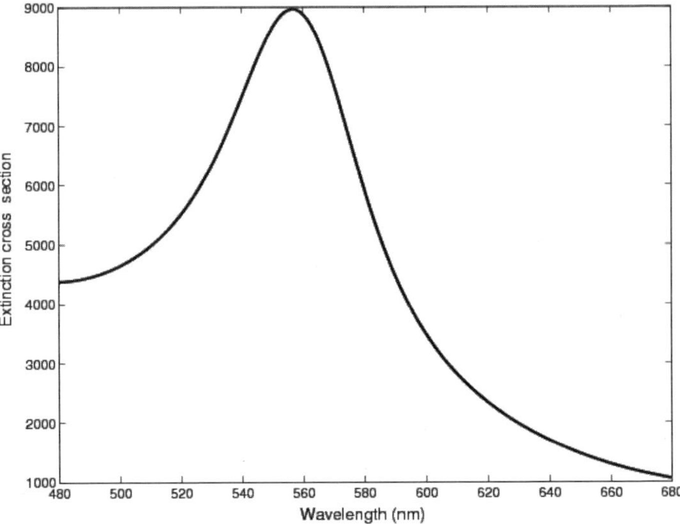

Figure 4.22

This eigenvalue problem can be used to compute resonance values of dielectric permittivity and resonance plasmon modes in multiple nanoparticle systems. It is apparent that this approach accounts for retardation effects through the nature of the kernels in integrals over S_m.

In this section, the perturbation technique has been extensively used for the solution of the scattering problem illustrated in Figure 4.6. It can be noted that other techniques can be used for this purpose as well. One such technique is the finite-difference time-domain (FDTD) method which has been discussed in section 5 of Chapter 3 and which is indeed widely used in the analysis of plasmon resonances in nanoparticles. Another technique which is attractive for the solution of the electromagnetic scattering problem is based on the finite element discretization of Maxwell equations. It has been recognized, however, that this discretization encounters two intrinsic difficulties. First, the conventional (node-based) finite elements prescribe the continuity of all three components of electric as well as all components of magnetic fields across media interfaces, whereas the continuity of only tangential components of field vectors occurs in reality at these interfaces. Second, the far-field computations require the knowledge of both electric and magnetic fields on some closed surfaces within the discretized region (see the Stratton-Chu formula (3.437)). This necessitates the differentiation of finite element solutions, which is inevitably associated with amplification of numerical errors. As far as the first difficulty is concerned, it can be circumvented by using special "edge elements" (see [13]). However, this substantially increases the number of unknowns and, besides, it does nothing to overcome the second difficulty mentioned above.

It turns out that these difficulties can be completely circumvented by using special "symmetric" boundary value problem formulation for the electromagnetic scattering problem [14]. The central idea of this formulation is to use different "state variables" (i.e., different unknowns) in different spatial regions. To demonstrate this idea, we shall use the total magnetic field $\mathbf{H}^+$ as a state variable inside the particle, i.e., in the region V^+, while we shall use the scattered electric field $\mathbf{E}_s^-$ in the exterior region V^-. Now, by using formulas (4.159), equations (4.160)-(4.161) and (4.176)-(4.177) as well as the boundary conditions (4.168)-(4.169), we arrive at the following boundary value problem:

$$\operatorname{curl}\operatorname{curl}\mathbf{H}^+ - k_+^2\mathbf{H}^+ = 0, \tag{4.274}$$

$$\operatorname{curl}\operatorname{curl}\mathbf{E}_s^- - k^2\mathbf{E}_s^- = 0, \tag{4.275}$$

$$\mathbf{n} \times \left(-\frac{j}{\omega\varepsilon}\operatorname{curl}\mathbf{H}^+ - \mathbf{E}_s^- \right) = \mathbf{n} \times \mathbf{E}_i, \tag{4.276}$$

$$\mathbf{n} \times \left(\mathbf{H}^+ - \frac{j}{\omega\mu_0}\operatorname{curl}\mathbf{E}_s^- \right) = \mathbf{n} \times \mathbf{H}_i. \tag{4.277}$$

Here, $k_+^2 = \omega^2 \mu_0 \varepsilon$, $k^2 = \omega^2 \mu_0 \varepsilon_0$ and appropriate outgoing radiation conditions at infinity are tacitly assumed.

The symmetry of the boundary value problem (4.274)-(4.277) is revealed by the fact that the mathematical structure of this boundary value problem is formally invariant with respect to permutation of $\mathbf{H}^+$ and $\mathbf{E}_s^-$. It is also worthwhile to note that by choosing different state variables for regions V^+ and V^-, the difficulty with node-based finite element implementations of interface boundary conditions on S is simply avoided.

The next step is to reduce the boundary value problem (4.274)-(4.277) to the weak Galerkin form. This can be accomplished by using the vectorial Green formula

$$\int_V (\mathbf{b} \cdot \operatorname{curl}\mathbf{a} - \mathbf{a} \cdot \operatorname{curl}\mathbf{b})\, dv = \pm \oint_S \mathbf{n} \cdot (\mathbf{a} \times \mathbf{b})\, ds, \tag{4.278}$$

which follows from the "divergence theorem" applied to vector $\mathbf{a} \times \mathbf{b}$. In the last formula, $\mathbf{n}$ is the unit vector of outward normal to S, while signs "+" and "−" correspond to the cases when V stands for interior and exterior regions, respectively.

By setting

$$\mathbf{a} = \operatorname{curl}\mathbf{H}^+ \tag{4.279}$$

and by using equation (4.274), from formula (4.278) we derive

$$\int_{V^+} \left(k_+^2 \mathbf{b}^+ \cdot \mathbf{H}^+ - \operatorname{curl}\mathbf{H}^+ \cdot \operatorname{curl}\mathbf{b}^+ \right) dv = \oint_S \mathbf{b}^+ \cdot \left(\mathbf{n} \times \operatorname{curl}\mathbf{H}^+ \right) ds. \tag{4.280}$$

Similarly, by setting

$$\mathbf{a} = \operatorname{curl}\mathbf{E}_s^- \tag{4.281}$$

and by using equation (4.275), from formula (4.278) we derive

$$\int_{V^-} \left(k^2 \mathbf{b}^- \cdot \mathbf{E}_s^- - \operatorname{curl}\mathbf{E}_s^- \cdot \operatorname{curl}\mathbf{b}^- \right) dv = -\oint_S \mathbf{b}^- \cdot \left(\mathbf{n} \times \operatorname{curl}\mathbf{E}_s^- \right) ds. \tag{4.282}$$

Now, by using boundary conditions (4.276) and (4.277) to transform the surface integrals in formulas (4.280) and (4.282), respectively, we end up with the following coupled weak Galerkin forms:

$$\int_{V^+} \left(k_+^2 \mathbf{b}^+ \cdot \mathbf{H}^+ - \operatorname{curl} \mathbf{H}^+ \cdot \operatorname{curl} \mathbf{b}^+ \right) dv - j\omega\varepsilon \oint_S \mathbf{b}^+ \cdot \left(\mathbf{n} \times \mathbf{E}_s^- \right) ds$$

$$= j\omega\varepsilon \oint_S \mathbf{b}^+ \cdot \left(\mathbf{n} \times \mathbf{E}_i \right) ds, \tag{4.283}$$

$$\int_{V^-} \left(k^2 \mathbf{b}^- \cdot \mathbf{E}_s^- - \operatorname{curl} \mathbf{E}_s^- \cdot \operatorname{curl} \mathbf{b}^- \right) dv - j\omega\mu_0 \oint_S \mathbf{b}^- \cdot \left(\mathbf{n} \times \mathbf{H}^+ \right) ds$$

$$= j\omega\mu_0 \oint_S \mathbf{b}^- \cdot \left(\mathbf{n} \times \mathbf{H}_i \right) ds. \tag{4.284}$$

The above coupled Galerkin forms are valid for any differentiable vectorial test functions $\mathbf{b}^+$ and $\mathbf{b}^-$. It can be shown that it is this fact that makes these Galerkin forms equivalent to the boundary value problem (4.274)-(4.277).

The above Galerkin forms are the foundation for finite element discretization of the scattering problem. To perform this discretization, the finite element mesh is first generated in V^+ and V^- (or its truncated version), then the approximate solution is looked for in the form

$$\mathbf{H}^+ \approx \sum_{m=1}^{N_+} \sum_{\nu=1}^{3} \mathbf{a}_\nu H_{\nu m} \psi_m^+, \tag{4.285}$$

$$\mathbf{E}_s^- \approx \sum_{n=1}^{N_-} \sum_{\nu=1}^{3} \mathbf{a}_\nu E_{\nu n} \psi_n^-, \tag{4.286}$$

where $\mathbf{a}_1 = \mathbf{a}_x$, $\mathbf{a}_2 = \mathbf{a}_y$ and $\mathbf{a}_3 = \mathbf{a}_z$, ψ_m^+ and ψ_n^- are finite element basis functions with local support in V^+ and V^-, respectively, which are equal to 1 at nodes number m and n, correspondingly, and zero at all other nodes, and N_+ and N_- are the total number of nodes in closed domains V^+ and V^-.

By substituting finite element expansions (4.285) and (4.286) into Galerkin forms (4.283) and (4.284) and by using for $\mathbf{b}^+$ and $\mathbf{b}^-$ in these forms the following test functions

$$\mathbf{b}_{\nu m}^+ = \mathbf{a}_\nu \psi_m^+, \qquad (\nu = 1, 2, 3; \quad m = 1, 2, ..., N_+), \tag{4.287}$$

$$\mathbf{b}_{\nu n}^- = \mathbf{a}_\nu \psi_n^-, \qquad (\nu = 1, 2, 3; \quad n = 1, 2, ..., N_-), \tag{4.288}$$

we end up with $3(N_+ + N_-)$ linear algebraic equations for $H_{\nu m}$ and $E_{\nu n}$. The solution of these algebraic equations will yield (among other things) the approximate values of $\mathbf{H}^+$ and $\mathbf{E}_s^-$ on the interface S. Then, by using these values, the Stratton-Chu formula (3.437) can be used to smooth the finite element solution (4.285)-(4.286) as well as to compute forward scattered far fields which are needed for the calculation of the extinction cross section. It is worthwhile to mention that the described technique is directly applicable to the case of multi-particle systems. In this case, V^+ and S in formulas (4.283) and (4.284) must be understood as the unions of regions occupied by nanoparticles and their boundaries, respectively.

4.4   Coupling of Plasmon Modes to Incident Radiation, Time-Dynamics of Their Excitation and Dephasing

It has been demonstrated in this book that the analysis of plasmon modes in nanoparticles can be framed as the eigenvalue problems for the following boundary integral equations:

$$\sigma_k(Q) = \frac{\lambda_k}{2\pi} \oint_S \sigma_k(M) \frac{\mathbf{r}_{MQ} \cdot \mathbf{n}_Q}{r_{MQ}^3} ds_M, \tag{4.289}$$

$$\tau_k(Q) = \frac{\lambda_k}{2\pi} \oint_S \tau_k(M) \frac{\mathbf{r}_{QM} \cdot \mathbf{n}_M}{r_{MQ}^3} ds_M, \tag{4.290}$$

where

$$\lambda_k = \frac{\varepsilon_k - \varepsilon_0}{\varepsilon_k + \varepsilon_0}. \tag{4.291}$$

As soon as the eigenvalue problem (4.289) (or (4.290)) is solved and eigenvalues λ_k are found, formula (4.291) can be used to compute the resonance values ε_k of nanoparticle permittivity, and then the resonance frequencies can be determined by using the dispersion relation

$$\varepsilon_k = \varepsilon'(\omega_k) = \mathrm{Re}\left[\varepsilon\left(\omega_k\right)\right]. \tag{4.292}$$

The electric fields of the corresponding plasmon modes can be computed by using surface single-layer charges $\sigma_k(M)$ or double-layer charges $\tau_k(M)$. It has been demonstrated in Chapter 2 (see formula (2.140)) that the following biorthogonality conditions are valid:

$$\oint_S \sigma_k(M)\tau_m(M)ds_M = \delta_{km},\qquad(4.293)$$

where δ_{km} is the Kronecker symbol. It is worthwhile to point out here that the solutions of eigenvalue problems (4.289) and (4.290) are defined up to some multiplicative constants. This freedom of scaling of eigenfunctions σ_k and τ_k can be used to enforce the normalization condition

$$\oint_S \sigma_k(M)\tau_k(M)ds_M = 1\qquad(4.294)$$

as well as the condition

$$\mathbf{p}_k = \mathbf{p}_{\sigma_k} = \mathbf{p}_{\tau_k}.\qquad(4.295)$$

Here, $\mathbf{p}_{\sigma_k}$ and $\mathbf{p}_{\tau_k}$ are dipole moments of the k-th plasmon mode defined by formulas (2.154) and (2.168), respectively. It is apparent that the normalization conditions (4.294) and (4.295) uniquely define dipole plasmon mode solutions of integral equations (4.289) and (4.290). It is also clear that for plasmon modes whose dipole moments are equal to zero, the equality of nonzero multipole moments can be used for normalization instead of equality (4.295). Finally, as will be shown later in this section, some simplification in mathematical formulas can be achieved by using the following normalization condition

$$\mathbf{p}_k = \mathbf{p}_{\sigma_k} = (\varepsilon_0 - \varepsilon_k)\,\mathbf{p}_{\tau_k}\qquad(4.296)$$

instead of the condition (4.295).

Now, we turn to the discussion of the problem of excitation of specific plasmon modes [15]. This problem includes the issues of the coupling of specific plasmon modes to incident radiation as well as the time dynamics of their excitation and dephasing. These issues can be studied by using time-domain formulation of the excitation problem.

Since the conducting media of metallic nanoparticles are dispersive, the constitutive relation between electric displacement $\boldsymbol{\mathcal{D}}(t)$ and electric field $\boldsymbol{\mathcal{E}}(t)$ is nonlocal in time,

$$\boldsymbol{\mathcal{D}}(t) = \int_0^t \tilde{\varepsilon}\,(t - t')\,\boldsymbol{\mathcal{E}}\,(t')\,dt',\qquad(4.297)$$

where $\tilde{\varepsilon}(t)$ and $\varepsilon(\omega)$ are related by the Fourier transform

$$\varepsilon(\omega) = \frac{1}{\sqrt{2\pi}} \int_{-\infty}^{\infty} \tilde{\varepsilon}(t) e^{j\omega t} dt, \tag{4.298}$$

$$\tilde{\varepsilon}(t) = \frac{1}{\sqrt{2\pi}} \int_{-\infty}^{\infty} \varepsilon(\omega) e^{-j\omega t} d\omega. \tag{4.299}$$

Since $\tilde{\varepsilon}(t)$ is a real-valued function of time, it is easy to conclude from formula (4.298) that

$$\varepsilon'(-\omega) = \varepsilon'(\omega), \tag{4.300}$$

$$\varepsilon''(-\omega) = -\varepsilon''(\omega). \tag{4.301}$$

In the absence of volume electric charges, we have

$$\operatorname{div} \boldsymbol{D} = 0. \tag{4.302}$$

According to formula (4.297), this implies that

$$\int_0^t \tilde{\varepsilon}\left(t - t'\right) \operatorname{div} \boldsymbol{\mathcal{E}}\left(t'\right) dt' = 0. \tag{4.303}$$

By using the Fourier transform of (4.303) and the fact that $\varepsilon(\omega)$ can be zero only for special and isolated values of ω, it can be established that

$$\operatorname{div} \boldsymbol{\mathcal{E}} = 0. \tag{4.304}$$

The total electric field $\boldsymbol{\mathcal{E}}(t)$ can be decomposed into two distinct components: incident $\boldsymbol{\mathcal{E}}_i(t)$ and scattered $\boldsymbol{\mathcal{E}}_s(t)$ fields,

$$\boldsymbol{\mathcal{E}}(t) = \boldsymbol{\mathcal{E}}_s(t) + \boldsymbol{\mathcal{E}}_i(t). \tag{4.305}$$

Since

$$\operatorname{div} \boldsymbol{\mathcal{E}}_i(t) = 0, \tag{4.306}$$

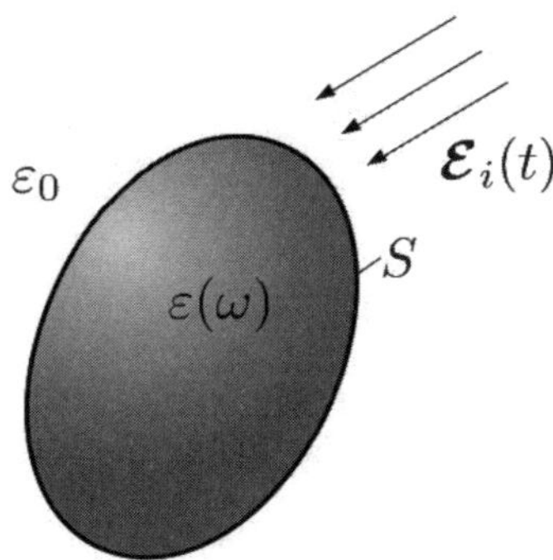

Figure 4.23

from formulas (4.304)-(4.306) we find

$$\operatorname{div} \boldsymbol{\mathcal{E}}_s(t) = 0. \tag{4.307}$$

Now, the excitation problem (see Figure 4.23) can be stated as the following boundary value problem:

$$\operatorname{curl} \boldsymbol{\mathcal{E}}_s^{\pm} = 0, \tag{4.308}$$

$$\operatorname{div} \boldsymbol{\mathcal{E}}_s^{\pm} = 0, \tag{4.309}$$

$$\mathbf{n}_Q \times \left[\boldsymbol{\mathcal{E}}_s^{+}(Q,t) - \boldsymbol{\mathcal{E}}_s^{-}(Q,t) \right] = 0, \tag{4.310}$$

$$\mathbf{n}_Q \cdot \left[\int_0^t \tilde{\varepsilon}(t-t')\, \boldsymbol{\mathcal{E}}_s^{+}(Q,t')\, dt' - \varepsilon_0 \boldsymbol{\mathcal{E}}_s^{-}(Q,t) \right]$$
$$= \mathbf{n}_Q \cdot \left[\varepsilon_0 \boldsymbol{\mathcal{E}}_i(Q,t) - \int_0^t \tilde{\varepsilon}(t-t')\, \boldsymbol{\mathcal{E}}_i(Q,t')\, dt' \right]. \tag{4.311}$$

We shall look for the solution of the boundary value problem (4.308)-(4.311) in the form

$$\boldsymbol{\mathcal{E}}_s(Q,t) = \frac{1}{4\pi\varepsilon_0} \oint_S \sigma(M,t) \frac{\mathbf{r}_{MQ}}{r_{MQ}^3}\, ds_M. \tag{4.312}$$

It is apparent that equations (4.308) and (4.309) as well as the boundary condition (4.310) are satisfied for any $\sigma(M,t)$. Then, by using formulas (2.30) and (2.31) from Chapter 2, we find that the boundary condition (4.311) is fulfilled if $\sigma(M, t)$ satisfies the equation

$$\int_0^t \frac{\tilde{\varepsilon}(t-t')}{\varepsilon_0} \sigma(Q, t') dt' - \frac{1}{2\pi} \oint_S \left[\int_0^t \frac{\tilde{\varepsilon}(t-t')}{\varepsilon_0} \sigma(M, t') dt' \right] \frac{\mathbf{r}_{MQ} \cdot \mathbf{n}_Q}{r_{MQ}^3} ds_M$$

$$+ \sigma(Q, t) + \frac{1}{2\pi} \oint_S \sigma(M, t) \frac{\mathbf{r}_{MQ} \cdot \mathbf{n}_Q}{r_{MQ}^3} ds_M$$

$$= 2 \int_0^t \tilde{\varepsilon}(t-t') \mathbf{n}_Q \cdot \boldsymbol{\mathcal{E}}_i(Q, t') dt' - \varepsilon_0 \mathbf{n}_Q \cdot \boldsymbol{\mathcal{E}}_i(Q, t). \tag{4.313}$$

Next, we shall use the following biorthogonal expansion of $\sigma(M, t)$:

$$\sigma(M, t) = \sum_{k=1}^{\infty} a_k(t) \sigma_k(M), \tag{4.314}$$

where, according to formula (4.293), the expansion coefficients $a_m(t)$ are given by the formula

$$a_m(t) = \oint_S \sigma(M, t) \tau_m(M) ds_M. \tag{4.315}$$

It is clear that the time evolution of the expansion coefficient $a_k(t)$ reveals the time-dynamics of the excitation of the plasmon mode corresponding to (and created by) $\sigma_k(M)$. For this reason, the time-dynamics of $a_k(t)$ is studied below. To derive the equation for $a_k(t)$, we substitute the expansion (4.314) into equation (4.313). Then, by using formula (4.289), we obtain

$$\sum_k \sigma_k(Q) \int_0^t \frac{\tilde{\varepsilon}(t-t')}{\varepsilon_0} a_k(t') dt' - \sum_k \frac{\sigma_k(Q)}{\lambda_k} \int_0^t \frac{\tilde{\varepsilon}(t-t')}{\varepsilon_0} a_k(t') dt'$$

$$+ \sum_k \sigma_k(Q) a_k(t) + \sum_k \frac{\sigma_k(Q)}{\lambda_k} a_k(t)$$

$$= 2 \int_0^t \tilde{\varepsilon}(t-t') \mathbf{n}_Q \cdot \boldsymbol{\mathcal{E}}_i(Q, t') dt' - \varepsilon_0 \mathbf{n}_Q \cdot \boldsymbol{\mathcal{E}}_i(Q, t). \tag{4.316}$$

By multiplying both sides of formula (4.316) by $\tau_m(Q)$ and then by integrating both sides over S, with the help of biorthogonality conditions (4.293) we derive the following equation for expansion coefficients $a_k(t)$:

$$\left(1 - \frac{1}{\lambda_k}\right) \int_0^t \frac{\tilde{\varepsilon}\left(t - t'\right)}{\varepsilon_0} a_k\left(t'\right) dt' + \left(1 + \frac{1}{\lambda_k}\right) a_k(t)$$

$$= 2 \int_0^t \tilde{\varepsilon}\left(t - t'\right) \left(\oint_S \mathbf{n}_Q \cdot \boldsymbol{\mathcal{E}}_i\left(Q, t'\right) \tau_k(Q) ds_Q \right) dt'$$

$$- \varepsilon_0 \oint_S \mathbf{n}_Q \cdot \boldsymbol{\mathcal{E}}_i(Q, t) \tau_k(Q) ds_Q. \tag{4.317}$$

The last equation can be appreciably simplified through its Fourier transform. This yields the relation

$$\left[\left(1 - \frac{1}{\lambda_k}\right) \frac{\varepsilon(\omega)}{\varepsilon_0} + \left(1 + \frac{1}{\lambda_k}\right)\right] A_k(\omega)$$

$$= 2\left[\varepsilon(\omega) - \varepsilon_0\right] \oint_S \mathbf{E}_i(Q, \omega) \cdot \mathbf{n}_Q \tau_k(Q) ds_Q, \tag{4.318}$$

where $A_k(\omega)$ and $\mathbf{E}_i(Q, \omega)$ are the Fourier transforms of $a_k(t)$ and $\boldsymbol{\mathcal{E}}_i(Q, t)$, respectively. Furthermore, by using formulas (see (4.291))

$$\lambda_k + 1 = \frac{2\varepsilon_k}{\varepsilon_k + \varepsilon_0}, \qquad \lambda_k - 1 = -\frac{2\varepsilon_0}{\varepsilon_k + \varepsilon_0}, \tag{4.319}$$

the equation (4.318) can be transformed as follows:

$$A_k(\omega) = \frac{(\varepsilon_k - \varepsilon_0)\left[\varepsilon(\omega) - \varepsilon_0\right]}{\varepsilon_k - \varepsilon(\omega)} \oint_S \mathbf{E}_i(Q, \omega) \cdot \mathbf{n}_Q \tau_k(Q) ds_Q. \tag{4.320}$$

In the case when the incident radiation is linearly polarized and its free-space wavelength is appreciably larger than the nanoparticle dimensions, it can be assumed with high accuracy that

$$\boldsymbol{\mathcal{E}}_i(Q, t) = \mathbf{E}_i(Q) f(t), \tag{4.321}$$

and

$$\mathbf{E}_i(Q, \omega) = \mathbf{E}_i(Q) F(\omega), \tag{4.322}$$

where $F(\omega)$ is the Fourier transform of $f(t)$. By using formula (4.322) in equation (4.320), we obtain

$$A_k(\omega) = \frac{\varepsilon(\omega) - \varepsilon_0}{\varepsilon_k - \varepsilon(\omega)} F(\omega) \left(\varepsilon_k - \varepsilon_0\right) \oint_S \mathbf{E}_i(Q) \cdot \mathbf{n}_Q \tau_k(Q) ds_Q. \tag{4.323}$$

It is apparent that the integral in the last formula accounts for the coupling of the incident radiation to a specific plasmon mode. This coupling can be made more explicit by assuming that $\mathbf{E}_i(Q)$ is uniform (constant) over the nanoparticle volume. The latter assumption is sufficiently accurate if the free-space wavelength of the incident radiation is appreciably larger than the nanoparticle dimensions and this radiation is produced by remote sources. The latter may not be the case in SERS-related research where plasmon resonances are utilized to enhance the Raman scattering of molecules localized on nanoparticle surfaces. Under the validity of the above uniformity assumption, the last formula can be transformed as follows:

$$A_k(\omega) = \frac{\varepsilon(\omega) - \varepsilon_0}{\varepsilon_k - \varepsilon(\omega)} F(\omega)(\varepsilon_k - \varepsilon_0)\mathbf{E}_i \cdot \oint_S \mathbf{n}_Q \tau_k(Q) ds_Q. \tag{4.324}$$

Now, by recalling formulas (2.168) and (4.296), we obtain

$$A_k(\omega) = \mathbf{E}_i \cdot \mathbf{p}_k \frac{\varepsilon(\omega) - \varepsilon_0}{\varepsilon_k - \varepsilon(\omega)} F(\omega). \tag{4.325}$$

The last formula clearly suggests that the incident radiation from remote sources is effectively coupled only to dipole plasmon modes and this coupling is controlled by the orientation of electric field of incident radiation with respect to the dipole moment of the plasmon mode. These facts, by the way, have been well understood in the area of SERS research.

It is instructive to write the last equation in the form

$$A_k(\omega) = (\mathbf{E}_i \cdot \mathbf{p}_k) R_k(\omega) F(\omega), \tag{4.326}$$

where

$$R_k(\omega) = \frac{\varepsilon(\omega) - \varepsilon_0}{\varepsilon_k - \varepsilon(\omega)} \tag{4.327}$$

can be construed as the transfer function for the k-th plasmon mode normalized by $\mathbf{E}_i \cdot \mathbf{p}_k$. The formula (4.327) reveals the resonance nature of excitation of the k-th plasmon mode at the frequency ω_k. Indeed, according to (4.292), at this frequency the denominator $\varepsilon_k - \varepsilon(\omega)$ becomes very small if $\varepsilon''(\omega_k)$ is very small. In the limiting case of vanishing ε'', $R_k(\omega)$ exhibits a pole at $\omega = \omega_k$.

By using formula (4.326), the following expression is derived for $a_k(t)$:

$$a_k(t) = \frac{1}{\sqrt{2\pi}} (\mathbf{E}_i \cdot \mathbf{p}_k) \int_{-\infty}^{\infty} r_k(t') f(t - t') dt', \tag{4.328}$$

where $r_k(t)$ is the inverse Fourier transform of $R_k(\omega)$.

Now, the algorithm of numerical analysis of time-dynamics of excitation of the k-th plasmon mode can be stated as follows. First, the eigenvalue problems (4.289) and (4.290) are solved and the proper normalization of $\sigma_k(M)$ and $\tau_k(M)$ defined by formulas (4.294) and (4.296) is performed. Then, $r_k(t)$ is determined through the inverse Fourier transform of $R_k(\omega)$. Finally, formula (4.328) is employed to evaluate the time evolution of $a_k(t)$, which reveals the time-dynamics of excitation (and/or dephasing) of the k-th plasmon mode. It is worthwhile to mention here that the outlined computations can be performed by using an actual, experimentally measured dispersion relation $\varepsilon(\omega)$.

Formulas (4.326)-(4.328) can be used for analytical calculations as well. We shall first demonstrate this by deriving the analytical expression for the steady state $a_k^{(ss)}(t)$ of $a_k(t)$ in the case of resonance excitation

$$f(t) = \sin \omega_k t. \tag{4.329}$$

In this case, we have

$$F(\omega) = j\sqrt{\frac{\pi}{2}}\left[\delta(\omega - \omega_k) - \delta(\omega + \omega_k)\right], \tag{4.330}$$

where δ stands for the Dirac delta function.

By substituting the last formula into equation (4.325) and by taking the inverse Fourier transform, we find

$$a_k^{(ss)}(t) = \frac{j}{2}(\mathbf{E}_i \cdot \mathbf{p}_k) \int_{-\infty}^{\infty} \frac{\varepsilon(\omega) - \varepsilon_0}{\varepsilon_k - \varepsilon(\omega)}\left[\delta(\omega - \omega_k) - \delta(\omega + \omega_k)\right]e^{-j\omega t}d\omega. \tag{4.331}$$

By performing the integration in the last formula and taking into account the relations (4.292), (4.300) and (4.301), we arrive at

$$
\begin{aligned}
a_k^{(ss)}(t) = -\frac{\mathbf{E}_i \cdot \mathbf{p}_k}{2\varepsilon''(\omega_k)} &\left\{\left[\varepsilon'(\omega_k) - \varepsilon_0 + j\varepsilon''(\omega_k)\right]e^{-j\omega_k t}\right. \\
&\left. + \left[\varepsilon'(\omega_k) - \varepsilon_0 - j\varepsilon''(\omega_k)\right]e^{j\omega_k t}\right\}.
\end{aligned}
\tag{4.332}
$$

Finally, the last expression can be simplified as follows:

$$\boxed{a_k^{(ss)}(t) = -(\mathbf{E}_i \cdot \mathbf{p}_k)\left[\frac{\varepsilon'(\omega_k) - \varepsilon_0}{\varepsilon''(\omega_k)}\cos \omega_k t + \sin \omega_k t\right],} \tag{4.333}$$

which can also be represented in the form

$$a_k^{(ss)}(t) = (\mathbf{E}_i \cdot \mathbf{p}_k) \frac{\sqrt{[\varepsilon'(\omega_k) - \varepsilon_0]^2 + [\varepsilon''(\omega_k)]^2}}{\varepsilon''(\omega_k)} \sin(\omega_k t + \varphi_k). \qquad (4.334)$$

In the case of strongly pronounced plasmon resonances, we have

$$|\varepsilon'(\omega_k)| \gg \varepsilon_0, \qquad (4.335)$$

$$|\varepsilon'(\omega_k)| \gg \varepsilon''(\omega_k), \qquad (4.336)$$

and formula (4.333) is transformed to

$$\boxed{a_k^{(ss)}(t) \approx -(\mathbf{E}_i \cdot \mathbf{p}_k) \frac{\varepsilon'(\omega_k)}{\varepsilon''(\omega_k)} \cos \omega_k t.} \qquad (4.337)$$

As typical for strong resonances, the steady state is shifted by almost 90° in time with respect to incident radiation. It is also apparent from the last formula that the strength of plasmon resonances is controlled by the ratio of the real part of permittivity to its imaginary part at the resonance frequency (resonance wavelength). For silver and gold, this ratio is most appreciable for wavelength ranges of 600-1400 nm and 650-1000 nm, respectively.

To evaluate the width of plasmon resonances, we consider the off-resonance excitation

$$f(t) = \sin \omega_0 t. \qquad (4.338)$$

Then,

$$F(\omega) = j\sqrt{\frac{\pi}{2}} \left[\delta(\omega - \omega_0) - \delta(\omega + \omega_0) \right]. \qquad (4.339)$$

By substituting the last formula into equation (4.325), taking the inverse Fourier transform and using relations (4.300)-(4.301), we derive

$$a_k^{(ss)}(t) = \frac{j}{2}(\mathbf{E}_i \cdot \mathbf{p}_k) \left[\frac{\varepsilon'(\omega_0) - \varepsilon_0 + j\varepsilon''(\omega_0)}{\varepsilon_k - \varepsilon'(\omega_0) - j\varepsilon''(\omega_0)} e^{-j\omega_0 t} \right.$$
$$\left. - \frac{\varepsilon'(\omega_0) - \varepsilon_0 - j\varepsilon''(\omega_0)}{\varepsilon_k - \varepsilon'(\omega_0) + j\varepsilon''(\omega_0)} e^{j\omega_0 t} \right]. \qquad (4.340)$$

Taking into account that

$$\frac{\varepsilon'(\omega_0) - \varepsilon_0 + j\varepsilon''(\omega_0)}{\varepsilon_k - \varepsilon'(\omega_0) - j\varepsilon''(\omega_0)} = \sqrt{\frac{[\varepsilon'(\omega_0) - \varepsilon_0]^2 + [\varepsilon''(\omega_0)]^2}{[\varepsilon_k - \varepsilon'(\omega_0)]^2 + [\varepsilon''(\omega_0)]^2}} e^{-j\varphi_0} \qquad (4.341)$$

and

$$\frac{\varepsilon'(\omega_0) - \varepsilon_0 - j\varepsilon''(\omega_0)}{\varepsilon_k - \varepsilon'(\omega_0) + j\varepsilon''(\omega_0)} = \sqrt{\frac{[\varepsilon'(\omega_0) - \varepsilon_0]^2 + [\varepsilon''(\omega_0)]^2}{[\varepsilon_k - \varepsilon'(\omega_0)]^2 + [\varepsilon''(\omega_0)]^2}} e^{j\varphi_0}, \qquad (4.342)$$

from formula (4.340) we derive

$$\boxed{a_k^{(ss)}(t) = (\mathbf{E}_i \cdot \mathbf{p}_k) C(\omega_0) \sin(\omega_0 t + \varphi_0),} \qquad (4.343)$$

where

$$\boxed{C(\omega_0) = \sqrt{\frac{[\varepsilon'(\omega_0) - \varepsilon_0]^2 + [\varepsilon''(\omega_0)]^2}{[\varepsilon_k - \varepsilon'(\omega_0)]^2 + [\varepsilon''(\omega_0)]^2}}.} \qquad (4.344)$$

By recalling the relation (4.292), it is easy to see that in the case of resonant excitation ($\omega_0 = \omega_k$) the last two formulas are reduced (as they must be) to the equation (4.334). It is also clear from the last two formulas that the sharpness of plasmon resonances is controlled by the nature of the dispersion relation around the resonance frequency. The examples of computations of $C(\omega_0)$ for silver (1) and gold (2) rectangular bar-type nanoparticles (length of 40 nm, width and height of 5.6 nm) are given in Figure 4.24. It is evident from this figure that the strength and sharpness of plasmon resonances in silver nanoparticles are much better than in gold nanoparticles. This is because the ratio of real to imaginary parts of dielectric permittivity is appreciably higher for silver than for gold (compare Figures 1.5 (gold) and 1.6 (silver) from Chapter 1). Thus, one may say that as far as the quality of plasmon resonances is concerned, silver is "gold" and gold is "silver". This is the reason why in the area of SERS research silver nanoparticles have been predominantly used in experiments.

Now, we proceed to the derivation of analytical expressions for $a_k(t)$ in the case of "rectangular" laser pulses of incident field. In these derivations, we shall use the Drude model (see Chapter 1) for $\varepsilon(\omega)$:

$$\varepsilon(\omega) = \varepsilon_0 \left[1 - \frac{\omega_p^2}{\omega(\omega + j\gamma)} \right], \qquad (4.345)$$

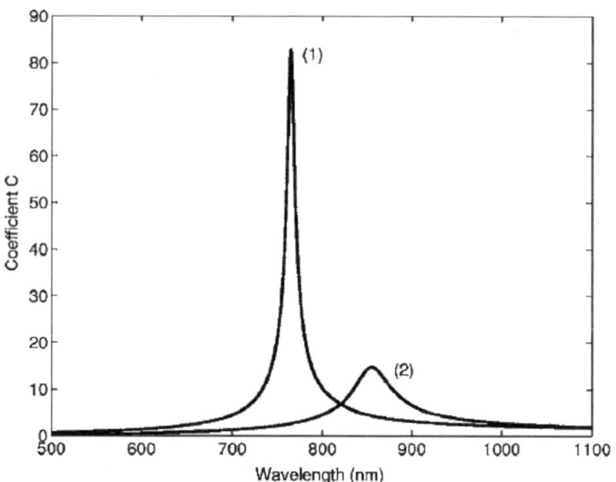

Figure 4.24

which implies that

$$\varepsilon'(\omega) = \varepsilon_0 \left[1 - \frac{\omega_p^2}{\omega^2 + \gamma^2} \right], \tag{4.346}$$

$$\varepsilon''(\omega) = \varepsilon_0 \frac{\omega_p^2 \gamma}{\omega \left(\omega^2 + \gamma^2 \right)}. \tag{4.347}$$

By using formulas (4.327), (4.346) and (4.347), we find

$$R_k(\omega) = -\frac{\omega_k^2 + \gamma^2}{(j\omega)^2 - \gamma(j\omega) + \omega_k^2 + \gamma^2}. \tag{4.348}$$

The quadratic equation (with respect to $j\omega$)

$$(j\omega)^2 - \gamma(j\omega) + \omega_k^2 + \gamma^2 = 0 \tag{4.349}$$

has the roots

$$\alpha_1 = \nu + j\beta \qquad \text{and} \qquad \alpha_2 = \nu - j\beta, \tag{4.350}$$

where

$$\nu = \frac{\gamma}{2}, \tag{4.351}$$

$$\beta = \frac{\sqrt{4\omega_k^2 + 3\gamma^2}}{2}. \tag{4.352}$$

As a result, the expression (4.348) can be transformed as follows:

$$R_k(\omega) = -\frac{\omega_k^2 + \gamma^2}{(\alpha_1 - j\omega)(\alpha_2 - j\omega)}, \tag{4.353}$$

which is equivalent to

$$R_k(\omega) = \frac{\omega_k^2 + \gamma^2}{\alpha_2 - \alpha_1}\left(\frac{1}{\alpha_2 - j\omega} - \frac{1}{\alpha_1 - j\omega}\right). \tag{4.354}$$

By performing the inverse Fourier transform of the last expression, we find

$$r_k(t) = \sqrt{2\pi}\frac{\omega_k^2 + \gamma^2}{\alpha_2 - \alpha_1}u(t)\left(e^{-\alpha_2 t} - e^{-\alpha_1 t}\right), \tag{4.355}$$

where $u(t)$ is the unit-step function

$$u(t) = \begin{cases} 1, & \text{if } t > 0, \\ 0, & \text{if } t < 0. \end{cases} \tag{4.356}$$

By using formulas (4.350)-(4.352), the relation (4.355) can be written in the form

$$r_k(t) = -\sqrt{2\pi}\frac{2\left(\omega_k^2 + \gamma^2\right)}{\sqrt{4\omega_k^2 + 3\gamma^2}}u(t)e^{-\frac{\gamma}{2}t}\sin\frac{\sqrt{4\omega_k^2 + 3\gamma^2}}{2}t. \tag{4.357}$$

By substituting the last formula into equation (4.328), we obtain

$$a_k(t) = -(\mathbf{E}_i \cdot \mathbf{p}_k)\frac{2\left(\omega_k^2 + \gamma^2\right)}{\sqrt{4\omega_k^2 + 3\gamma^2}}\int_{-\infty}^{t}e^{-\frac{\gamma}{2}(t-t')}\sin\frac{\sqrt{4\omega_k^2 + 3\gamma^2}}{2}(t - t')\, f\left(t'\right)dt'. \tag{4.358}$$

In the case of strong plasmon resonances as well as the "transparency" of dispersive media, the following inequality holds

$$\omega_k \gg \gamma, \tag{4.359}$$

which leads to the appreciable simplifications of formulas (4.357) and (4.358):

$$r_k(t) \simeq -\sqrt{2\pi}\omega_k u(t)e^{-\frac{\gamma}{2}t}\sin\omega_k t, \tag{4.360}$$

$$a_k(t) \simeq -(\mathbf{E}_i \cdot \mathbf{p}_k)\omega_k\int_{-\infty}^{t}e^{-\frac{\gamma}{2}(t-t')}\sin\omega_k\left(t - t'\right)f\left(t'\right)dt'. \tag{4.361}$$

 Plasmon Resonances in Nanoparticles

Next, we shall use formulas (4.358) and (4.361) for the analysis of time-dynamics of excitation and dephasing of plasmon modes. We start with the case of lossless media

$$\gamma = 0 \tag{4.362}$$

and "rectangular" time-step of incident radiation at resonance frequency:

$$f(t) = u(t) \sin \omega_k t. \tag{4.363}$$

By substituting formulas (4.362) and (4.363) into equation (4.361), we find

$$a_k(t) = -(\mathbf{E}_i \cdot \mathbf{p}_k)\omega_k \int_0^t \sin \omega_k (t - t') \sin \omega_k t' dt'. \tag{4.364}$$

By evaluating the integral in the last formula, we obtain that for $t > 0$

$$a_k(t) = \frac{\mathbf{E}_i \cdot \mathbf{p}_k}{2} \left[\omega_k t \cos \omega_k t - \sin \omega_k t \right]. \tag{4.365}$$

For sufficiently large time t, we have

$$a_k(t) \approx \frac{\mathbf{E}_i \cdot \mathbf{p}_k}{2} \omega_k t \cos \omega_k t. \tag{4.366}$$

The last formula reveals the linear in time growth of amplitude of the excited plasmon mode, which is typical for ideal (lossless) resonances.

It is interesting and instructive to consider the time-dynamics of the growth of lossless plasmon modes in the case of off-resonance excitation

$$f(t) = u(t) \sin \omega_0 t. \tag{4.367}$$

By substituting formulas (4.362) and (4.367) into equation (4.361), we find

$$a_k(t) = -(\mathbf{E}_i \cdot \mathbf{p}_k)\omega_k \int_0^t \sin \omega_k (t - t') \sin \omega_0 t' dt'. \tag{4.368}$$

By evaluating the integral in the last formula, we arrive at the following result:

$$a_k(t) = (\mathbf{E}_i \cdot \mathbf{p}_k) \left[\frac{\omega_k^2}{\omega_0 + \omega_k} \frac{2 \sin \frac{\omega_0 - \omega_k}{2} t}{\omega_0 - \omega_k} \cos \frac{\omega_0 + \omega_k}{2} t - \frac{\omega_k}{\omega_0 + \omega_k} \sin \omega_k t \right].$$

$$\tag{4.369}$$

It is apparent that in the limit of $\omega_0 \to \omega_k$, the last formula is reduced to the equation (4.365). In the case when ω_0 is close to ω_k, we can retain two terms in the Taylor expansion of $\sin \frac{\omega_0 - \omega_k}{2} t$, which leads to the formula

$$a_k(t) \approx \frac{\mathbf{E}_i \cdot \mathbf{p}_k}{2} \omega_k t \left[1 - \frac{[(\omega_0 - \omega_k)t]^2}{24} \right] \cos \omega_k t. \qquad (4.370)$$

The last formula reveals the correction to the linear growth of plasmon mode magnitude caused by the deviation of the excitation frequency ω_0 from the resonance frequency ω_k.

Now, we consider the time dynamics of $a_k(t)$ in the lossy case when $\gamma \neq 0$. In the case of resonance excitation (see formula (4.363)), the equation (4.358) can be written as follows:

$$a_k(t) = -(\mathbf{E}_i \cdot \mathbf{p}_k) \frac{2\left(\omega_k^2 + \gamma^2\right)}{\sqrt{4\omega_k^2 + 3\gamma^2}} \int_0^t e^{-\frac{\gamma}{2}(t-t')} \sin \frac{\sqrt{4\omega_k^2 + 3\gamma^2}}{2}(t - t') \sin \omega_k t' dt'.$$

$$(4.371)$$

By evaluating the integral in the last formula, after lengthy (but straightforward) transformations we arrive at the following result:

$$a_k(t) = -\ (\mathbf{E}_i \cdot \mathbf{p}_k)\omega_k e^{-\frac{\gamma}{2}t} \left(\frac{1}{\gamma} \cos \frac{\sqrt{4\omega_k^2 + 3\gamma^2}}{2} t \right.$$

$$\left. - \frac{4}{\sqrt{4\omega_k^2 + 3\gamma^2}} \sin \frac{\sqrt{4\omega_k^2 + 3\gamma^2}}{2} t \right) + a_k^{(ss)}(t), \qquad (4.372)$$

where

$$a_k^{(ss)}(t) = (\mathbf{E}_i \cdot \mathbf{p}_k) \left[\frac{\omega_k}{\gamma} \cos \omega_k t - \sin \omega_k t \right]. \qquad (4.373)$$

It is easy to see that $a_k^{(ss)}(t)$ given by the last formula is identical to $a_k^{(ss)}(t)$ given by the formula (4.333). This is because according to relations (4.346) and (4.347) we have

$$\frac{\omega_k}{\gamma} = -\frac{\varepsilon'(\omega_k) - \varepsilon_0}{\varepsilon''(\omega_k)}. \qquad (4.374)$$

In the important case of strong plasmon resonances when the inequality (4.359) holds, the formula (4.372) can be substantially simplified and written as follows:

$$\boxed{a_k(t) \approx (\mathbf{E}_i \cdot \mathbf{p}_k)\frac{\omega_k}{\gamma}\left(1 - e^{-\frac{\gamma}{2}t}\right)\cos\omega_k t.}$$

(4.375)

It is apparent that at the initial stage of the excitation process (i.e., for relatively small t), the last formula can be reduced to formula (4.366) by retaining two terms in the Taylor expansion of $e^{-\frac{\gamma}{2}t}$. This means that at the initial stage of the excitation process there is almost linear growth in time of the amplitude (i.e., peak value) of the plasmon mode. As the excitation proceeds, this linear growth is saturated due to ohmic losses accounted for by finite γ. In the limit of $t \to \infty$, $a_k(t)$ reaches its steady state

$$a_k^{(ss)}(t) = (\mathbf{E}_i \cdot \mathbf{p}_k)\frac{\omega_k}{\gamma}\cos\omega_k t,$$

(4.376)

which coincides with (4.373) if the second term, small in comparison with $\frac{\omega_k}{\gamma}$, is neglected.

In the case of off-resonance excitation (see formula (4.367)) and small γ (see inequality (4.359)), the relation (4.361) can be written as follows:

$$a_k(t) = -(\mathbf{E}_i \cdot \mathbf{p}_k)\omega_k \int_0^t e^{-\frac{\gamma}{2}(t-t')}\sin\omega_k(t - t')\sin\omega_0 t'dt'.$$

(4.377)

By evaluating the integral in the last formula and neglecting small terms, we arrive at the following expression:

$$\boxed{\begin{aligned}a_k(t) \approx (\mathbf{E}_i \cdot \mathbf{p}_k)\frac{\omega_k}{2}&\left[\frac{\frac{\gamma}{2}}{\frac{\gamma^2}{4} + (\omega_0 - \omega_k)^2}\left(\cos\omega_0 t - e^{-\frac{\gamma}{2}t}\cos\omega_k t\right)\right.\\ &\left.+ \frac{\omega_0 - \omega_k}{\frac{\gamma^2}{4} + (\omega_0 - \omega_k)^2}\left(\sin\omega_0 t - e^{-\frac{\gamma}{2}t}\sin\omega_k t\right)\right].\end{aligned}}$$

(4.378)

It is clear that in the limit of $\omega_0 \to \omega_k$, the last formula is reduced to the formula (4.375).

Finally, in order to study the time-dynamics of dephasing of plasmon modes, we consider the time-evolution of $a_k(t)$ in the case of excitation of plasmon modes by a "rectangular" pulse of incident optical radiation. In this case,

$$f(t) = [u(t) - u(t - T)] \sin \omega_k t, \qquad (4.379)$$

where T is the duration of the pulse. To simplify final formulas we consider the natural case when the inequality (4.359) holds. In this case, formula (4.375) is valid for $0 < t < T$. For $t > T$, we shall use the relation (4.361) which, after substitution of formula (4.379), has the form

$$a_k(t) = -(\mathbf{E}_i \cdot \mathbf{p}_k)\omega_k \int_0^T e^{-\frac{\gamma}{2}(t-t')} \sin \omega_k (t - t') \sin \omega_k t' dt'. \qquad (4.380)$$

By evaluating the integral in the last formula and neglecting the terms small in comparison with $\frac{\omega_k}{\gamma}$, we arrive at the following result:

$$\boxed{a_k(t) \approx (\mathbf{E}_i \cdot \mathbf{p}_k)\frac{\omega_k}{\gamma} e^{-\frac{\gamma}{2}t} \left(e^{\frac{\gamma T}{2}} - 1 \right) \cos \omega_k t.} \qquad (4.381)$$

In the case of rectangular pulse of optical radiation of "off-resonance" frequency, we have

$$f(t) = [u(t) - u(t - T)] \sin \omega_0 t. \qquad (4.382)$$

By substituting the last formula into equation (4.361), we find

$$a_k(t) = -(\mathbf{E}_i \cdot \mathbf{p}_k)\omega_k \int_0^T e^{-\frac{\gamma}{2}(t-t')} \sin \omega_k (t - t') \sin \omega_0 t' dt'. \qquad (4.383)$$

By evaluating the integral in formula (4.383) and neglecting small terms, we obtain

$$\boxed{\begin{aligned} a_k(t) = (\mathbf{E}_i \cdot \mathbf{p}_k)\frac{\omega_k e^{-\frac{\gamma}{2}t}}{\gamma^2 + 4(\omega_0 - \omega_k)^2} &\left\{ \gamma \left[e^{\frac{\gamma T}{2}} \cos[\omega_k t + (\omega_0 - \omega_k)T] - \cos \omega_k t \right] \right. \\ &\left. + 2(\omega_0 - \omega_k) \left[e^{\frac{\gamma T}{2}} \sin[\omega_k t + (\omega_0 - \omega_k)T] - \sin \omega_k t \right] \right\}. \end{aligned}}$$

$$(4.384)$$

It is apparent that in the limit of $\omega_0 \to \omega_k$, the last formula is reduced to the equation (4.381).

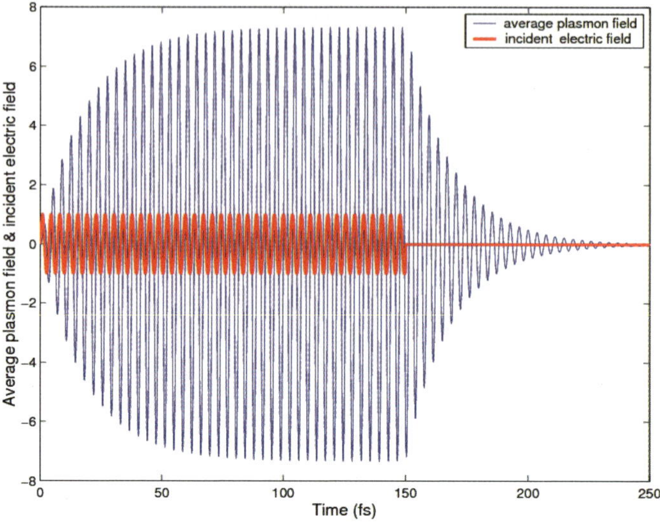

Figure 4.25

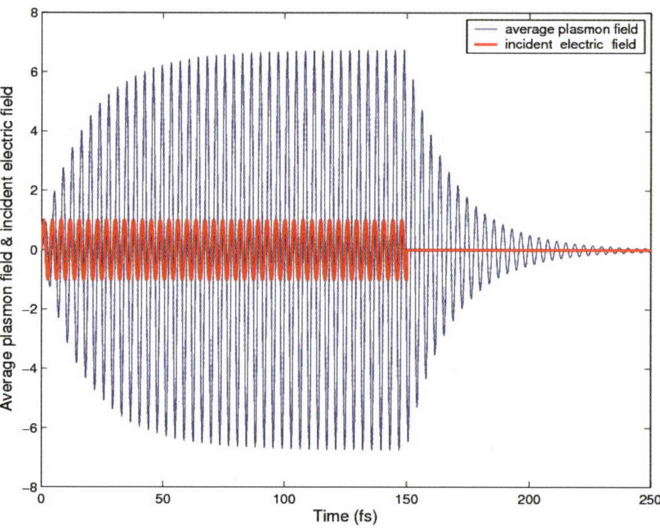

Figure 4.26

All the formulas of this section have been derived for nanoparticles in free space. However, these formulas are also valid for nanoparticles placed on dielectric substrates. Indeed, the derivation has been based on the biorthogonal expansion (4.314)-(4.315) and the specific mathematical forms of the kernels of integral equations (4.289) and (4.290) has not been essential in this derivation. This means that the presented derivation can be literally repeated for the kernels which account for the presence of dielectric substrates (see Chapter 2).

It is instructive to compare the computational results obtained by using the analytical formulas derived by employing the Drude model (4.345) with the computational results obtained by using the experimentally measured dispersion relation $\varepsilon(\omega)$. This comparison is presented by Figures 4.25 and 4.26 for gold rings on a glass substrate subject to rectangular pulses of optical radiation with resonance wavelength of 1102 nm. In computations, $\gamma = 1.075 \cdot 10^{14}$ s^{-1} and gold dispersion relation from [7] have been used. It is clear from these figures that the Drude model leads to quantitatively similar results as the use of the measured dispersion relation from [7]. It is worthwhile to point out that formulas (4.381) and (4.384) imply that $\frac{1}{\gamma}$ can be identified as the decay (dephasing) time for the light intensity (not electric field) of the excited plasmon modes. In accordance with the available experimental data for γ (see [7]), this suggests that the theoretically predicted decay (dephasing) time for plasmon modes excited in gold and/or silver nanoparticles is in the range of 5-12 fs. The latter is consistent with the experimental results reported in [16]-[21].

To conclude this section, we present the comparison between the computational results based on the presented theory and the experimental results reported in [16, 21]. By using equations (4.327) and (4.328), we have computed the time-dynamics of the 774 nm resonance wavelength plasmon mode for gold cylinders on a glass substrate [16]. Figure 4.27 presents the time variation of the incident electric field of the laser pulse (see bold line) used in experiments discussed in [16]. The computed corresponding time-dynamics of the average (over nanoparticle volume) electric field of the plasmon mode is present on this figure by the thin line. This plasmon time-dynamics has been used to compute the third-order autocorrelation function (ACF), which was compared with the experimentally measured ACF from [16]. The results of this comparison are presented in Figure 4.28, where the envelope of the calculated third-order ACF (filled circles) is superimposed on the measured third-order ACF (continuous line) from [16]. Another comparison with

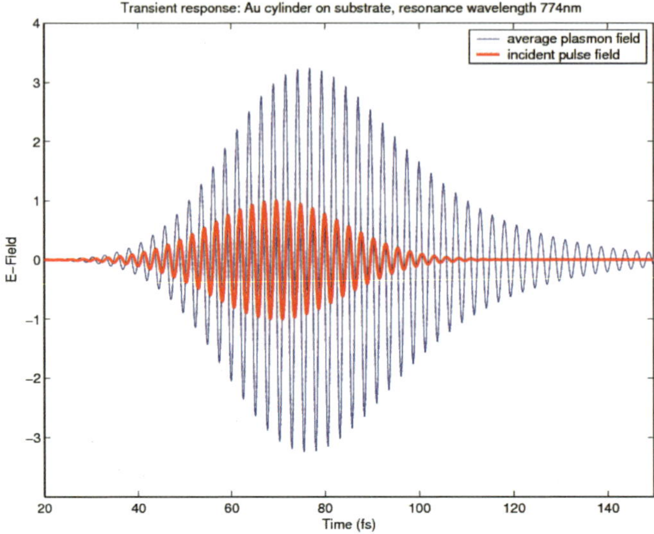

Figure 4.27

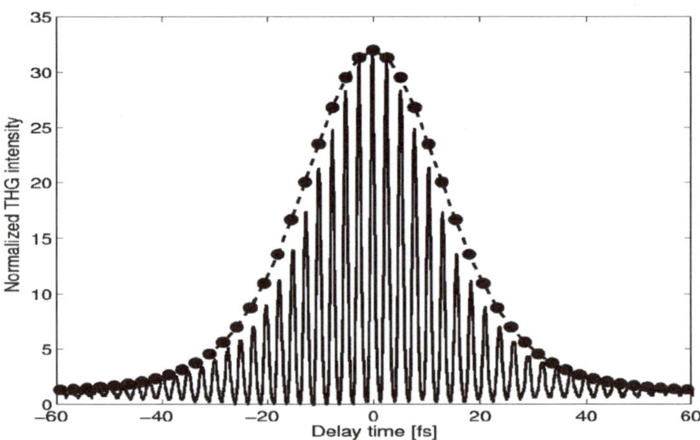

Figure 4.28

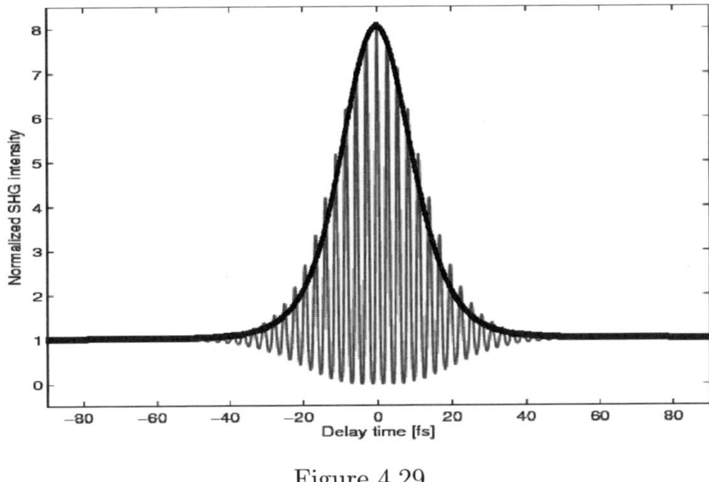

Figure 4.29

experiment is illustrated by Figure 4.29, which presents the calculated second-order ACF for L-shape gold nanoparticles from [21] and the envelope measured in [21] of the second-order ACF. The L-shape nanoparticles have height of about 21 nm, arm length of about 150 nm and arm width of about 75 nm. These nanoparticles were placed on the substrate with permittivity $\varepsilon_s = 4.33$ and resonated at the wavelength of 838 nm. Figures 4.28 and 4.29 suggest the agreement with experimental data within 10%.

4.5 Selective Applications of Plasmon Resonances

This section deals with a few sample applications of plasmon resonances in nanoparticles. The choice of these applications reflects research interests of the author and, for this reason, the list is not complete by any standards. Nevertheless, it is hoped that these applications will be of interest to a broad audience of scientists and engineers. Most of these applications are probably discussed for the first time in monographic literature. Some of these applications are maybe novel, although may not be fully developed yet.

4.5.1 Plasmon Resonance Enhancement of Faraday Rotation in Garnet Films

On the macroscopic level, garnets act as gyrotropic media that discriminate between right-handed and left-handed circular polarizations of light. This discrimination leads to the Faraday magneto-optic effect. However, on the microscopic level, magneto-optic effects are controlled by spin-orbit coupling. The spin-orbit coupling is a relativistic effect and its Hamiltonian is given by the formula [22, 23]

$$\hat{H}_{so} = \frac{\hbar}{4m^2c^2} \left(\nabla V \times \hat{\mathbf{p}} \right) \cdot \hat{\boldsymbol{\sigma}}, \tag{4.385}$$

where all notations have their usual meaning.

The last equation reveals that the spin-orbit coupling depends on local electric fields $(-\nabla V)$. These fields can be optically induced by exciting plasmon resonances in metallic nanoparticles embedded in garnet films, which may eventually result in the enhancement of magneto-optic effects. This suggests that plasmon resonances in embedded nanoparticles can be potentially utilized for the enhancement of the Faraday effect in garnets as well as for the probing of the origin of this effect on the fundamental microscopic level. The latter is of interest in a larger sense, because it may point the way for possible optical manipulation (and control) of the spin-orbit coupling through the excitation of plasmon resonances. This may find applications beyond the area of magneto-optic effects.

First experimental observations of plasmon resonance enhancement of magneto-optic effects have been reported in [24]-[27]. In these publications, garnet films sputtered over substrates populated by gold nanoparticles have been used in experiments. Below, the results of our work [28]-[30] on garnet films grown by liquid phase epitaxy (LPE) over (100)-oriented substituted gadolinium gallium garnet (SGGG) substrates partially populated with gold nanoparticles are reported. This work has been performed at the ECE Department of the University of Maryland and LPS.

Thin layers of gold, between 5 nm and 10 nm, have been evaporated in selected areas (see Figure 4.30) of (100)-oriented SGGG substrates. These gold layers have been subsequently annealed in air at temperatures varying from 700°C to 900°C. This annealing resulted in segregation of gold nano-layers and formation of dome-shaped (hemispherical) nanoparticles with average height of about 25-40 nm and diameters between 40 and 100 nm. Sample

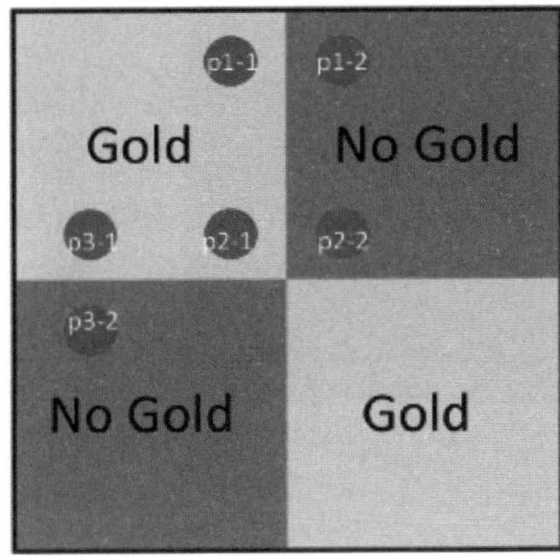

Figure 4.30

atomic force microscopy images of these gold nanoparticles are shown in Figures 4.31 and 4.32. By using an ellipsometer, light transmission experiments have been conducted to measure the transmission coefficients of the substrates in the areas with and without gold. Figure 4.33 presents the results of these measurements which reveal the appreciable decrease in transmission coefficients with a well defined minimum at about 600 nm. It is apparent from Figure 4.33 that this minimum can be attributed to the presence of gold nanoparticles and excitation of plasmon resonances in these nanoparticles. It has been repeatedly emphasized in this book that plasmon resonance wavelengths are scale invariant, i.e., these wavelengths depend on nanoparticle shapes rather than their dimensions, provided that these dimensions are small in comparison with resonance wavelengths. This implies that self-similar gold nanoparticles may all resonate at the same wavelength. Deviation from self-similarity may manifest itself in the broadening of resonance (as well as transmission coefficient) curves which represent collective effects of plasmon resonances in all the gold nanoparticles.

The experimental results in Figure 4.33 are consistent with computations performed for a dome-shaped nanoparticle on a substrate (see Figure 4.34) by using integral equation (2.87). The results of computations for various height-to-radius ratios are shown in Figure 4.35. It is apparent from this figure that for hemispherical nanoparticles (i.e., when the height and radius are approximately the same) the resonance wavelength is about 600 nm. The computations have also been performed for two and four adjacent dome-shaped

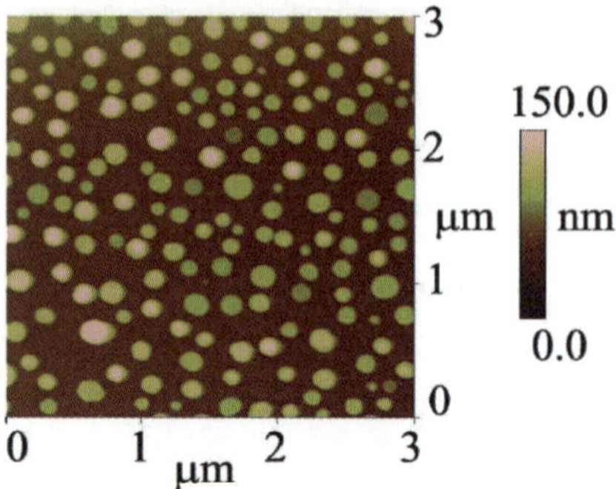

Figure 4.31

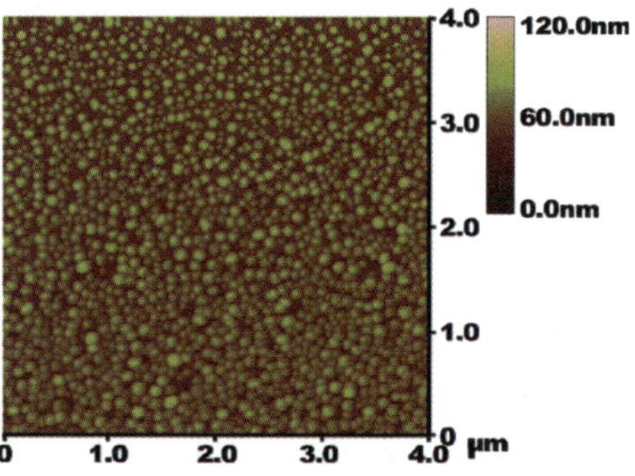

Figure 4.32

nanoparticles to examine if the plasmon resonance wavelength is affected by their proximity. The results of computations are shown in Figure 4.36a and 4.36b, respectively, and they clearly demonstrate that the plasmon resonance wavelength is more affected by the aspect ratio h/r of the nanoparticles than by their proximity.

By using the liquid phase epitaxy technique, garnet films of (Bi, Pr, Y, Gd)$_3$(Fe, Ga)$_5$O$_{12}$ composition and thickness of about 500 nm have been grown on the annealed substrates. These garnet films have in-plane magnetization. It is clear that gold nanoparticles got embedded in the epitaxially grown garnet films. It is most likely that the shape of the gold nanoparticles and especially their aspect ratios have been appreciably modified by

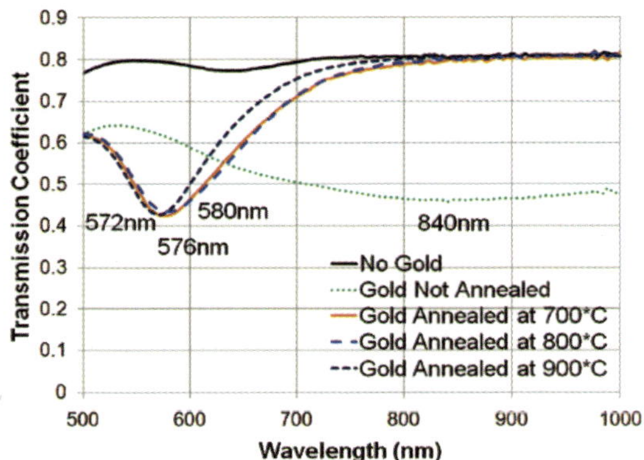

Figure 4.33

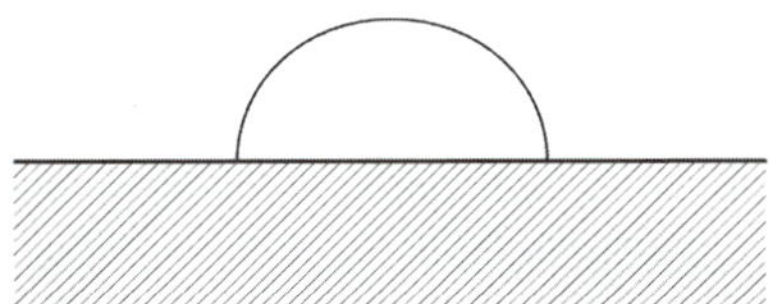

Figure 4.34

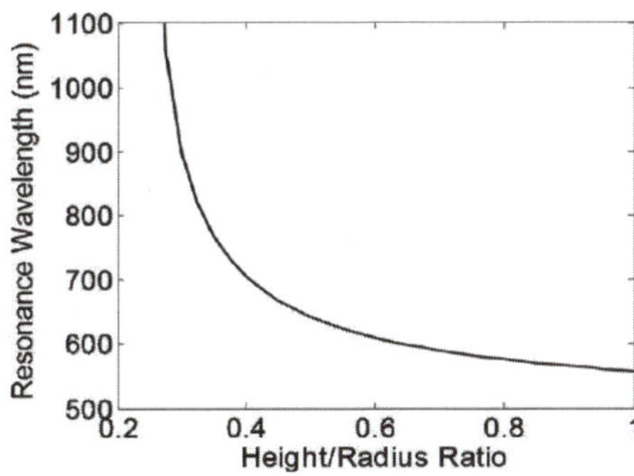

Figure 4.35

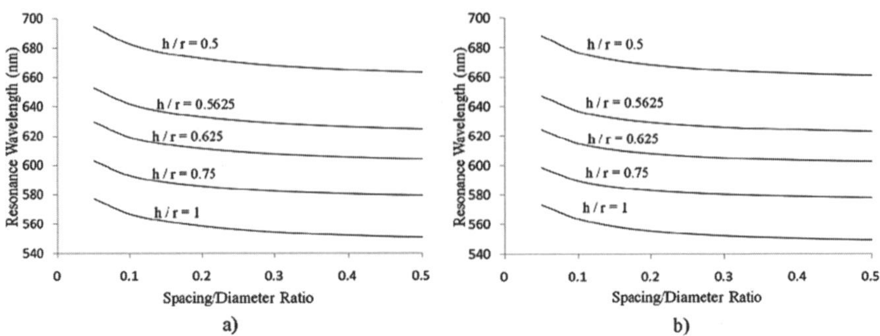

Figure 4.36

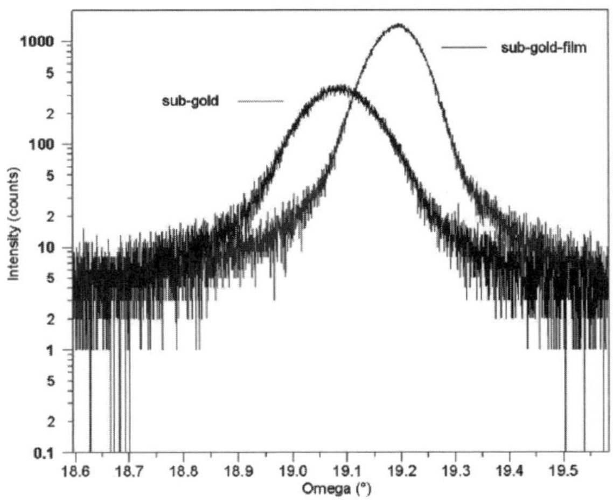

Figure 4.37

the epitaxial growth process. However, these gold nanoparticles survived the growth process in some crystalline form. The last assertion is supported by x-ray diffraction measurements which revealed the presence of gold (111) peaks near 19.1°, both prior and after the epitaxial growth of garnet film (see Figure 4.37). Light transmission experiments have been performed to measure the transmission coefficients in the areas with and without embedded nanoparticles. To avoid (or minimize) the ambiguity caused by possible variations in the thickness and composition of epitaxially grown garnet films,

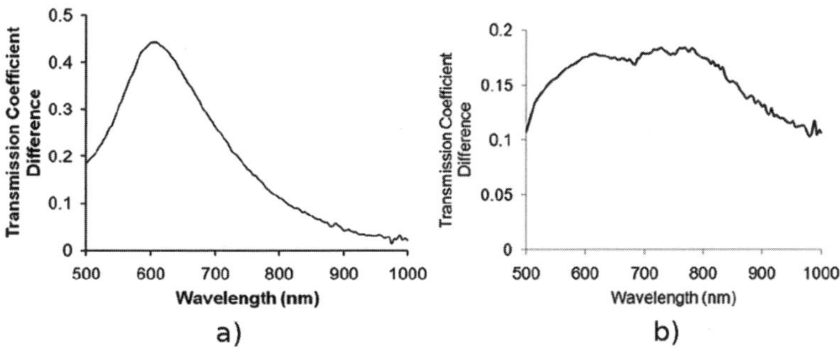

Figure 4.38

the measurements have been performed for pairs of adjacent circular areas located on opposite sides of the boundary between "Gold" and "No Gold" regions (see Figure 4.30). Figure 4.38b presents the measured difference between transmission coefficients in adjacent "p1-1" and "p1-2" areas, while Figure 4.38a presents the same difference prior to the epitaxial garnet film growth. It is conceivable that the epitaxial growth of garnet films over substrates populated with gold nanoparticles results in some appreciable spread of aspect ratios of embedded nanoparticles which leads to the broadening of the curve in Figure 4.38b in comparison with the curve in Figure 4.38a. This interpretation is supported by the computational results shown in Figure 4.36. The "broad" plasmon resonances exhibited by the curve shown in Figure 4.38b may be beneficial because they may eventually lead to the Faraday rotation enhancement in a wider wavelength range than may be the case otherwise.

The Faraday rotation enhancement has been measured for pairs of adjacent "circular" regions shown in Figure 4.30. The measurements have been performed for perpendicular to the film plane applied magnetic field by using stabilized 532 nm and 633 nm lasers. Some results of these measurements are summarized in the table below, where normalized Faraday rotations per unit length of film thickness are reported. It is evident from this table that the Faraday enhancement between 9% and 13% has been observed in the "Gold" regions p1-1, p2-1 and p3-1. It is also apparent that the Faraday rotation enhancement reported in the table occurs in the wide wavelength range which is consistent with the "broad" plasmon resonance curve presented in Figure 4.38b. The relatively small Faraday rotation enhancement

Table 4.5

	FR/thick (increase) at 633 nm	FR/thick (increase) at 532 nm
p1-1 (Gold)	0.74°/μm (13.6%)	2.85°/μm (9.0%)
p1-2 (No Gold)	0.65°/μm	2.62°/μm
p2-1 (Gold)	0.85°/μm (12.8%)	2.95°/μm (9.9%)
p2-2 (No Gold)	0.75°/μm	2.68°/μm
p3-1 (Gold)	0.89°/μm (13.2%)	2.97°/μm (11.1%)
p3-2 (No Gold)	0.78°/μm	2.67°/μm

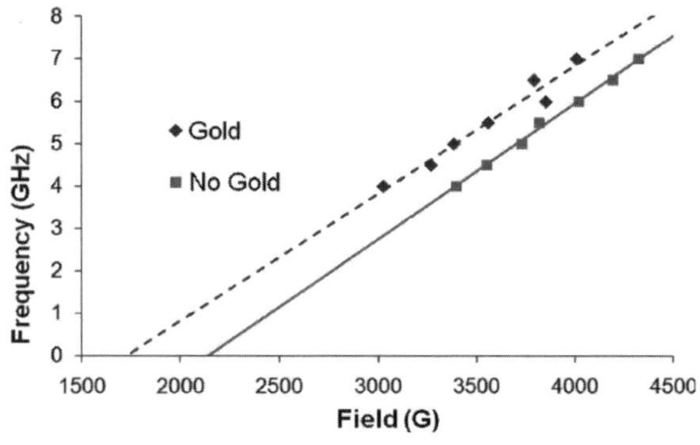

Figure 4.39

reported in the table can be attributed to the fact that the electric field of excited plasmon modes is concentrated near the gold nanoparticles and does not extend over the garnet film thickness. In other words, only relatively thin layers of epitaxially grown garnet films are affected by these electric fields and contribute to the overall Faraday rotation enhancement. This suggests that epitaxially grown multilayer structures with gold nanoparticles embedded between thin layers may be beneficial to the increase in plasmon resonance-induced Faraday rotation enhancement.

Ferromagnetic resonance (FMR) experiments were performed for the epitaxially grown garnet films using a microwave (46 GHz to 76 GHz) stripline. The results of these measurements reveal appreciably different effective magnetic fields in the "Gold" and "No Gold" regions. The results of these FMR experiments are shown in Figure 4.39. These results are consistent with the

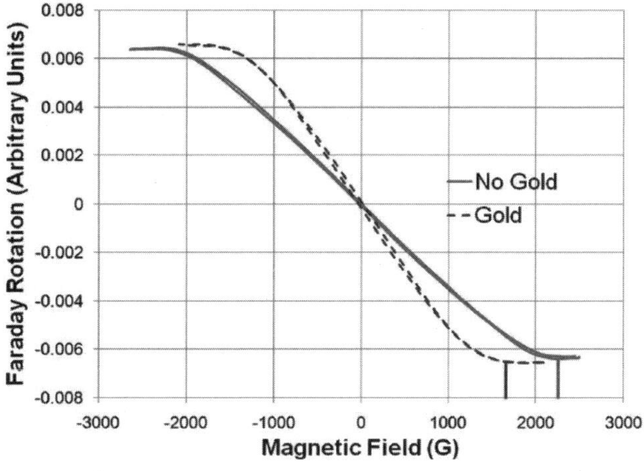

Figure 4.40

optical loop measurements shown in Figure 4.40. These measurements were performed for applied magnetic fields perpendicular to the film plane, and they reveal that the epitaxially grown garnet films with in-plane magnetization have a reverse ("negative") Faraday rotation. It is clear from Figure 4.40 that the regions with and without gold nanoparticles have saturation fields of 1.6 kG and 2.2 kG, respectively. It can be observed that these saturation fields are very close to the "intercept" fields from Figure 4.39 for FMR experiments. This implies the consistency between FMR and optical loop measurements.

The presented FMR and optical loop measurement results clearly suggest that there are different magnetic properties of epitaxially grown garnet films in adjacent regions with and without gold nanoparticles. Because these variations of magnetic properties are exhibited within the same garnet film, they can be clearly attributed to the presence of embedded gold nanoparticles. The most plausible explanation of the observed modification of magnetic properties of epitaxially grown garnet films is that the presence of gold nanoparticles affects the growth condition of LPE-grown garnet films and may result in different anisotropy properties of these films. This phenomenon is of interest in its own right because by selectively populating different areas of substrates with gold nanoparticles local control of anisotropy of epitaxially grown garnet films can be attempted. This phenomenon has been further

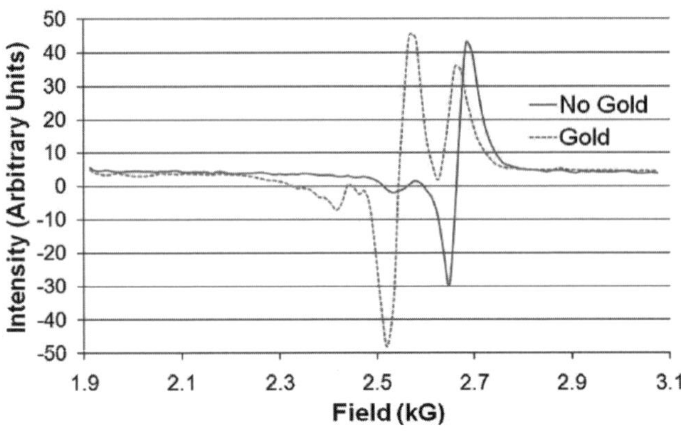

Figure 4.41

studied for garnet films with out-of-plane magnetization orientation. These films have been epitaxially grown over (100)-oriented SGGG substrates partially populated with gold nanoparticles. Their composition is described as $(Bi, Gd, Lu)_3(Fe, Ga)_5O_{12}$. To enhance the effect of gold nanoparticles on the magnetic properties of epitaxially grown garnet films, special efforts have been made to grow submicron garnet films of thickness about 280 nm. The FMR and optical loop measurements of these garnet films have been performed. Figure 4.41 presents the results of FMR measurements that most strongly exhibit the difference between magnetic properties of epitaxially grown garnet films in the regions with and without gold nanoparticles. Two ferromagnetic resonances can be observed for the region with gold nanoparticles. The appearance of two resonances may be attributed to the variation of the effective anisotropy field over the film thickness. It also cannot be precluded that the observed multiple FMR peaks are due to spin waves which arise from the abovementioned variations of effective anisotropy fields and the possible pinning of magnetization by surface anisotropy [31]. However, the latter explanation is somewhat difficult to reconcile with the observation of the larger FMR amplitude at lower magnetic fields. An example of optical loop measurements is shown in Figure 4.42, which reveals noticeable broadening of optical loops in the regions populated with gold nanoparticles. This broadening can also be attributed to the increase in anisotropy in those regions.

Finally, it is known that garnet films with out-of-plane magnetization orientation exhibit serpentine domains whose structure depends on magnetic properties of the garnet films. Figures 4.43a and 4.43b present optical images of serpentine domain structures in the "No Gold" and "Gold" regions, respectively, which clearly reveal the appreciable differences. These images are suggestive that magnetic anisotropy may be different in those regions.

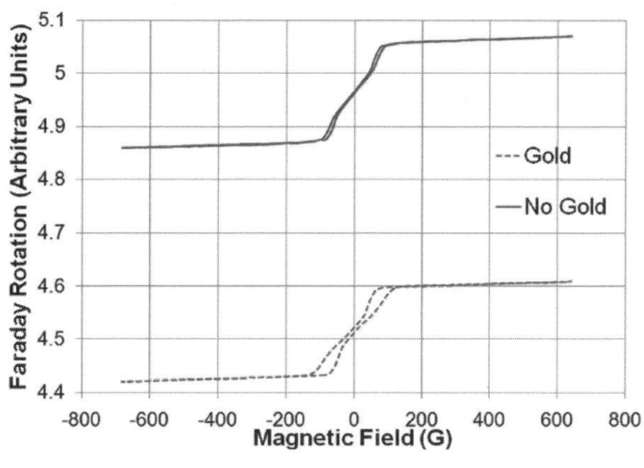

Figure 4.42

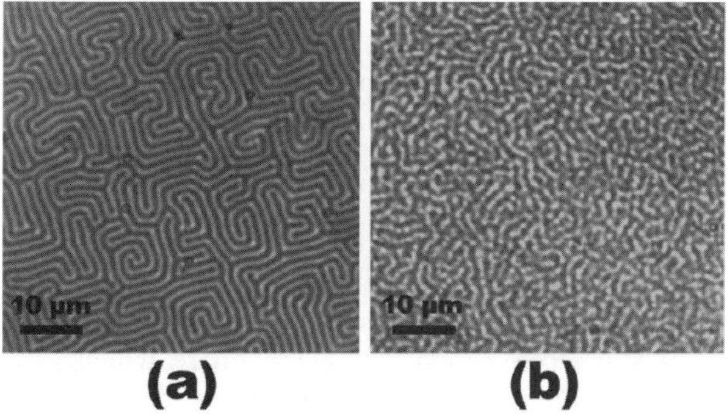

Figure 4.43

4.5.2 Application of Plasmon Resonances to Heat-Assisted Magnetic Recording

Magnetic recording (in particular, hard disk data storage) technology has been evolving at a remarkable pace that has few equals. This evolution is driven by relentless increase in areal storage density, increase in rate of data transfer and decrease in cost [32, 33]. Increase in storage density results in decrease in bit dimensions and this inevitably reduces thermal stability of recorded information and its long-time reliability. The only way to overcome this problem is to increase the coercivity of recording media. This requires very strong magnetic fields for recording and makes purely magnetic writing process practically impossible. For this reason, novel techniques for recording of digital information are currently being developed. One of them is heat-assisted magnetic recording (HAMR). In this technique, a nanoscale spot of recording medium is optically heated close to Curie temperature. This substantially reduces the coercivity of the heated local spot and makes the magnetic writing of desired magnetization pattern in this spot possible by using relatively small magnetic fields. It is apparent that the central issue in the implementation of heat-assisted magnetic recording is the development of optical sources of nanoscale resolution and high light intensity. The proper profiling of optical spots is also an important practical issue because it may reduce the detrimental effect of the collateral heating of adjacent recorded bits. It has been demonstrated [34, 35] that plasmon resonances in metallic nanoparticles and perforated metallic nano-films hold unique promise for the development of such optical sources.

Below, we numerically study some promising designs and identify specific plasmon modes that create strong and well localized optical fields. It is worthwhile to stress again that plasmon resonance modes are scale invariant and this is very beneficial for scaling of these designs to higher storage densities.

First, we consider plasmon modes in rectangular bar-type metallic nanoparticles of different aspect ratios deposited on dielectric substrates with $\varepsilon = 2.25\varepsilon_0$ (see Figure 4.44). By using the eigenvalue technique developed in the book, the specific plasmon resonance mode which generates strong optical field has been identified. This plasmon mode occurs in rectangular bar nanoparticles with 5:0.7:0.7 side aspect ratio. The resonance value of permittivity is $\varepsilon = -28.7\varepsilon_0$ and the dipole moment of the mode is oriented along the longest side of the bar-type nanoparticle. By using the dispersion

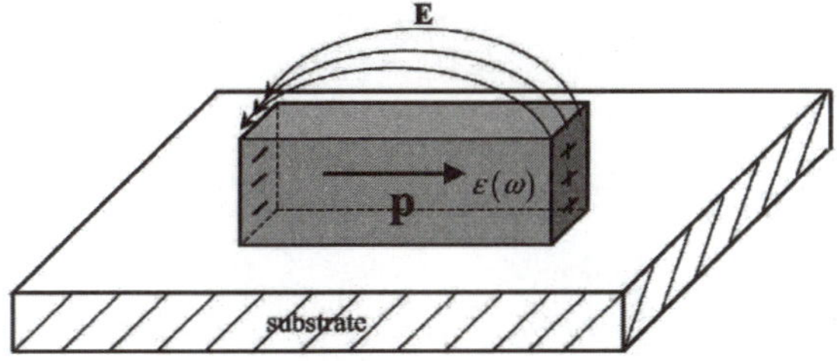

Figure 4.44

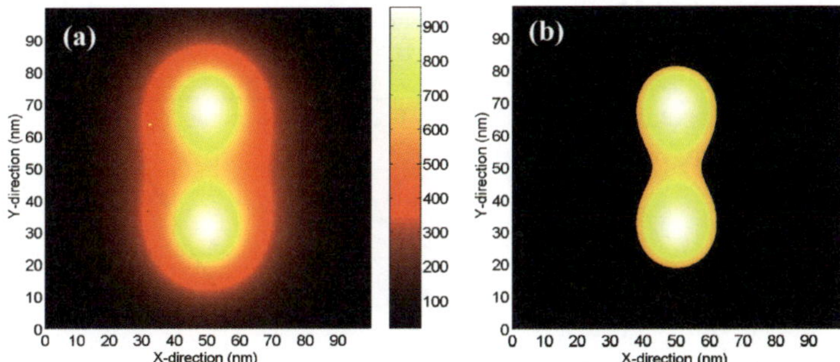

Figure 4.45

relations of gold and silver [7], it has been found that this plasmon mode can be excited by incident laser radiation with resonance wavelengths of 853 nm and 770 nm, respectively. The computed surface electric charge density of this plasmon mode has been used to calculate the mode light intensity. This plasmon mode light intensity and full width at its half maximum for bar-type silver nanoparticle (length 40 nm, width 5.6 nm, height 5.6 nm) are shown in Figures 4.45a and 4.45b, respectively, at the distance of 12 nm from the nanoparticle in the direction normal to the substrate. For a gold nanoparticle of the same dimensions, a similar plot is presented in Figure 4.46. Color bars on these figures indicate the ratios of optical near-field intensities to the intensity of the incident light. It is apparent that silver nanoparticles generate higher light intensity and this is due to more favorable ratio of real to imaginary parts of permittivity at respective resonance wavelengths. It is also clear that the optical spots shown in Figures 4.45 and 4.46 are

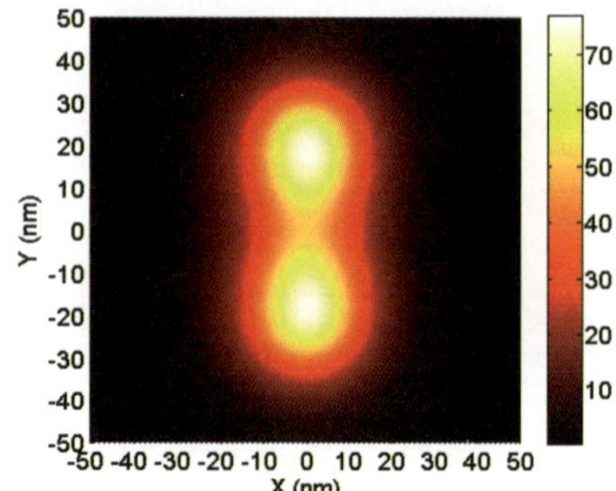

Figure 4.46

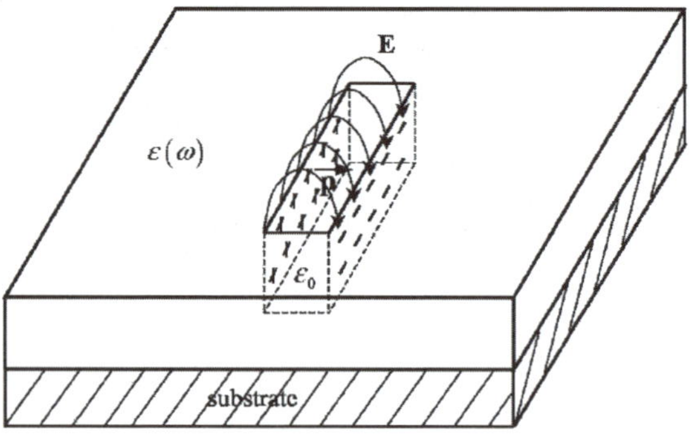

Figure 4.47

almost rectangular in shape with 4:1 aspect ratio, which may be desirable in recording applications.

Another example of a possible design that has been numerically studied is presented in Figure 4.47, which shows a perforated metallic nano-film deposited on a dielectric substrate ($\varepsilon = 2.25\varepsilon_0$). This design can be accomplished by using the ion beam milling technique. In this design, the strongest optical spots are generated by plasmon modes localized around the aperture. By using the eigenvalue technique, the extensive numerical study of plasmon modes localized around rectangular apertures with different aspect ratios has been performed. As an example, for the aspect ratio of 4:0.6:2, the specific localized plasmon mode which generates strong optical field has been

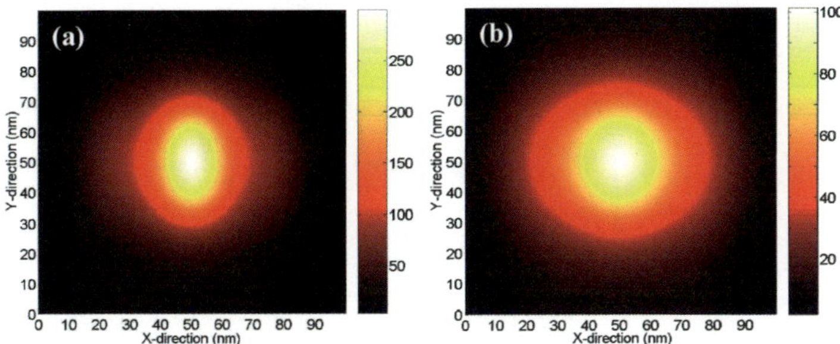

Figure 4.48

identified. This mode occurs at the resonance value of dielectric permittivity $\varepsilon = -17.2\varepsilon_0$ and its dipole moment is normal to the longest side of the aperture. This mode can be excited by the incident laser radiation with resonance wavelengths of 710 nm for gold nano-films and 617 nm for silver nano-films. The optical spots generated by this plasmon mode at the vertical distance of 10 nm and 15 nm from the silver film with aperture of 40 nm in length, 6 nm in width and 20 nm in height are shown in Figure 4.48a and 4.48b. As before, color bars indicate the ratios of optical near-field intensity to the intensity of incident light.

It is worthwhile to point out that the ratios of $\varepsilon'(\omega)$ and $\varepsilon''(\omega)$ are equal to 16.25 and 82.95 for the gold and silver bar-type nanoparticles, respectively, while these ratios are equal to 15.40 and 34.44 for gold and silver perforated nano-films, respectively. These ratios suggest that the corresponding plasmon resonances are strongly pronounced as far as the enhancement of local electric field is concerned. This enhancement is even more pronounced for the light intensity, which is proportional to the square of the electric field. Furthermore, it can be mentioned that the dielectric (glass) substrate is not only a structural part of the design, but it also results in the downward (red) shift of plasmon resonance frequencies. This makes the ratio of $\varepsilon'(\omega)$ to $\varepsilon''(\omega)$ more favorable, especially for gold.

The plasmon electric fields and local light intensities can be enhanced not only due to the large ratios of $\varepsilon'(\omega)$ to $\varepsilon''(\omega)$ at resonance frequencies, but also by proper choice of geometry of apertures. To demonstrate this, consider a T-shaped aperture in metallic nano-film (see Figure 4.49) placed on a glass ($\varepsilon = 2.25\varepsilon_0$) substrate. In this case, the corners formed by horizontal and vertical legs of the aperture substantially contribute to the local field and light

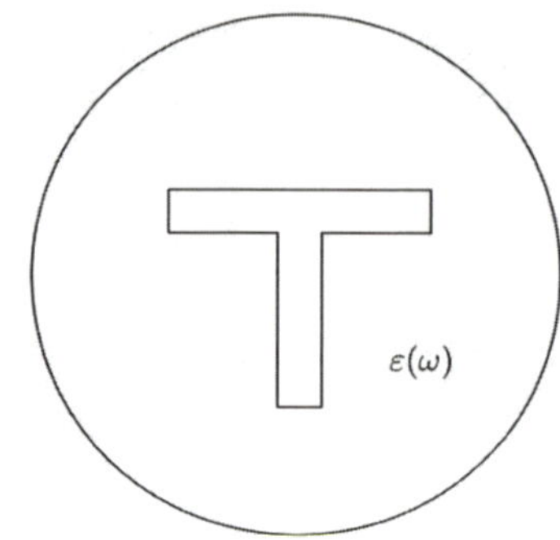

Figure 4.49

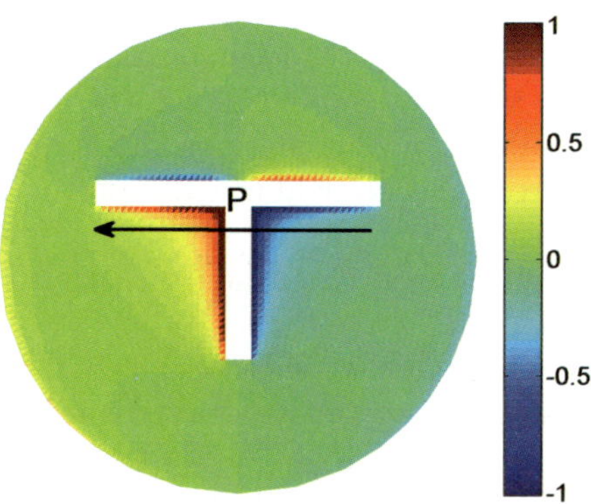

Figure 4.50

intensity enhancements. To demonstrate this, we present the results of our numerical study of plasmon modes localized around the T-shaped aperture with 110 nm length of the horizontal leg, 60 nm length of the vertical leg, the leg width of 10 nm and the film thickness of 30 nm. Through numerical calculations, we have identified the localized plasmon mode with resonance value of permittivity $\varepsilon = -34.58\varepsilon_0$ and with the dipole moment oriented along the horizontal leg of the aperture. The resonance wavelengths of this mode are 919 nm and 838 nm for gold and silver nano-films, respectively. The surface electric charge distribution for this plasmon mode is shown in Figure 4.50. The light intensity of optical spot at the vertical distance of

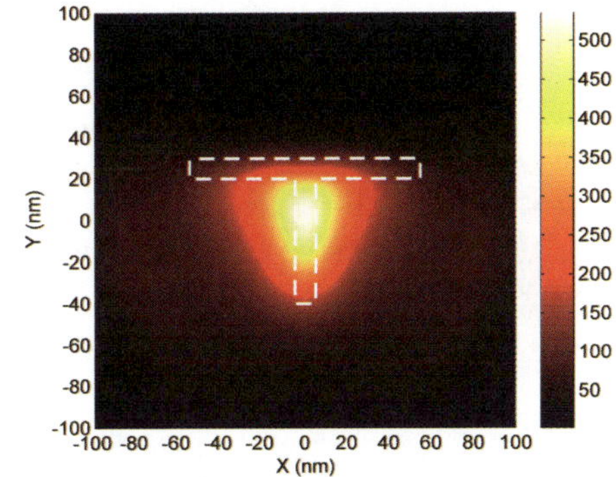

Figure 4.51

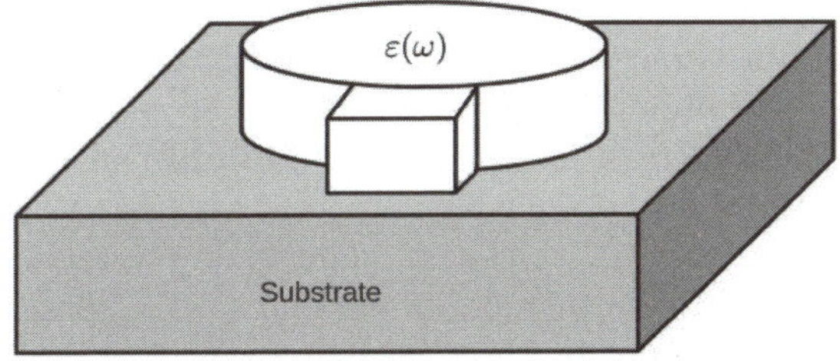

Figure 4.52

15 nm from the gold film has been computed. This light intensity is presented in Figure 4.51, which reveals that the highest enhancement of light intensity indeed occurs in the vicinity of the corners formed by the horizontal and vertical legs of the T-shaped aperture.

To further demonstrate the local field and light enhancement engineered by the proper choice of nanostructure geometry, consider plasmon modes in gold "lollipop" nanodisks deposited on a dielectric (Ta_2O_5, $\varepsilon = 4.6\varepsilon_0$) substrate (see Figure 4.52). This structure has recently been proposed and extensively studied in [35]. Furthermore, this "lollipop" structure has been integrated into the design of a recording head structure, which has been successfully used to demonstrate the actual heat-assisted magnetic recording of tracks of about 70 nm width [35]. Hence, the plasmon mode analysis of this structure is of special interest.

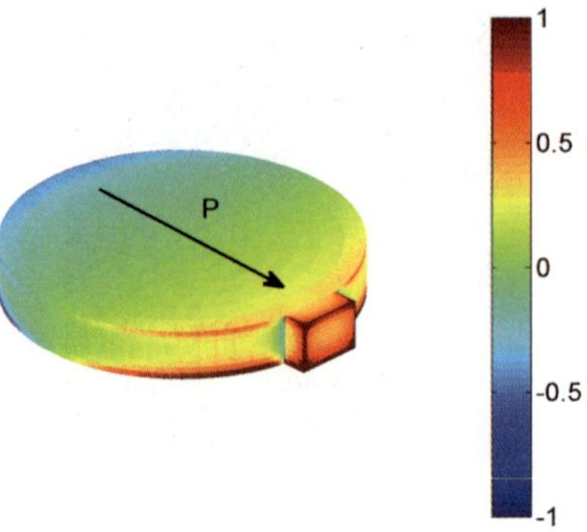

Figure 4.53

We have analyzed the "lollipop" disk studied in [35] by using the following dimensions: 200 nm diameter, 15 nm length of the peg, 40 nm width of the peg. Two different lollipop thicknesses (22 and 30 nm) have been used in calculations. Our numerical (eigenvalue) analysis of this structure suggests that the above "lollipop" design has superior resonance properties at the thickness of 30 nm. At this thickness, the dipole plasmon mode exists at $\varepsilon = -25.52\varepsilon_0$ and its dipole moment is parallel to the sides of the peg. The electric charge distribution for this plasmon mode is shown in Figure 4.53 and it exhibits the concentration of charges at the end of the peg (along its edges). This concentration is the primary reason why the peg is needed, and this concentration results in appreciable enhancement of local electric fields and local light intensity. At 30 nm disk and peg thickness, the resonance wavelength is 817 nm, which is beneficial for the excitation by the 830 nm laser. Furthermore, at 817 nm the ratio of $\varepsilon'(\omega)$ to $\varepsilon''(\omega)$ for gold is equal to 16.9 and is closer (or may be equal) to its maximum value. This makes the plasmon resonances strongly pronounced. The latter is demonstrated by the computed extinction cross sections shown in Figure 4.54.

It should be mentioned that there are many engineering challenges in the HAMR area. They include thermal management of the head, lubricant and recording media as well as the nonlinear dynamics of magnetization at elevated temperatures [36]-[38] during the recording process. However, the proper design of plasmonic structures for focused heating of recording media is central to the eventual success of HAMR.

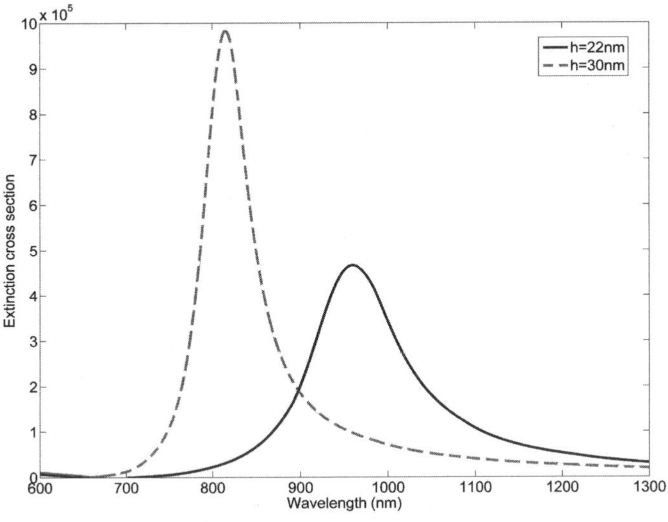

Figure 4.54

4.5.3   Application of Plasmon Resonances to All-Optical Magnetic Recording

It has been recently demonstrated [39] that magnetization reversal can be achieved by using circularly polarized laser pulses, that is, without applying any external magnetic fields. The direction of magnetization reversals is controlled by the helicity of circularly polarized light. The plausible explanation of this experimental phenomenon is based on the notion that magnetization reversals occur due to the combination of the local laser heating of magnetic media close to the Curie temperature and the simultaneous action of circularly polarized light as an effective magnetic field. The direction of this magnetic field is parallel to the light wave vector and right- and left-handed circularly polarized waves act as magnetic fields of opposite directions. The latter may be connected to the fact that the magnetic field and magnetization are not true vectors but rather axial (or pseudo-) vectors whose directions are reversed as a result of the change from right- to left-handed coordinate systems and the other way around.

The performed experiments reported in [39] have revealed femtosecond magnetization reversals of about 100 micrometer spots on magnetic media. It is apparent that this all-optical magnetization switching will be technologically feasible for commercial data storage if this switching is realized at the nanoscale. The latter is only possible if the techniques of delivery of nanoscale-focused circularly polarized light are developed. It is demonstrated below through theoretical and numerical analysis that plasmon resonances in uniaxially symmetric metallic nanostructures can be efficiently used for dramatic enhancement of circularly polarized light intensity at the nanoscale [40]-[42]. It is conceivable that this nanoscale enhancement of circularly polarized light may find other applications far beyond the area of all-optical recording, for instance, in biology for light probing of helical molecular structures.

The starting and central point of our discussion is the fact that in the case of uniaxially symmetric nanostructures there are two-fold degenerate eigenvalues λ_k (and corresponding negative values of ε_k) of integral equations (4.289) and/or (4.290). For these eigenvalues, two identical (up to ninety-degree rotation in space) plasmon modes (eigenfunctions) exist. Such two plasmon modes have the same resonance frequency (wavelength) and can be simultaneously excited by the incident circularly polarized light. These two simultaneously excited plasmon modes form a *circularly polarized plasmon mode*, and this may result in dramatic nanoscale enhancement (focusing) of the incident circularly polarized light, which can be eventually used in all-optical nanoscale switching of magnetization. The existence of such *circularly polarized plasmon modes* is a consequence of rotational symmetry of uniaxial nanostructures. It is worthwhile to point out that each of the two plasmon modes mentioned above is not rotationally invariant, while their combination as a *circularly polarized plasmon mode* is rotationally invariant.

Next, we discuss several promising designs of uniaxial nanostructures that can be potentially used for nanoscale enhancement of circularly polarized light. We begin with the example of metallic nanorings. The surface charge distribution for one of the two identical (but ninety-degree shifted in space) dipole plasmon modes is shown in Figure 4.55 for a ring with outer radius of 30 nm, height of 20 nm and ring wall thickness of 7 nm placed on a glass substrate. The color bar indicates the normalized surface charge density. The resonance dielectric permittivity of this circularly polarized plasmon mode is equal to $-38.45\varepsilon_0$, the corresponding resonance wavelength for a silver nanoring is 878 nm and the ratio of $\varepsilon'(\omega)$ to $\varepsilon''(\omega)$ at the resonance wavelength

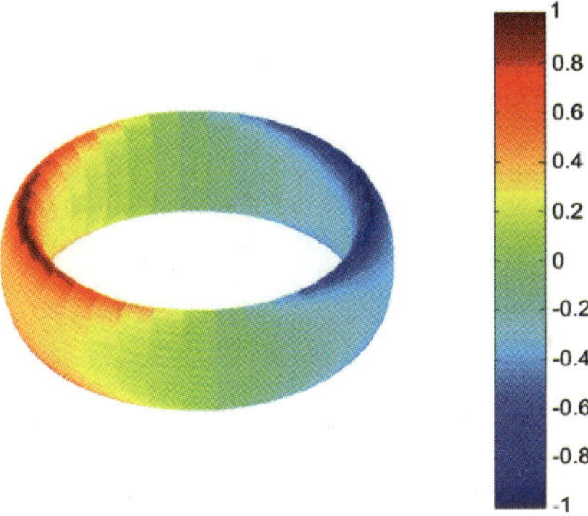

Figure 4.55

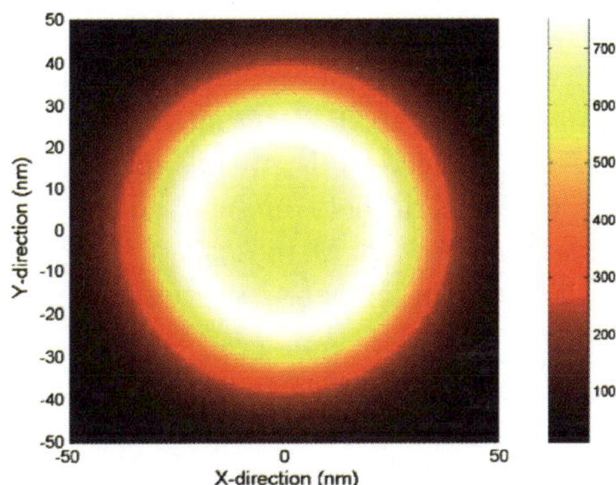

Figure 4.56

is about 84. The latter means that the plasmon resonance is strongly pronounced. This also can be inferred from Figure 4.56, which presents the light intensity of circularly polarized light (normalized by intensity of incident light) at the vertical distance of 10 nm from the ring. This figure demonstrates that the intensity of the incident circularly polarized light has been enhanced up to 750 times.

Another interesting design is a circular aperture in a metallic nanofilm (see Figure 4.57). The circularly polarized plasmon modes localized around this type of aperture have been numerically studied. It has been found that

Figure 4.57

the plasmon resonance is strongly pronounced for circular apertures in silver nanofilms with the film thickness to aperture diameter ratio of 1:6. At this aspect ratio, there are two identical and ninety-degree shifted in space plasmon modes for resonance permittivity of $-37.06\varepsilon_0$. The corresponding resonance wavelength for silver nanofilm is 864 nm and the ratio of $\varepsilon'(\omega)$ to $\varepsilon''(\omega)$ at this wavelength is about 76. The latter suggests that the plasmon resonance is strongly pronounced and this is supported by Figure 4.58 (see [53]), which presents the electric field intensity of circularly polarized light (normalized by incident electric field intensity) of the optical spot at the vertical distance of 10 nm from the circular aperture.

Finally, we consider the use of spherical and spheroidal nanoshells for the enhancement of circularly polarized incident light at the nanoscale. The analytical study of plasmon resonances in such nanoshells is presented in section 1 of Chapter 3. Nanoshell structures are very attractive because the resonance values of shell permittivity and resonance wavelengths can be effectively controlled by changing the thickness of the nanoshells. To illustrate this, we have computed by using formulas (3.151)-(3.155) the resonance values of permittivity $\varepsilon_1^-/\varepsilon_0$ of circularly polarized dipole plasmon modes in spherical nanoshells as a function of the ratio δ/a of its thickness to its inner radius for different values of core permittivity $\tilde{\varepsilon}$. The results of computations are presented in Figure 4.59. Similar computations have been performed for spheroidal nanoshells by using formulas (3.207)-(3.209). The results of these computations are presented in Figure 4.60 for the aspect ratio $a_1/b_1 = 3$. It is apparent from the above figures that the desired resonance values of shell permittivities can be achieved by the proper choice of shell thickness. For the sake of comparison, Figures 4.61a and 4.61b present the intensities of

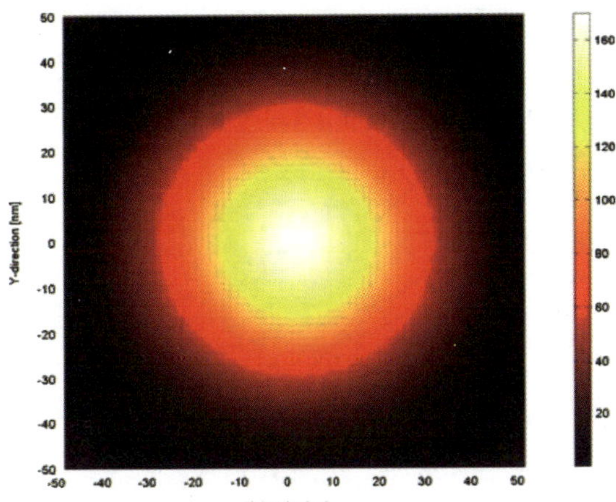

Figure 4.58

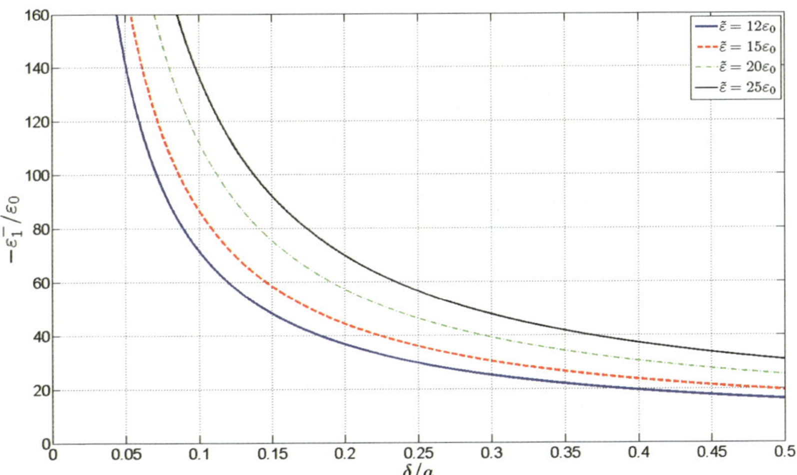

Figure 4.59

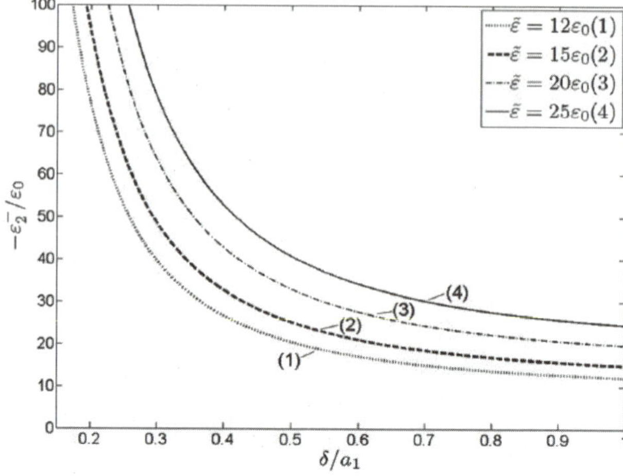

Figure 4.60

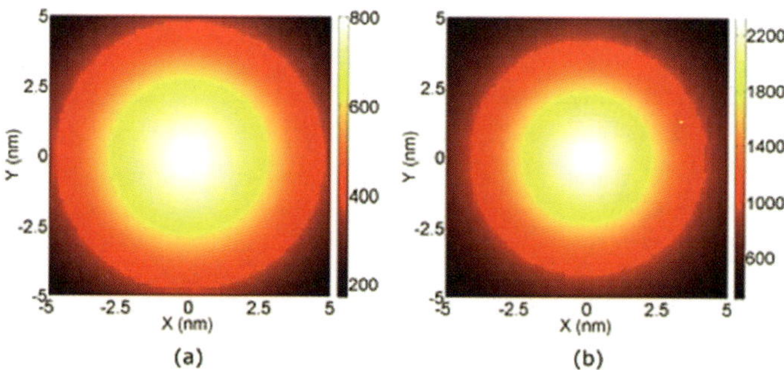

Figure 4.61

circularly polarized light normalized by the intensity of incident light at the distance of 5 nm from the spherical and spheroidal nanoshells, respectively. It is apparent from the computational results presented in the last figures that spheroidal nanoshells are superior to the spherical nanoshells as far as the enhancement of circularly polarized light intensity and its nanoscale focusing are concerned, although fabrication of spheroidal nanoshells can be quite challenging.

4.5.4 SERS and Plasmon Resonances

SERS is the abbreviation which stands for surface enhanced Raman scattering (or spectroscopy). SERS has been a very active area of research for the past 30 years. The reason is that SERS provides high sensitivity and high specificity for molecular spectroscopy [43, 44]. It has been long understood that the primary and dominant mechanism of enhancement in SERS is electromagnetic in nature, and it is due to the excitation of plasmon resonances in metallic (silver) nanoparticles adjacent (or attached) to adsorbed molecules. In this well-accepted electromagnetic (EM) mechanism of SERS, the incoming optical radiation excites a desired (dipole) plasmon resonance mode in a nanoparticle or a cluster of nanoparticles. This results in strong optical electric field on nanoparticle boundaries which usually leads to the strong enhancement of the incident electric field at the location of the adsorbed molecule. The strong (plasmon resonance generated) electric field is scattered by a molecule adsorbed at the metal nanoparticle surface at a Raman-shifted frequency. This scattered field may in turn excite the resonance plasmon mode in the nanoparticle (or cluster of nanoparticles) at the (slightly shifted from resonance) Raman frequency. As a result, enormous enhancement of overall Raman scattering may be achieved. Indeed, the SERS enhancement factors as large as 10^{10} to 10^{11} have been reported. It is clear that the contribution of plasmon resonances to SERS is two-fold: 1) the enhancement of the incident optical radiation and 2) the enhancement of the Raman-scattered signal.

It is apparent that the theory of plasmon resonances presented in this book can be used for the detailed (quantitative) study of the electromagnetic mechanism of SERS. It follows from the above description of this mechanism that fine-tuning of the following conditions is desirable for the achievement of very strong SERS activity:

a) The incoming light closely matches the resonance frequency and polarization of the desired dipole plasmon mode;

b) The adsorbed molecule is in the vicinity where the maximum of the plasmon mode electric field occurs;

c) The Raman-scattered field is strongly coupled to the same (or another) dipole plasmon mode by properly matching its resonance frequency as well as its polarization (i.e., its spatial field distribution).

Next, we shall discuss these three conditions. It is clear that by using the eigenvalue formulation of plasmon resonances presented in the book the resonance frequencies (wavelengths) of plasmon modes can be computed along with their dipole moments. This information is important for proper choice of incident laser radiation and its polarization. It is also clear from formula (4.337) that the plasmon resonance enhancement of optical electric fields in the vicinity of nanoparticles is controlled by the ratio of the real to imaginary parts of dielectric permittivity at the resonance frequency. This ratio is more favorable for silver than for gold, and it is most appreciable in the 600-1400 nm wavelength range. For this reason, it is desirable to design nanoparticles (or their clusters) that will have dipole modes in this wavelength range. It can be accomplished by increasing the geometric aspect ratio of nanoparticles (for instance, by choosing ellipsoidal nanoparticles) or by using spherical nanoshells.

It is important to note that the mathematical eigenvalue formalism for computing resonance plasmon modes in nanoparticles may require some modification to account for interaction between nanoparticles and adsorbed molecules. In other words, plasmon modes of a system consisting of interacting nanoparticles and adsorbed molecules should be examined. These computations will help to examine if the nanoparticle-molecule interaction may lead to appreciable shifts in resonance frequencies (wavelengths) as well as in appreciable changes in electric fields in molecule locations. It is conceivable that these computations may (at least partially) account for so-called "chemical enhancement." Below, we shall briefly describe the modification of the eigenvalue formalism for the case of a single nanoparticle and a single adsorbed molecule (see Figure 4.62). The generalization of this modification to the case of nanoparticle clusters on substrates and multiple adsorbed molecules is straightforward.

By using the same line of reasoning as before, it can be shown that the distribution of surface electric charges on the nanoparticle boundary S satisfies the following integral equation:

$$\sigma(Q) - \frac{\lambda}{2\pi} \oint_S \sigma(M) \frac{\mathbf{r}_{MQ} \cdot \mathbf{n}_Q}{r_{MQ}^3} ds_M - 2\varepsilon_0 \lambda \mathbf{n}_Q \cdot \mathbf{E}^{(m)}(Q) = 0, \qquad (4.386)$$

where

$$\lambda = \frac{\varepsilon - \varepsilon_0}{\varepsilon + \varepsilon_0} \qquad (4.387)$$

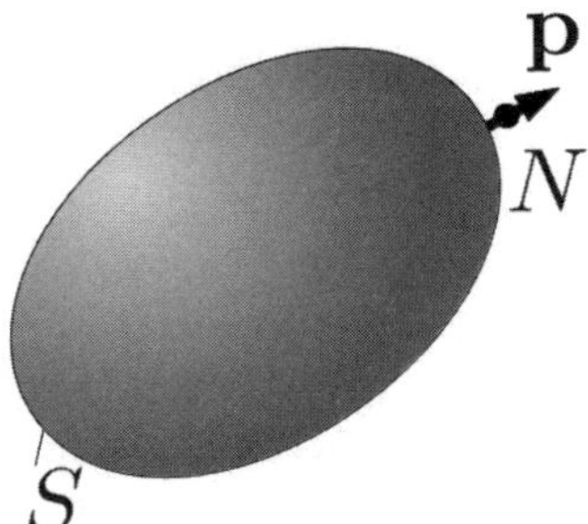

Figure 4.62

and $\mathbf{E}^{(m)}(Q)$ is the electric field due to the dipole moment of the adsorbed molecule. This field is given by the formula

$$\mathbf{E}^{(m)}(Q) = \frac{1}{4\pi\varepsilon_0}\left[\frac{3\left(\mathbf{p}\cdot\mathbf{r}_{NQ}\right)\mathbf{r}_{NQ}}{r_{NQ}^5} - \frac{\mathbf{p}}{r_{NQ}^3}\right],\tag{4.388}$$

where $\mathbf{p}$ is the dipole moment of the molecule, while point N is its (electric) center.

This dipole moment is related to the electric field $\mathbf{E}^{(\sigma)}(N)$ created by surface charges $\sigma(M)$ at point N by the formula

$$\mathbf{p} = \hat{\alpha}_0\mathbf{E}^{(\sigma)}(N),\tag{4.389}$$

where $\hat{\alpha}_0$ is the time-independent (Rayleigh) part of the polarizability tensor of the adsorbed molecule and the electric field $\mathbf{E}^{(\sigma)}(N)$ is defined by the formula

$$\mathbf{E}^{(\sigma)}(N) = \frac{1}{4\pi\varepsilon_0}\oint_S \sigma(M)\frac{\mathbf{r}_{MN}}{r_{MN}^3}ds_M.\tag{4.390}$$

By using sequential substitutions of formula (4.390) into relation (4.389), then relation (4.389) into formula (4.388) and, finally, formula (4.388) into equation (4.386), after simple transformation we obtain the following eigenvalue problem:

$$\sigma(Q) = \frac{\lambda}{2\pi} \oint_S \sigma(M) \left\{ \frac{\mathbf{r}_{MQ} \cdot \mathbf{n}_Q}{r_{MQ}^3} \right.$$

$$\left. - \frac{\hat{\alpha}_0 \mathbf{r}_{MN}}{4\pi\varepsilon_0 r_{MN}^3} \cdot \left[\frac{3\mathbf{r}_{NQ}(\mathbf{r}_{NQ} \cdot \mathbf{n}_Q)}{r_{NQ}^5} - \frac{\mathbf{n}_Q}{r_{NQ}^3} \right] \right\} ds_M. \qquad (4.391)$$

After solving this eigenvalue problem, the electric field at the adsorbed molecule location can be computed by using formula (4.390). This field can then be used for the calculation of the molecule dipole moment responsible for Raman scattering, which is given by the formula

$$\mathbf{p}_{\text{Raman}} = \hat{\beta}\mathbf{E}^{(\sigma)}(N). \qquad (4.392)$$

Here, $\hat{\beta}$ is the time-dependent (Raman) part of the polarizability tensor which is due to the internal rotational and vibrational dynamics of the molecule [45].

Our computations based on the modified eigenvalue formulation (4.391) have not produced any evidence of appreciable changes in plasmon resonance wavelengths or in plasmon electric fields at the location of the adsorbed molecule. The reason is that $\hat{\alpha}_0$ is very small. However, if the so-called "chemical enhancement" mechanism may result in substantial increase in components of $\hat{\alpha}_0$, then the eigenvalue formulation (4.391) may be useful to translate the "chemical" enhancement into the electromagnetic enhancement, which can be experimentally observed.

The plasmon electric field at the location of the adsorbed molecule can be enhanced by the proper choice of nanoparticle shape (with edges and corners) as well as by using dimers or clusters of nanoparticles. In the case of dimers, the excitation of the dipole plasmon modes with dipole moments "normal" to dimer gaps is most desirable. The plasmon electric field of such dipole modes is most enhanced in the gap region, and this is the most favorable location for adsorbed molecules. The use of dimers has other advantages. First, for small gaps the dipole mode resonance wavelengths of dimers are upward shifted in comparison with resonance wavelengths of single nanoparticles. These shifts occur due to the overall increase in aspect ratio of dimer nanostructures, and they may result in larger ratios of real to imaginary parts of particle permittivities at resonance conditions. The latter, as mentioned before, is beneficial to the strength of plasmon resonances. Second, the redistribution of electric charges on boundaries of dimer nanoparticles due to nanoparticle interactions may result in the conversion of multipole modes

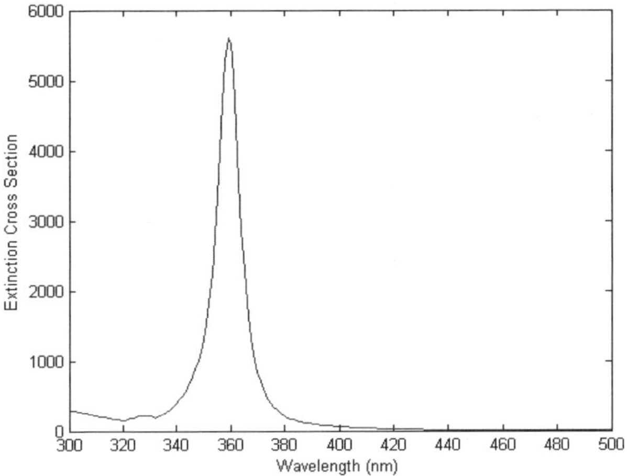

Figure 4.63

of single nanoparticles into dipole modes for dimers. We shall demonstrate this effect for spherical nanoparticle dimers. Figures 4.63 and 4.64 present the computed extinction cross sections for dimers of (diameter D) spherical nanoparticles with large $(d = 0.5D)$ and small $(d = 0.1D)$ separation d, respectively. It is apparent from these figures that an additional peak in the extinction cross section appears at small nanoparticle separation which can be attributed to the conversion of the specific multipole mode into the dipole mode. The latter interpretation is supported by the distribution of surface charges shown in Figures 4.65 and 4.66 for $d = 0.5D$ and $d = 0.1D$, respectively. It is also apparent from Figures 4.63 and 4.64 that for small separation $(d = 0.1D)$ there is upward shift of the resonance wavelength from 360 nm to 380 nm. The effect of multipole-to-dipole mode conversion can be quite pronounced when dimer nanoparticles are placed on a substrate. This is illustrated by Figure 4.67 for a spherical dimer on a silicon substrate with $d = 0.5D$ and $h = 0.01D$, where h is the separation from the substrate. It is most likely that this mode conversion is due to the "interaction" of spherical nanoparticles with their images due to the substrate presence. The existence of two adjacent (in resonance wavelength) dipole modes may be beneficial for SERS because it facilitates the simultaneous matching of plasmon mode resonance wavelengths by the incident radiation and the Raman-scattered radiation.

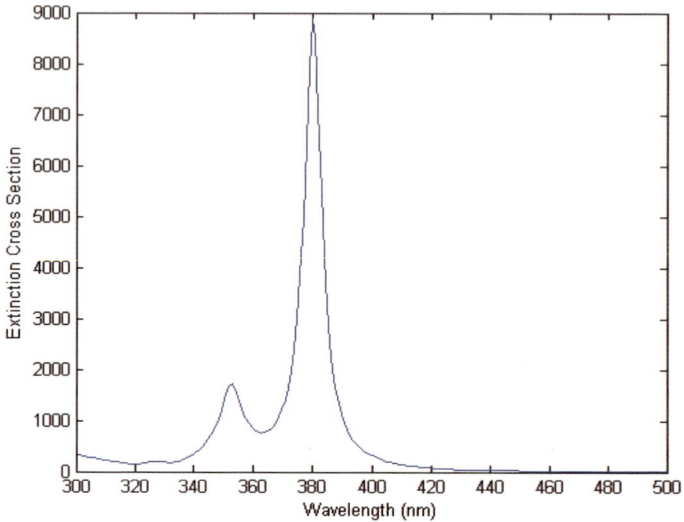

Figure 4.64

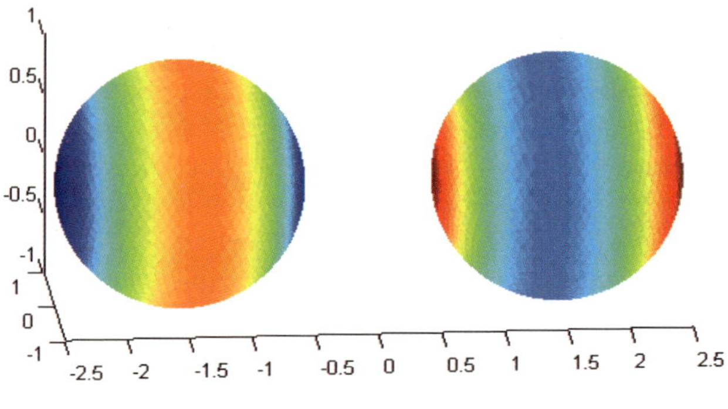

Figure 4.65

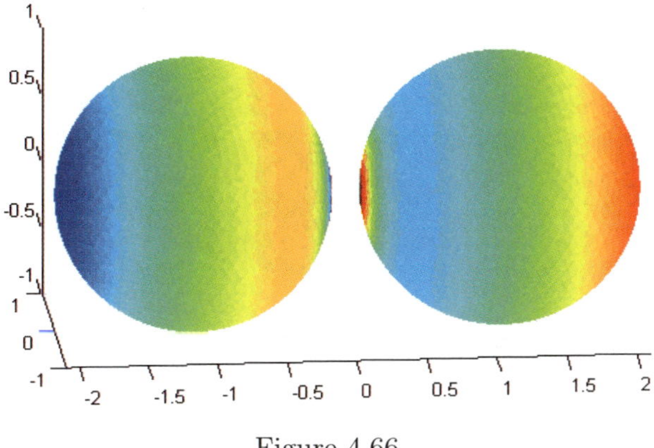

Figure 4.66

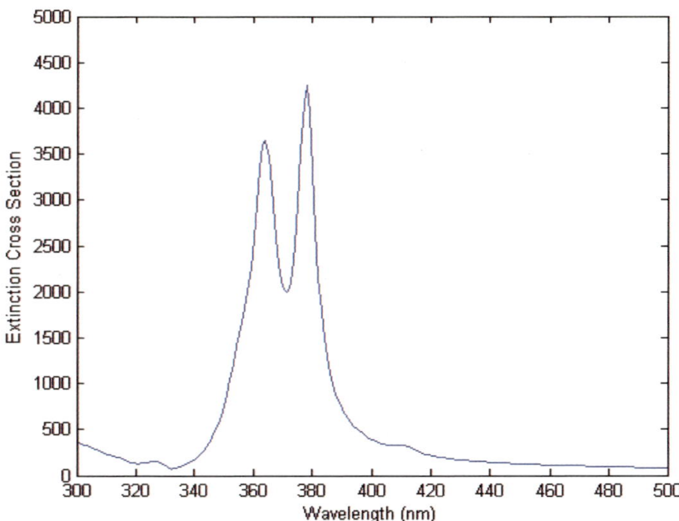

Figure 4.67

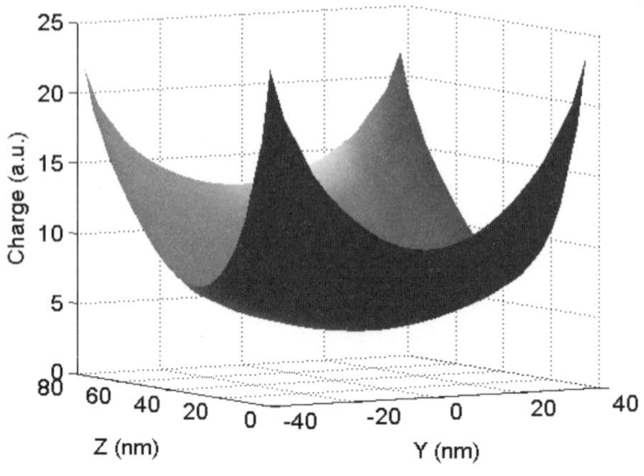

Figure 4.68

Dimers of nanocubes are very attractive [46]-[48], because corners and edges of cubes result in appreciable enhancement of local electric fields. This is illustrated by Figure 4.68 where the distribution of electric charges for the face-to-face nanocube dimer on the silicon substrate with the separations $d = 0.025L$, $h = 0.01L$ (L being the edge length) is presented for one of the opposite faces. The computed extinction cross section of this dimer is shown in Figure 4.69. It is apparent from this figure that the resonance wavelength is about 600 nm and appreciably shifted upwards in comparison with dimers of spherical nanoparticles (see Figure 4.67). This is another advantage of nanocube dimers. Furthermore, there is also the appreciable broadening of the extinction cross section of nanocube dimers, which is probably the result of the existence of two dipole modes with approximately the same resonance wavelengths. This broadening may be beneficial for SERS because it facilitates tuning the laser wavelength in such a way that near-resonance excitations are simultaneously achieved by the laser radiation and the Raman-shifted radiation, respectively. The proper assembly of nanocube dimers on substrates is a challenging problem. This problem has been addressed in [48] by using patterned substrates with cylindrical cavities (pores) and by employing the vertical deposition technique from a colloidal solution. The scanning electron microscopy (SEM) images of two isolated dimers trapped in cylindrical cavities of the patterned substrate are shown in Figure 4.70.

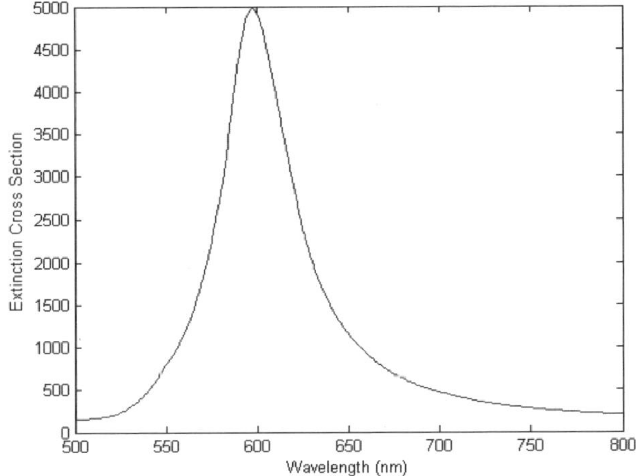

Figure 4.69

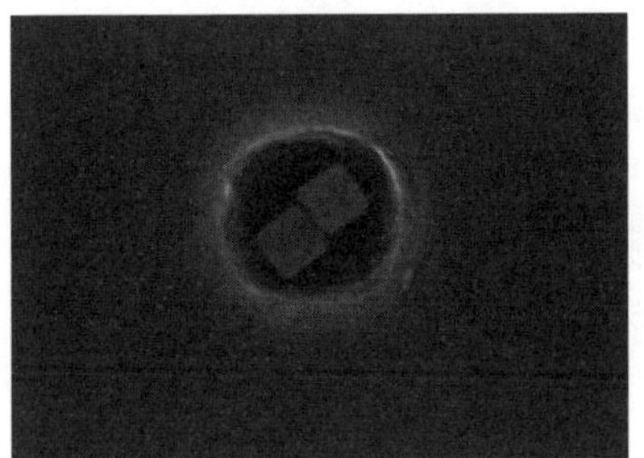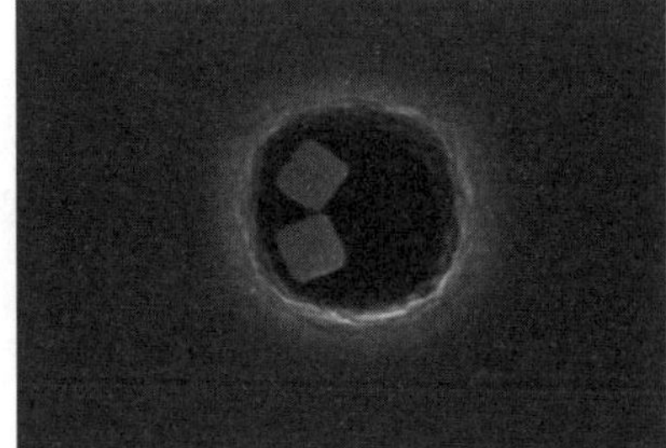

Figure 4.70

Finally, it is important to stress that the coupling between the Raman radiation of adsorbed molecules and plasmon dipole modes is quite different in nature than the coupling between the incident laser radiation and the same dipole modes. The reason is that incident (primary) radiation is produced by remote sources. For this reason, its electric field is practically uniform over the nanoparticle regions if the free-space wavelength of the incident radiation is appreciably larger than the nanoparticle dimensions. In this situation, the coupling between dipole plasmon modes and the incident radiation can be fully characterized by the dot-product $\mathbf{E}_i \cdot \mathbf{p}_k$ (see formulas (4.333) and

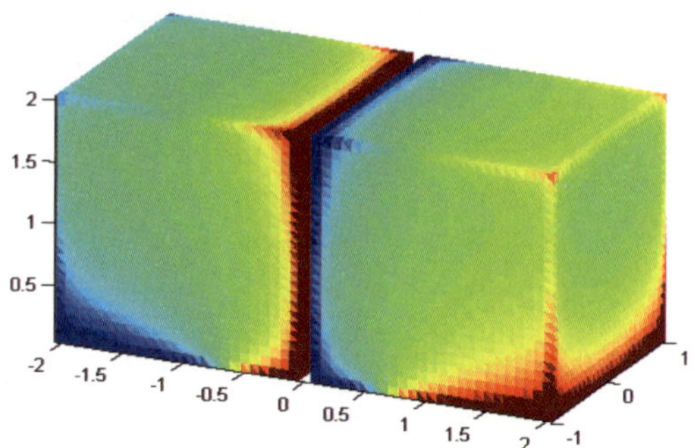

Figure 4.71

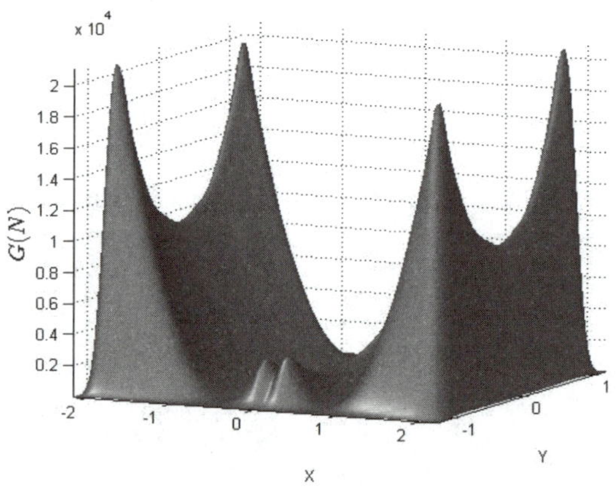

Figure 4.72

(4.343)). This is not the case for the Raman-shifted radiation produced by the adsorbed molecule. The electric field $\mathbf{E}_R$ of the Raman radiation is highly nonuniform over the nanoparticle region and its coupling with the dipole modes is described by the integral (see formula (4.323))

$$G(N) = \oint_S \tau_k(Q)\mathbf{E}_R(Q) \cdot \mathbf{n}_Q ds_Q. \tag{4.393}$$

It is clear from the last formula that this coupling may depend on the location (and orientation) of the adsorbed molecule. Indeed, $\mathbf{E}_R(Q)$ can be computed by using formula (4.388), and the integral (4.393) is then the

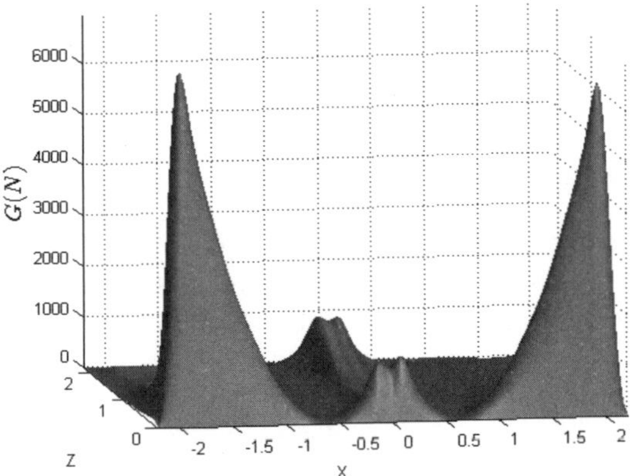

Figure 4.73

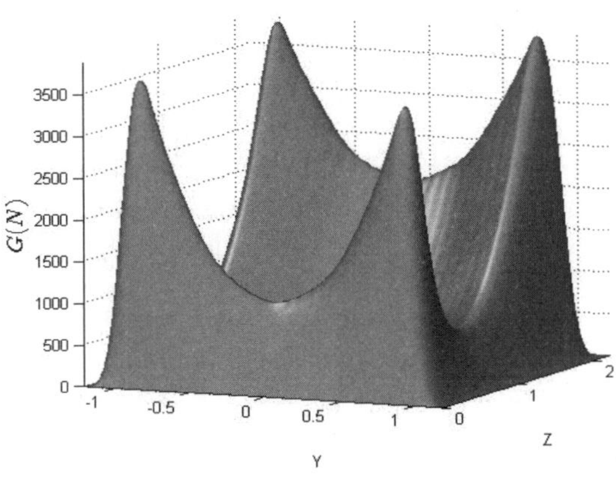

Figure 4.74

function of the location (point N) of the adsorbed molecule and the orientation of its dipole moment **p**. To illustrate the location-dependent coupling, the "location coupling factor" $G(N)$ given by formula (4.393) has been computed for the nanocube dimer (Figure 4.71) on a silicon substrate with $d = 0.025L$ and $h = 0.01L$. The computations have been performed under the assumption that the orientation of the molecule dipole is normal to the dimer boundaries. The results of computations are shown in Figures 4.72, 4.73 and 4.74 for "horizontal," "in-gap" and "side" planes, respectively. It is apparent from these figures that "hot" coupling spots coincide, by and large, with "hot" electric field "spots."

4.5.5 Ball Lightning

The enigmatic natural phenomenon of ball lightning usually occurs during or after thunderstorms. Its manifestation is the appearance of a small luminous fireball of a few inches in diameter. This ball is usually red, orange or yellow and it lasts for about five seconds. This natural phenomenon is quite rare and quite accidental. For this reason, it resists "pre-planned" observation. Nevertheless, it has been observed independently many times. Thus, its existence is beyond any doubt.

The physical understanding of the ball lightning phenomenon requires the explanations of its nucleation and growth, the mechanism of energy accumulation inside this lightning as well as why this lightning is usually of spherical shape. It is briefly discussed below how the theory of plasmon resonances may provide a plausible explanation for these three essential features of ball lightning [1, 49].

A lightning appearance is usually preceded by plasma formation, for instance, along numerous paths of so-called "step leaders." A subsequent lightning strike may also serve as a source of considerable electromagnetic radiation. If the frequency spectrum of this radiation is such that the dielectric permittivity of the formed plasma is negative, then plasmon resonances may occur. The nucleation of plasmon resonance and the spatial growth of resonance regions may be facilitated by the scale invariance of plasmon resonance frequencies. Indeed, if a plasmon resonance occurs initially in a small region, then it may result in strong electric fields on its boundary, which in turn may produce ionization in layers of air adjacent to this boundary. If this expansion of ionized region is geometrically self-similar in nature, it does not affect plasmon resonance conditions due to the scale invariance of plasmon resonances. This growth of plasmon resonance regions may be aborted for two reasons: a) reduction of surface electric fields on the resonance region boundary due to the reduction of its curvature as a result of its (self-similar) expansion and b) increase in radiation losses when the geometric dimensions of the resonance region become comparable with the resonance free-space wavelength. By the way, this resonance wavelength is naturally much larger than in the case of plasmon resonance in metallic nanoparticles because the electron density of atmospheric plasma is many orders of magnitude smaller than the density of conduction electrons in silver or gold.

These plasmon resonances in atmospheric plasma are more likely to occur for regions with spherical boundaries. The reason is that for such regions the coupling of atmospheric radiation to dipole modes occurs for any polarization of radiation. This is the consequence of rotational symmetry, which results in three identical dipole modes with mutually orthogonal dipole moments.

Plasmon resonances may produce a considerable accumulation of electromagnetic energy in a spherical resonance region that may visually manifest itself as a ball lightning. Thus, the notion of plasmon resonances may provide a plausible explanation for the energy accumulation in the ball lightning, its nucleation and growth as well as for its spherical shape. Recent laboratory "fireball" experiments [50] support the plausibility of the outlined explanation.

4.5.6   Optical Controllability of Plasmon Resonances

Next, we proceed to the brief discussion of optical controllability of plasmon resonances in nanoparticles. It has been repeatedly emphasized in the book that the occurrence of these resonances depends on the dispersion relation $\varepsilon(\omega)$. In the simplest (but reasonably accurate) case when electron collisions and energy losses are neglected (see Chapter 1), this dispersion relation can be written as follows:

$$\varepsilon(\omega) = \varepsilon_0 \left(1 - \frac{\omega_p^2}{\omega^2} \right), \tag{4.394}$$

$$\omega_p = \frac{e^2 N}{\varepsilon_0 m}, \tag{4.395}$$

where all the notations have their usual meaning.

The last two formulas clearly suggest that the dispersion relation $\varepsilon(\omega)$, and, consequently, plasmon resonances can be controlled through the manipulation of conduction electron density N. Semiconductor nanoparticles are especially convenient for this purpose. First, semiconductors, like metals, may exhibit dispersion of the dielectric permittivity and its real part may assume negative values in the optical frequency range below the plasma frequency ω_p [51]. Second, the manipulation of conduction electron density can be accomplished by doping or depletion means. Indeed, by appropriate doping, the wide range of controllability of ω_p can be achieved and, in this way, plasmon resonances can be tuned to desirable frequencies (wavelengths). The doping and depletion means can be especially convenient for tuning localized plasmon resonances in thin films with apertures. Third, it may be most attractive to optically control conduction electron density N (and, consequently, $\varepsilon(\omega)$). In this case, one light beam can be used to generate conduction electrons and, in this way, to drive semiconductor nanoparticles into conditions when plasmon resonances can be excited by another light beam. If this light gating of plasmon resonances can be realized, then it may open the opportunity for the development of nanoscale light switches. In this way, semiconductors may well play a role in nanophotonics similar to what they do in electronics.

Finally, there may also be a possibility for optical control of plasmon resonances in metallic nanoparticles. This possibility is based not on the optical manipulation of electron density, but rather on the optical manipulation of polarization of incident radiation. Indeed, according to formula (4.333), the excitation of plasmon resonances is controlled by proper polarization of electric field $\mathbf{E}_i$ of incident radiation with respect to the dipole moment $\mathbf{p}_k$ of the plasmon mode. By controlling in time the light polarization, the optical gating of plasmon resonances in metallic nanoparticles can be in principle made possible.

References

[1] I.D. Mayergoyz, D.R. Fredkin, and Z. Zhang, Physical Review B 72, 155412 (2005).

[2] C.F. Bohren and D.R. Huffman, Absorption and Scattering of Light by Small Particles, John Wiley, New York (1983).

[3] J. Aizpurua, P. Hanarp, D.S. Sutherland, M. Käll, G.W. Bryant, and F.J. García de Abajo, Physical Review Letters 90, 057401 (2003).

[4] L.J. Sherry, S.-H. Chang, B.J. Wiley, Y. Xia, G.C. Schatz, and R.P. Van Duyne, Nano Letters 5, 2034 (2005).

[5] J.D Jackson, Classical Electrodynamics, John Wiley, New York (1999).

[6] A. Ishimaru, Electromagnetic Wave Propagation, Radiation and Scattering, Prentice Hall, New Jersey (1999).

[7] P.B. Johnson and R.W. Christy, Physical Review B 6, 4370 (1972).

[8] P. Royer, J.P. Goudonnet, R.J. Warmack, and T.L. Ferrell, Physical Review B 35, 3753 (1986).

[9] S.A. Maier, P.G. Kik, and H.A. Atwater, Applied Physics Letters 81, 1714 (2002).

[10] S.A. Maier, M.L. Brongersma, P.G. Kik, and H.A. Atwater, Physical Review B 65, 193408 (2002).

[11] S.A. Maier and H.A. Atwater, Journal of Applied Physics 67, 205402 (2003).

[12] I.D. Mayergoyz and Z. Zhang, IEEE Transactions on Magnetics 43, 1685 (2007).

[13] A. Bossavit and I. Mayergoyz, IEEE Transactions on Magnetics 25, 2816 (1989).

[14] I.D. Mayergoyz and J.D. D'Angelo, IEEE Transactions on Magnetics 27, 3967 (1991).

[15] I.D. Mayergoyz, Z. Zhang, and G. Miano, Physical Review Letters 98, 147401 (2007).

[16] B. Lamprecht, J.R. Krenn, A. Leitner, and F.R. Aussenegg, Physical Review Letters 83, 4421 (1999).

[17] J. Lehmann, M. Merschdorf, W. Pfeiffer, A. Thon, S. Voll, and G. Gerber, Physical Review Letters 85, 2921 (2000).

[18] A. Kubo, K. Onda, H. Petek, Z. Sun, Y.S. Jung, and H.K. Kim, Nano Letters 5, 1123 (2005).

[19] O.L. Muskens, N.D. Fatti, and F. Valee, Nano Letters 6, 552 (2006).

[20] G. Gay, O. Alloschery, B. Viaris de Lesegno, C. O'Dwyer, J. Weiner, and H.J. Lezec, Nature Physics 2, 262 (2006).

[21] T. Zentgraf, A. Christ, J. Kuhl, and H. Giessen, Physical Review Letters 93, 243901 (2004).

[22] R.J. Elliott, Physical Review 96, 266 (1954).

[23] G. Dresselhaus, Physical Review 100, 580 (1955).

[24] S. Tomita, T. Kato, S. Tsunashima, S. Iwata, M. Fujii, and S. Hayashi, Physical Review Letters 96, 167402 (2006).

[25] R. Fujikawa, A.V. Baryshev, J. Kim, H. Uchida, and M. Inoue, Journal of Applied Physics 103, 07D301 (2008).

[26] H. Uchida, Y. Masuda, R. Fujikawa, A.V. Baryshev, and M. Inoue, Journal of Magnetism and Magnetic Materials 231, 843 (2009).

[27] F.E. Moolekamp and K.L. Stokes, IEEE Transactions on Magnetics 45, 4888 (2009).

[28] I.D. Mayergoyz, G. Lang, L. Hung, S. Tkachuk, C. Krafft, and O. Rabin, Journal of Applied Physics 107, 09A925 (2010).

[29] S. Tkachuk, G. Lang, C. Krafft, O. Rabin, and I. Mayergoyz, Journal of Applied Physics 109, 07B717 (2011).

[30] G. Lang, D. Bowen, L. Hung, C. Krafft, and I. Mayergoyz, Journal of Applied Physics 111, 07A505 (2012).

[31] C. Kittel, Introduction to Solid State Physics, Wiley and Sons, New York (2005).

[32] S.X. Wang and A.M. Taratorin, Magnetic Information Storage Technology, Academic Press, New York (1999).

[33] I. Mayergoyz and C. Tse, Spin-Stand Microscopy of Hard Disk Data, Elsevier, Amsterdam (2007).

[34] Z. Zhang and I.D. Mayergoyz, Journal of Applied Physics 103, 07F510 (2008).

[35] W.A. Challener, C. Peng, A.V. Itagi, D. Karns, W. Peng, Y. Peng, X.-M. Yang, X. Zhu, N.J. Gokemeijer, Y.-T. Hsia, G. Ju, R.E. Rottmayer, M.A. Seigler, and E.C. Gage, Nature Photonics 3, 220 (2009).

[36] G. Bertotti, I. Mayergoyz, and C. Serpico, Nonlinear Magnetization Dynamics in Nanosystems, Elsevier, Amsterdam (2009).

[37] I. Mayergoyz, G. Bertotti, C. Serpico, Z. Liu, and A. Lee, Journal of Applied Physics 111, 07D501 (2012).

[38] I. Mayergoyz, G. Bertotti, and C. Serpico, Physical Review B (Rapid Communications) 83, 020402 (2011).

[39] C.D. Stanciu, F. Hansteen, A.V. Kimel, A. Kirilyuk, A. Tsukamoto, A. Itoh, and Th. Rasing, Physical Review Letters 99, 047601 (2007).

[40] I. Mayergoyz, Z. Zhang, P. McAvoy, D. Bowen, and C. Krafft, IEEE Transactions on Magnetics 44, 3372 (2008).

[41] I. Mayergoyz, P. McAvoy, G. Lang, D. Bowen, and C. Krafft, Journal of Applied Physics 105, 07B904 (2009).

[42] L. Hung, G. Lang, P. McAvoy, C. Krafft, and I. Mayergoyz, Journal of Applied Physics 111, 07A915 (2012).

[43] J.P. Camden, J.A. Dieringer, J. Zhao, and R.P. Van Duyne, Accounts of Chemical Research 41, 1653 (2008).

[44] M. Moskovits, Journal of Raman Spectroscopy 36, 485 (2005).

[45] N.B. Colthup, L.H. Daly, and S.E. Wiberley, Introduction to Infrared and Raman Spectroscopy, Academic Press, New York (1990).

[46] P.H.C. Camargo, M. Rycenga, L. Au, and Y. Xia, Angewandte Chemie International Edition 48, 2180 (2009).

[47] D.-S. Kim, J. Heo, S.-H. Ahn, S.W. Han, W.S. Yun, and Z.H. Kim, Nano Letters 9, 3619 (2009).

[48] S.Y. Lee, L. Hung, G.S. Lang, J.E. Cornett, I.D. Mayergoyz, and O. Rabin, ACS Nano 4, 5763 (2010).

[49] D.R. Fredkin and I.D. Mayergoyz, Physical Review Letters 91, 253902 (2003).

[50] V. Dikhtyar and E. Jerby, Physical Review Letters 96, 045002 (2006).

[51] Handbook of Optical Constants of Solids (edited by E.D. Palik), Academic Press, New York (1985).

[52] I.D. Mayergoyz and Z. Zhang, IEEE Transactions on Magnetics **43**, 1681 (2007).

[53] I. Mayergoyz, Z. Zhang, P. McAvoy, D. Bowen, and C. Krafft, IEEE Transactions on Magnetics **45**, 1634 (2009).

Index